INORGANIC SYNTHESES

Volume 27

Editor-in-Chief
ALVIN P. GINSBERG

●●●●●●●●●●●●●●●●●●●●●●●●●●●●●●●●●

INORGANIC
SYNTHESES

Volume 27

A Wiley-Interscience Publication
JOHN WILEY & SONS

New York   Chichester   Brisbane   Toronto   Singapore

Published by John Wiley & Sons, Inc.

Copyright © 1990 Inorganic Syntheses, Inc.

Library of Congress Catalog Number: 39-23015

ISBN 0-471-50976-0

Printed in the United States of America

10 9 8 7 6 5 4 3 2 1

To John C. Bailar, Jr.,
Associate Editor of Volumes I–III and
Editor-in-Chief of Volume IV,
in appreciation of his contributions to
the development of *Inorganic Syntheses*.

PREFACE

Inorganic Syntheses is 50 years old. The first volume, edited by Harold S. Booth, appeared in 1939 and contained 90 checked procedures. Booth, Professor of Chemistry at Western Reserve University, had seen a need for a series of volumes in which would appear detailed and tested procedures for the synthesis of inorganic compounds. His efforts to bring *Inorganic Syntheses* into being dated back to 1933 when he was joined in the venture by L. F. Audrieth, W. C. Fernelius, W. C. Johnson, and R. E. Kirk. To quote one of the cofounders: "It would be fair to say that the inauguration of *Inorganic Syntheses* was not a product of the times but more a reaction against the times. In the late 20's and early 30's of this century there were in this country only a few inorganic chemists and little research in the field" [W. C. Fernelius, "History of *Inorganic Syntheses*," October 1983 (unpublished)].

Over the 50 years since Volume 1 appeared, research in inorganic chemistry has undergone an enormous expansion, and the *Inorganic Syntheses* series has grown to a total of 27 volumes containing more than 3000 checked procedures. Although it would be incorrect to claim that all of these procedures are foolproof [witness *Inorg. Synth.*, **23**, 199 (1985)], they do comprise the largest available collection of *reliable* preparations of inorganic compounds. The series has established itself as the reference of first resort for anyone with the need to synthesize an inorganic compound. The procedures in *Inorganic Syntheses* owe their reliability to the requirements that they be checked in a laboratory other than that of the submitters and that they be presented in detailed and unambiguous form. Both of these requirements frequently result in substantial modification of submitted manuscripts.

The present volume contains over 200 syntheses, presented in 68 numbered sections grouped into nine chapters that correspond for the most part to

---

Previous volumes of *Inorganic Syntheses* are available. Many of the volumes originally published by McGraw Hill, Inc. are available from R. E. Krieger Publishing Co., Inc., P.O. Box 9542, Melbourne, FL 32901. Please write this publisher for a current list. Volumes out of print with John Wiley & Sons, Inc. are also available from Krieger Publishing. Recent back volumes can be obtained from John Wiley & Sons, Inc., 605 Third Avenue, New York, NY 10158. Please write the publisher for a current list of available volumes.

currently active areas of research. Although all of the chapters contain important and timely compounds, I would single out for special mention Chapter 3 on the early transition metal polyoxoanions. This chapter, organized and solicited by Walter G. Klemperer, addresses the synthesis of a class of compounds that are notably difficult to prepare in pure form. The successful preparation of many of these compounds requires close attention to the details of the procedure, and in several cases critical details taken for granted by the submitters were brought out by the checkers.

An *Inorganic Syntheses* volume is the product of the efforts of many people. There are, of course, the submitters and checkers who produce the checked syntheses. In addition to these contributors, I would like to thank the following people for their efforts on behalf of Volume 27: Walter Klemperer for organizing Chapter 3 and providing an introduction to it; Andrea Wayda for organizing Chapter 4; Thomas E. Sloan and William Powell for advice on chemical nomenclature and for compiling the index; and the members of the Inorganic Syntheses Corporation who provided reviews of the submitted manuscripts, especially Bob Angelici, John Bailar, Dimitri Coucouvanis, and Therald Moeller. John Bailar and Therald Moeller also helped with checking the revised manuscripts. Finally, I thank Fred Basolo for efficiently resolving an awkward checking problem and Duward Shriver for his "Guide for Editors of *Inorganic Syntheses*", which facilitated my task as Editor-in-Chief.

ALVIN P. GINSBERG

Berkeley Heights, New Jersey
February, 1990

NOTICE TO CONTRIBUTORS
AND CHECKERS

The *Inorganic Syntheses* series is published to provide all users of inorganic substances with detailed and foolproof procedures for the preparation of important and timely compounds. Thus the series is the concern of the entire scientific community. The Editorial Board hopes that all chemists will share in the responsibility of producing *Inorganic Syntheses* by offering their advice and assistance in both the formulation of and the laboratory evaluation of outstanding syntheses. Help of this kind will be invaluable in achieving excellence and pertinence to current scientific interests.

There is no rigid definition of what constitutes a suitable synthesis. The major criterion by which syntheses are judged is the potential value to the scientific community. For example, starting materials or intermediates that are useful for synthetic chemistry are appropriate. The synthesis also should represent the best available procedure, and new or improved syntheses are particularly appropriate. Syntheses of compounds that are available commercially at reasonable prices are not acceptable. We do not encourage the submission of compounds that are unreasonably hazardous, and in this connection, authors are requested to avoid procedures involving perchlorate salts due to the high risk of explosion in combination with organic or organometallic substances. Authors are also requested to avoid the use of solvents known to be carcinogenic.

The Editorial Board lists the following criteria of content for submitted manuscripts. Style should conform with that of previous volumes of *Inorganic Syntheses*. The introductory section should include a concise and critical summary of the available procedures for synthesis of the product in question. It should also include an estimate of the time required for the synthesis, an indication of the importance and utility of the product, and an admonition if any potential hazards are associated with the procedure. The Procedure should present detailed and unambiguous laboratory directions and be written so that it anticipates possible mistakes and misunderstandings on the part of the person who attempts to duplicate the procedure. Any unusual equipment or procedure should be clearly described. Line drawings should be included when they can be helpful. All safety measures should be stated clearly. Sources of unusual starting materials must be given, and, if possible, minimal standards of purity of reagents and solvents should be stated. The scale should be reasonable for normal laboratory operation, and any problems involved in scaling the procedure either up or down should be discussed. The criteria for judging the purity of the final product should be delineated clearly. The section on Properties should supply and discuss those physical and chemical characteristics that are relevant to judging the purity of the product and to permitting its handling and use in an intelligent manner. Under References, all pertinent literature citations should be listed in order. A style sheet is available from the Secretary of the Editorial Board.

The Editorial Board determines whether submitted syntheses meet the general specifications outlined above, and the Editor-in-Chief sends the manuscript to an independent laboratory where the procedure must be satisfactorily reproduced.

Each manuscript should be submitted in duplicate to the Secretary of the Editorial Board, Professor Jay H. Worrell, Department of Chemistry, University of South Florida, Tampa, FL 33620. The manuscript should be typewritten in English. Nomenclature should be consistent and should follow the recommendations presented in *Nomenclature of Inorganic Chemistry*, 2nd ed., Butterworths & Co., London, 1970, and in *Pure and Applied Chemistry*, Volume 28, No. 1 (1971). Abbreviations should conform to those used in publications of the American Chemical Society, particularly *Inorganic Chemistry*.

Chemists willing to check syntheses should contact the editor of a future volume or make this information known to Professor Worrell.

TOXIC SUBSTANCES AND LABORATORY HAZARDS

Chemicals and chemistry are by their very nature hazardous. Chemical reactivity implies that reagents have the ability to combine. This process can be sufficiently vigorous as to cause flame, an explosion, or, often less immediately obvious, a toxic reaction.

The obvious hazards in the syntheses reported in this volume are delineated, where appropriate, in the experimental procedure. It is impossible, however, to foresee every eventuality, such as a new biological effect of a common laboratory reagent. As a consequence, *all* chemicals used and *all* reactions described in this volume should be viewed as potentially hazardous. Care should be taken to avoid inhalation or other physical contact with all reagents and solvents used in procedures described in this volume. In addition, particular attention should be paid to avoiding sparks, open flames, or other potential sources that could set fire to combustible vapors or gases.

A list of 400 toxic substances may be found in the *Federal Register*, Vol. 40, No. 23072, May 28, 1975. An abbreviated list may be obtained from *Inorganic Syntheses*, Volume 18, p. xv, 1978. A current assessment of the hazards associated with a particular chemical is available in the most recent edition of *Threshold Limit Values for Chemical Substances and Physical Agents in the Workroom Environment* published by the American Conference of Governmental Industrial Hygienists.

The drying of impure ethers can produce a violent explosion. Further information about this hazard may be found in *Inorganic Syntheses*, Volume 12, p. 317. A hazard associated with the synthesis of tetramethyl-diphosphine disulfide [*Inorg. Synth.*, **15**, 186 (1974)] is cited in *Inorganic Syntheses*, Volume 23, p. 199.

CONTENTS

**Chapter Two   TRANSITION METAL CHALCOGENIDE
COMPLEXES**

**Chapter Three EARLY TRANSITION METAL
POLYOXOANIONS**

Chapter Four   LANTHANIDE AND ACTINIDE COMPLEXES

Chapter Five TRANSITION METAL CLUSTER COMPLEXES

Chapter Eight   MISCELLANEOUS TRANSITION METAL COMPLEXES

Chapter Nine   LIGANDS AND OTHER MAIN GROUP COMPOUNDS

INORGANIC SYNTHESES

Volume 27

Chapter One

TRANSITION METAL POLYHYDRIDE COMPLEXES

1. MOLECULAR HYDROGEN COMPLEXES OF Mo AND W

Submitted by GREGORY J. KUBAS*
Checked by CARL HOFF[†]

The complexes $M(CO)_3(PR_3)_2(\eta^2\text{-}H_2)$ [M = Mo or W; R = cyclohexyl (cy) or isopropyl (i-Pr)] are the first recognized examples of molecular hydrogen complexes.[1] The side-on bonding of H_2 represents the first example of stable intermolecular coordination of a σ bond to a metal, much as η^2-ethylene coordination was the first interaction of a π bond with a metal center. The H_2 complexes also represent arrested intermediates in the oxidative addition of hydrogen to metal centers and, more importantly, may possess a unique chemistry of their own, including direct reactivity in catalysis. One of the remarkable features of molecular hydrogen coordination is that it has not been recognized earlier, since several known polyhydride complexes, for example, $RuH_4(PPh_3)_3$, which was first prepared over 20 years ago, are now being shown to contain one or more H_2 ligands.[1d,2] Part of the reason for this is that complexes containing hydrogen ligands (whether H_2, H^-, or

*Los Alamos National Laboratory, University of California, Los Alamos, NM 87545. This work was performed under the auspices of the U.S. Department of Energy, Division of Chemical Sciences, Office of Basic Energy Sciences.
†Department of Chemistry, University of Miami, Coral Gables, FL33124.

1

both) are generally fluxional and can possess significantly different structural features in solution than in the solid state. Indeed the complexes described here, which were shown by neutron diffraction to possess only H_2 ligands in the solid state, contain, *in solution*, up to 30% of the seven-coordinate *dihydride* form, $MH_2(CO)_3(PR_3)_2$, in dynamic equilibrium with the six-coordinate H_2 complex:

Thus, in the same sense, many known polyhydride complexes, including those shown to contain only hydrides in the solid state, may in fact contain H_2 species in solution.

The syntheses of $M(CO)_3(PR_3)_2(H_2)$ are quite straightforward and facile, similar to those for coordinatively unsaturated $M(CO)_3(PR_3)_2$, from which the H_2 complexes can be prepared by H_2 addition.[3] The reactions involve displacement of the triolefin in $M(CO)_3(\eta^6\text{-cycloheptriene})$ by two equivalents of phosphine, which leaves an "open" sixth-coordination site to which H_2 can bind [the H_2 is present in the reaction atmosphere and effectively displaces an intramolecularly bound "agostic" phosphine C—H group from the sixth site of $M(CO)_3(PR_3)_2$].[3a] Because ligand redistribution to $M(CO)_3(PR_3)_3$ occurs if commonly used smaller phosphines are employed (including PPh_3), the preparations succeed only for R = cy, *i*-Pr, or similar* *bulky* phosphines, which make tris-phosphine species too unstable on account of steric crowding. The complex $W(CO)_3[P(i\text{-Pr})_3]_2(H_2)$ is perhaps the most useful analog because of its high solubility and its function as a precursor to $W(CO)_3[P(i\text{-Pr})_3]_2$, which, unlike the Pcy_3 analog, cannot be prepared directly from $W(CO)_3(\eta^6\text{-cycloheptatriene})$. The complex $Mo(CO)_3[P(i\text{-Pr})_3]_2(H_2)$ is not isolable as a solid, and $Cr(CO)_3(Pcy_3)_2(H_2)$ has been found to be stable only under 300 psi of H_2.[4] The primary difficulty in the synthesis of the tungsten complexes is the commercial unavailability of $W(CO)_3(\eta^6\text{-cycloheptatriene})$. An improved synthesis of the latter has been reported,[5] but preparation still requires several days. The detailed synthesis is presented below, in slightly modified form.

* Analogs with tricyclopentylphosphine[6] and $Pcy_2(i\text{-Pr})^{1c}$ have been prepared but the syntheses failed for $P(t\text{-Bu})_3$ [*tert*-butyl (t-Bu)] or the other usual bulky phosphines reported in the literature.

A. TRICARBONYL(DIHYDROGEN)BIS(TRICYCLOHEXYL-PHOSPHINE)MOLYBDENUM

$$Mo(CO)_3(\eta^6\text{-cycloheptatriene}) + 2Pcy_3 \xrightarrow{H_2} Mo(CO)_3(Pcy_3)_2(H_2) + C_7H_8$$

Procedure

- **Caution.** *The syntheses described below should be carried out in a well-ventilated hood because of the flammability of hydrogen and the toxicity of phosphines.*

The following manipulations are carried out on a Schlenk system[7] using hydrogen (99.9% minimum purity) atmospheres and reagent grade deoxygenated (by bubbling argon or H_2, *but not* N_2, through) solvents without further purification. The complex $Mo(CO)_3(\eta^6\text{-cycloheptatriene})$ and the phosphine were obtained from Strem Chemicals.

A mixture of $Mo(CO)_3(\eta^6\text{-cycloheptatriene})$ (1.88 g, 6.94 mmol), Pcy_3 (4.0 g, 14.3 mmol), and toluene (25 mL) in a 100-mL Schlenk flask is vigorously stirred magnetically under hydrogen for ~3 h. Precipitation of yellow microcrystalline $Mo(CO)_3(Pcy_3)_2(H_2)$ begins and is completed by addition of 60 mL of heptane (or hexane) and further stirring (2–3 h). The product is collected on a frit and washed with H_2-saturated 2:1 heptane–toluene (~25 mL). The washings are normally slightly purple in color because of the $Mo(CO)_3(Pcy_3)_2$ formed by dissociation of H_2. The bright yellow complex is dried in a stream of H_2 (taking care to maintain ~1 atm of hydrogen pressure over the solid), then briefly (~1 min) *in vacuo*. The solid becomes darker as a result of H_2 loss *in vacuo*, but the color immediately brightens when an H_2 atmosphere is restored. This procedure is repeated until the complex is completely dry. The complex should *always* be kept under an H_2 or argon–H_2 (~10:1) atmosphere when transferred or stored. Nitrogen cannot be used because it displaces the H_2 ligand. Yield: 3.62 g (84%). Yields for larger scale reactions can approach 95%. The D_2 isotopomer is prepared in exactly the same fashion.

Anal. Calcd. for $C_{39}H_{68}O_3P_2Mo$: C, 63.1; H, 9.2; P, 8.3. Found: C, 62.8; H, 9.0; P, 8.0.

Properties

The solid complex slowly dissociates H_2 and must be kept in an H_2-enriched atmosphere as discussed previously. It is sparingly soluble in aromatic hydrocarbons, imparting a purple color resulting from the presence of small equilibrium amounts of $Mo(CO)_3(Pcy_3)_2$ (even under 1 atm of H_2). The latter can readily be prepared from the H_2 complex.[1c] Decomposition occurs in most polar solvents, including halogenated hydrocarbons, although tetrahydrofuran (THF) and other ethers do not readily displace the H_2. The microcrystalline yellow solid slowly decomposes in air with only slight darkening, although larger crystals can be exposed for several hours before significant decomposition occurs. It is also slightly photosensitive, slowly (weeks) becoming orange on the surface in room light, and it should be stored in darkness.

The complexes can be characterized by IR spectroscopy by loading samples in a glove bag initially filled with argon, to which is then added $\sim$ 10 to 20% H_2. Alternatively, solution samples can be transferred by syringe directly into holders using Schlenk techniques. Unlike the spectra of the tungsten analogs (see below), the IR spectrum of the Mo complex shows no readily observable bands due to the H_2 ligand (a partially obscured band at $885 \, cm^{-1}$ may be discerned). Carbonyl peaks occur at 1966 and $1853 \, cm^{-1}$ (Nujol mull) and 1960 and $1842 \, cm^{-1}$ (toluene). A band near $1865 \, cm^{-1}$ is normally also present as a result of a very intense mode of a minor amount ($< 5\%$) of $Mo(CO)_4(Pcy_3)_2$ impurity. The latter is difficult to avoid because it is readily formed upon minor air oxidation, photochemical decomposition, and spontaneous ligand redistribution of the H_2 complex. Because of low solubility of the complex, ^1H NMR is not useful for characterization. In solution, the complex exists in dynamic equilibrium with the seven-coordinate *dihydride* form, $MoH_2(CO)_3(Pcy_3)_2$ ($\sim$ 20–30%), formed by oxidative addition of the H_2.[1b,c] The latter gives a weak carbonyl band at $1998 \, cm^{-1}$ in the solution IR.

B.   TRICARBONYL(η^6-CYCLOHEPTATRIENE)TUNGSTEN

$$W(CO)_6 + 3NCEt \longrightarrow W(CO)_3(NCEt)_3 + 3CO$$

$$W(CO)_3(NCEt)_3 + cyclopheptatriene$$

$$\longrightarrow W(CO)_3(\eta^6\text{-cycloheptatriene}) + 3NCEt$$

Procedure

Into a 500-mL round-bottomed Schlenk flask equipped with a magnetic stirring bar is placed 220 mL of propionitrile (■ **Caution.** *Highly toxic; may be fatal if absorbed through skin. Gloves should be worn.*), which is degassed by pumping *in vacuo* and refilling with nitrogen. With the stirrer in motion and a heating mantle in place, 25 g (0.1 mol) of $W(CO)_6$ is added (slowly, in order to keep the stirrer moving). A reflux condenser with a nitrogen line attached is then added so that a positive pressure of N_2 is maintained in the system. It is very important that CO evolved from the reaction is allowed to escape freely. ■ **Caution.** *W(CO)$_6$ and CO are toxic by inhalation. An efficient fume hood must be used.* The solution is initially heated to *mild* reflux so that the condensing solvent is just below the condenser [in order to avoid clogging of the condenser with W(CO)$_6$]. After ~ 1 day, or when most of the W(CO)$_6$ has reacted, the reflux rate is increased. The CO groups are displaced stepwise and all three W(CO)$_x$(NCEt)$_{6-x}$ ($x = 1$–3) species are present in solution midway through the reaction. After a few days, the reaction should be monitored by solution IR of an aliquot diluted with propionitrile by ~ 10:1. The complex W(CO)$_3$(NCEt)$_3$ in NCEt has strong CO bands at 1909 and 1790 cm^{-1}, whereas those for the bis species are at 2021, 1898, and 1840 cm^{-1}. When the latter nearly disappear (~ 6 days or less depending on reaction conditions such as reflux temperatures, which are lower at higher altitudes, and escape of CO from the reaction), the volume of the now red solution is reduced to 60–80 mL *in vacuo*. Precipitation of fine light yellow needles of W(CO)$_3$(NCEt)$_3$ suddenly occurs during this process. The precipitation is completed by addition of diethyl ether (150 mL), and the voluminous product is collected on a frit, washed with diethyl ether (2 × 40 mL), and dried *in vacuo* (~ 5 min, extended pumping may give decomposition). Yield: 25 g (81%). The complex slowly decomposes in air and is sometimes tinged with green because of minor surface oxidation.

Anal. Calcd. for $C_{12}H_{15}N_3O_3W$: C, 33.3; H, 3.5; N, 9.7. Found: C, 32.5; H, 3.3; N, 9.4.

    Heptane (670 mL), 1,4-dioxane (40 mL), and cycloheptatriene (60 mL) are placed into a 1-L Schlenk flask and degassed as above. ■ **Caution.** *1,4-Dioxane is harmful if inhaled or absorbed through the skin. Cycloheptatriene has a very strong odor and is harmful if inhaled. Wear gloves and work in a well-ventilated hood.* Finely divided W(CO)$_3$(NCEt)$_3$ (19.2 g, 44.3 mmol) is added while the suspension is vigorously stirred magnetically, and the mixture is heated to reflux under N_2 for 16 h. [*Note*: The dioxane increases the reaction rate and yield, but can be omitted if it is unavailable (the reflux time should then be increased to 63 h)]. The solution becomes deep red and is cooled to ~ 40 °C and decanted from a fine brown precipitate [the last ~ 50 mL is filtered (can be slow) through a medium frit; the presence of small amounts of precipitate in the filtrate is not detrimental]. ■ **Caution.** *When dry, the precipitate as well as the unsublimed residue from the sublimation below decompose in air with considerable heat evolution and smoking.* The filtrate is rotoevaporated to ~ 150 mL and cooled to − 20 °C for a few hours. The resulting deep red crystalline precipitate is collected on a coarse frit and dried in air. Any residues of product remaining in the flask or other equipment can readily be extracted (in air) with small amounts of solvent, for example, dichloromethane, and recovered by solvent removal. The complex contains impurities but can readily be purified by sublimation at 100 °C, using a dynamic vacuum (0.01 torr). The impurities occasionally interfere with the sublimation in that they disproportionate to $W(CO)_6$ on heating (the latter sublimes as colorless crystals). However, the $W(CO)_6$ sublimes first and can be separated from the ruby red crystals of $W(CO)_3(\eta^6$-cycloheptatriene). Also, allowing the crude product to stand overnight in air aids in decomposing the air-sensitive impurities to nonvolatile materials. The sublimation is slow because of buildup of fine solid on the surface of the material in the sublimer (a cheesecloth layer should be used to prevent this lightweight solid from flying up onto the probe). The residue in the sublimer must be checked carefully to determine if all the $W(CO)_3(\eta^6$-cycloheptatriene) has sublimed. Normally the sublimation is complete within 1–2 days. Yield: 12.22 g (73%).

Properties

The compound $W(CO)_3(\eta^6$-cycloheptatriene) is an air-stable, ruby red solid that is soluble in most organic solvents. The complex should be kept in a freezer for long-term storage (months). It displays strong, sharp, carbonyl IR bands at 1991, 1924, and 1898 cm^{-1} in cyclohexane solution. Other properties and uses in synthesis have been previously described.[5,8]

C. TRICARBONYL(DIHYDROGEN)BIS(TRICYCLOHEXYL-PHOSPHINE)TUNGSTEN

$$W(CO)_3(\eta^6\text{-cycloheptatriene}) + 2Pcy_3 \xrightarrow{\text{H}_2} W(CO)_3(Pcy_3)_2(H_2) + C_7H_8$$

Procedure

The synthesis follows that for the Mo analog in Section A, using 2.50 g of $W(CO)_3(\eta^6$-cycloheptatriene). The reaction time is much shorter (30 min for the initial reaction and 30 min to complete precipitation). Yield: 5.06 g (88%).

Anal. Calcd. for $C_{39}H_{68}O_3P_2W$: C, 56.4; H, 8.3; P, 7.5. Found: C, 56.9; H, 8.5; P, 7.5.

Properties

The properties of $W(CO)_3(Pcy_3)_2(H_2)$ are very similar to those of the Mo analog, except that it is somewhat more stable to H_2 loss and to atmospheric oxidation. The dissociation pressure of hydrogen over the solid complex is ~ 1 torr. Nujol mull IR spectra show bands resulting from coordinated H_2 at $2690\,cm^{-1}$ (w, br, ν_{HH}), $1575\,cm^{-1}$ (w, br, ν_{WH}), and $953\,cm^{-1}$ (m, ν_{WH}) and ν_{CO} at 1963 and $1843\,cm^{-1}$.

D. TRICARBONYL(DIHYDROGEN)BIS(TRIISOPROPYLPHOS-PHINE)TUNGSTEN

$$W(CO)_3(\eta^6\text{-cycloheptatriene}) + 2P(i\text{-Pr})_3$$

$$\xrightarrow{H_2} W(CO)_3[P(i\text{-Pr})_3]_2(H_2) + C_7H_8$$

Procedure

The reaction conditions are similar to those for the Pcy_3 complexes. A mixture of 4.757 g (13.21 mmol) of $W(CO)_3(\eta^6\text{-cycloheptatriene})$ (powder or microcrystalline particle size), 5.5 mL (4.5 g, 28 mmol) of $P(i\text{-Pr})_3$, and 6 mL of hexane is stirred for 4 h in a 25-mL Schlenk flask. A yellow microcrystalline precipitate forms (may appear to look orange until washed), with the solution phase usually retaining a reddish color. The flask is cooled to $-20\,°C$ in a freezer (or ice–HCl bath) for several hours, and the product is collected on a frit. Because of the high solubility of the product, a somewhat better yield is obtained if the frit is cooled (e.g., by wrapping with a material capable of absorbing liquid N_2) during filtration and washing. The solid is washed with cold H_2-saturated hexane ($2 \times 4\,mL$) until the reddish color of the mother liquor has been removed. Drying, transfer, and storage of the complex *must* be done under an H_2-enriched atmosphere in the absence of nitrogen (see synthesis of the $Mo(CO)_3(Pcy_3)_2(H_2)$ analog). Yield: 4.6 g (59%).

Properties

Unlike the Pcy_3 complexes, the i-Pr complex is very soluble, even in hexane. It is less stable to H_2 loss (dissociation pressure is ~ 10 torr) and air, rapidly reacting with N_2 to form sparingly soluble red-orange $[W(CO)_3[P(i\text{-}Pr)_3]_2]_2(\mu\text{-}N_2)$.[3] The latter and $W(CO)_4[P(i\text{-Pr})_3]_2$ [see properties of $Mo(CO)_3(Pcy_3)_2(H_2)$] are often minor impurities in the H_2 complex, showing IR bands at 1948 and $1870\,cm^{-1}$, respectively. The complex $W(CO)_3(P\text{-}i\text{-}Pr_3)_2(H_2)$ displays IR bands (Nujol mull) similar to those for the Pcy_3 analog at 2695, 1965, 1852, 1567, and $953\,cm^{-1}$. Because of the partial solubility of the complex in Nujol, a band at $1993\,cm^{-1}$ due to the solution equilibrium dihydride species (see above) is usually visible also. The hexane solution IR shows ν_{CO} due to the H_2 complex at 1969 and 1856 cm^{-1} and somewhat weaker ν_{CO} due to the dihydride form at 1993, 1913, 1867, and 1828 cm^{-1}. Thus up to eight carbonyl bands can be seen, but all are accountable (see published spectrum).[1c] Proton NMR (200 MHz, 25 °C, methylcyclohexane-d_{14}) shows a broad singlet resulting from the H_2 ligand near -4.5 ppm and a field- and temperature-dependent weaker signal resulting from the equilibrium dihydride at -3.7 ppm (triplet, $J_{PH} = 38$ Hz) with intensity ratio $\sim 4:1$.

References

1. (a) G. J. Kubas, R. R. Ryan, B. I. Swanson, P. J. Vergamini, and H. J. Wasserman, *J. Am. Chem. Soc.*, **106**, 451 (1984). (b) G. J. Kubas, R. R. Ryan, and D. Wrobleski, *J. Am. Chem. Soc.*, **108**, 1339 (1986). (c) G. J. Kubas, C. J. Unkefer, B. I. Swanson, and E. Fukushima, *J. Am. Chem. Soc.*, **108**, 7000 (1986). (d) G. J. Kubas, *Acc. Chem. Res.*, **21**, 120 (1988).

2. R. H. Crabtree and D. G. Hamilton, *J. Am. Chem. Soc.*, **108**, 3124 (1986).

3. (a) H. J. Wasserman, G. J. Kubas, and R. R. Ryan, *J. Am. Chem. Soc.*, **108**, 2294 (1986). (b) G. J. Kubas, *Organometallic Syntheses*, Vol. 3, R. B. King and J. J. Eisch (*eds.*), Elsevier, New York, 1986, p. 254.

4. A. A. Gonzalez, S. L. Mukerjee, S.-J. Chou, Z. Kai, and C. D. Hoff, *J. Am. Chem. Soc.*, **110**, 4419 (1989).

5. G. J. Kubas, *Inorg. Chem.*, **22**, 692 (1983).

6. K. A. Kubat-Martin, G. J. Kubas, G. R. K. Khalsa, L. S. Stepan Van Der Sluys, and R. R. Ryan, submitted.

7. D. F. Shriver, *The Manipulation of Air-Sensitive Compounds*, 2nd Ed., Wiley-Interscience, New York, 1986.

8. R. B. King and A. Fronzaglia, *Inorg. Chem.*, **5**, 1837 (1966).

2. MOLYBDENUM AND TUNGSTEN PHOSPHINE POLYHYDRIDES

Submitted by GREGORY G. HLATKY[†§] and ROBERT H. CRABTREE[*§]
Checked by KIMBERLY A. KUBAT-MARTIN[‡] and G. J. KUBAS[‡]

Transition metal polyhydrides have several features that are currently attracting considerable interest.[1] For example, some examples have been found to have hydrides bound in a nonclassical η^2-H_2 fashion, *e.g.*,

*Department of Chemistry, Yale University, 225 Prospect Street, New Haven, CT 06520.
†Polymers Technology Division, Exxon Chemical Co., P.O. Box 5200, Baytown, TX 77522.
‡Los Alamos National Laboratory, Los Alamos, NM 87545.
§The submitters thank the NSF for support.

$[IrH_2(H_2)\{P(C_6H_{11})_3\}_2]^+$ (ref. 2) and $FeH_2(H_2)(PEtPh_2)_3$ (ref. 3). Other polyhydrides such as $IrH_5\{P(C_6H_{11})_3\}_2$ are classical,[4] having only terminal M—H bonds. The routes to the molybdenum and tungsten phosphine polyhydrides described here[5] have certain advantages over published syntheses in being more direct, or giving higher yields of product. The use of the soluble hydride donors $Li[BEt_3H]$ and $Na[AlH_2(OCH_2CH_2OCH_3)_2]$ has advantages over commonly used $Na[BH_4]$ and $Li[AlH_4]$. With the aluminum compound, an aqueous NaOH work-up is used to avoid the formation of a gelatinous precipitate. The tungsten tetrahydride is formed in a single step from WCl_6, avoiding the two-step synthesis of $WCl_4(PR_3)_2$.[6]

■ **Caution.** *Alkyl phosphines are foul-smelling and toxic. Hydrogen gas is evolved in the hydrolysis step. All reactions should be carried out in a well-ventilated fume hood. Benzene is a suspected human carcinogen. It should be used only in a well-ventilated fume hood, and protective gloves should be worn. Tetrahydrofuran (THF = solvent, thf = ligand) forms explosive peroxides: only fresh, peroxide-free material should be used. The compound $Li[BEt_3H]$ (Super Hydride, 1.0 M in THF) and $Na[AlH_2(OCH_2CH_2-OCH_3)_2]$ (3.4 M solution in toluene) can ignite upon contact with water, alcohols, or air. This reagent should be handled only under an inert atmosphere.*

A. TETRAHYDRIDOTETRAKIS(METHYLDIPHENYL-PHOSPHINE)MOLYBDENUM(IV)

$$MoCl_4(C_4H_8O)_2 + 4PMePh_2 + 4Li[BEt_3H]$$
$$\longrightarrow MoH_4(PMePh_2)_4 + 4LiCl + 4BEt_3 + 2C_4H_8O$$

Procedure

All solvents in the preparations in Sections A to D are dried by distillation over $CaH_2(CH_2Cl_2)$ or $Na/Ph_2CO(thf)$ or Mg (MeOH, EtOH) and degassed before use by purging with argon for 5 min. A 250-mL three-necked round-bottomed flask is equipped with a magnetic stirring bar and reflux condenser connected to an argon bubbler (a nitrogen atmosphere can also be used, but the yields are poorer). One neck of the three-necked flask contains the condenser, or contains the argon inlet, and the other is stopped. After purging with dry argon, $MoCl_4(thf)_2$ (1.91 g, 5 mmol)[62] is added to the flask, followed by THF (50 mL, freshly distilled from the deep purple sodium benzophenone and purged with argon) and $PMePh_2$ (5 mL, 25 mmol). Addition of the phosphine causes a color change from brownish-orange to red. The mixture is heated at reflux for 30 min. On cooling at 20 °C, an orange-red solid is deposited. A solution of $Li[BEt_3H]$ (1 M in THF; 65 mL,

65 mmol) (Aldrich) is added slowly with a syringe over 10 min. The deep red homogeneous mixture is stirred at room temperature for 12 h.

The flask is then cooled in an ice bath, and 10 mL of absolute ethanol is carefully added dropwise by syringe over a period of 5 min. A vigorous effervescence of hydrogen and evolution of heat take place. The solvent is evaporated *in vacuo* to give a deep red oil. Addition of methanol (100 mL) and stirring precipitate an orange-yellow solid, which is filtered using a Schlenk tube fitted with a sintered glass frit under argon, and the filtered precipitate is washed with methanol (3 × 10 mL). Recrystallization from a minimum amount of benzene by slow addition of methanol gives 3.0 g (67%) of yellow powder.

Anal. Calcd. for $C_{52}H_{56}P_4Mo$: C, 69.33; H, 6.27. Found: C, 69.49; H, 6.30.

Properties

The bright yellow solid can be handled in air without decomposition, but is somewhat photosensitive. It is best stored in an inert atmosphere in the dark. Solutions are much more air sensitive. The compound is soluble in aromatic solvents, THF, dichloromethane (with decomposition), slightly soluble in diethyl ether, and insoluble in methanol and hexane. The solid state structure shows a dodecahedral arrangement of the eight ligands.[7] In C_6D_6 at 20 °C a binomial quintet is observed at $\delta - 2.05$ ($^2J_{PH} = 33$ Hz). The long T_1 for the Mo—H protons (165 ms at 250 K, toluene-d_8 at 250 MHz) confirms that a classical structure is retained in solution.[8] The dynamic behavior of the complex has been fully studied by NMR.[6d] The compound $MoH_4(PMePh_2)_4$ reacts with $H[BF_4]$ in THF to give the unusual fluoride-bridged complex $[\{MoH_2(PMePh_2)_3\}_2(\mu\text{-}F)_3][BF_4]$.[5c]

Analogous Complexes

The 1, 2-bis(diphenylphosphino)ethane complex can be prepared in an almost identical way, but it is recrystallized from a minimum amout of hot benzene by precipitation with ethanol in 50% yield.

B.   TETRAHYDRIDOTETRAKIS(METHYLDIPHENYL-PHOSPHINE)TUNGSTEN(IV)

$$WCl_6 + 4PMePh_2 + 6Li[BEt_3H]$$

$$\longrightarrow WH_4(PMePh_2)_4 + 6LiCl + H_2 + 6BEt_3$$

Procedure

To a three-necked round-bottomed flask, equipped as described earlier, are added dry, argon-saturated THF (40 mL) and $PMePh_2$ (4 mL, 20 mmol), followed by WCl_6 (1.6 g, 4 mmol). The mixtue is heated at reflux under argon for 1 h. The stirred mixture is cooled to room temperature, and $Li[BEt_3H]$ (50 mL of a 1 M solution in THF, 50 mmol) is added by syringe over 10 min to give a deep red homogeneous solution. After stirring at 20 °C for 12 h, ethanol (10 mL) is *cautiously* added with cooling as described previously. The solvents are removed *in vacuo* to give an orange-red oil. The crude product is precipitated by slow addition of ethanol (100 mL) by syringe with stirring. Recrystallization form a minimum amount of benzene and ethanol gives 2.78 g (70%) of a yellow microcrystalline product.

Anal. Calcd. for $C_{52}H_{56}P_4W$: C, 63.16; H, 5.70. Found: C, 63.01; H, 5.81.

Properties

The physical and solubility properties of the tungsten complex are similar to those of the molybdenum analog, except that it is less photosensitive. The ^1H NMR spectrum at room temperature (toluene-d_8) shows this complex to be stereochemically rigid, but at 60 °C, a binomial quintet is observed at $\delta - 1.72$ for the W—H protons ($^2J_{PH} = 31$ Hz).

Noncoordinating acids react with this complex in THF to give the unusual pentahydrido cation $[WH_5(PMePh_2)_4]^+$.[9] T_1 data (179 ms at 240 K in CD_2Cl_2 at 250 MHz) suggest that this complex has a classical formulation.[8]

C. HEXAHYDRIDOTRIS(DIMETHYLPHENYLPHOSPINE)-TUNGSTEN(VI)

$$WCl_4(PMe_2Ph)_3 + 4Na[AlH_2(OCH_2CH_2OCH_3)_2] + 12H_2O + 4NaOH$$

$$\longrightarrow WH_6(PMe_2Ph)_3 + 4NaCl + 4Na[Al(OH)_4]$$

$$+ 8HOCH_2CH_2OCH_3 + 3H_2$$

Procedure

A 250-mL round-bottomed flask is equipped as above, but an argon inlet connected to a bubbler is substituted for the reflux condenser. Dry THF (50 mL) and $Na[AlH_2(OCH_2CH_2OCH_3)_2]$ (10 mL of a 3.4 M solution in toluene, 34 mmol) (Aldrich) are added. The solution is cooled using a CO_2–acetone or CO_2–2-propanol bath. The bath is at $\sim - 78$ °C (during the reaction, the contents of the flask are probably warmer than this). Solid $WCl_4(PMe_2Ph)_3$ (2.16 g, 2.92 mmol)[10],* is slowly added to the stirred reaction mixture over 20 min. The mixture is allowed to warm to room temperature and is stirred overnight, turning from deep red to pale yellow.[†] The solvent is evaporated *in vacuo* to give a yellow oil, to which benzene (50 mL) is added. To this mixture degassed 10% aqueous NaOH solution (10 mL) is *carefully* added dropwise. After hydrolysis, the benzene layer is decanted by a cannula to an argon-purged 100-mL Schlenk tube, dried with anhydrous $MgSO_4$, and filtered. The solvent is evaporated *in vacuo* and diethyl ether (20 mL), and then hexanes (50 mL) are added. The precipitated crude product is removed by filtering on a Schlenk frit, washed with hexanes, and dried *in vacuo*. The grayish crude product is purified by dissolution in diethyl ether (100 mL) to which anhydrous $MgSO_4$ (1 g) is added (this decolorizes the mixture and may remove excess phosphine). After stirring for 30 min, the mixture is filtered using a Schlenk frit, and the pale yellow solution is slowly evaporated *in vacuo* until the onset of precipitation at room temperature. Cooling to $-20\,°C$ overnight yields colorless needles. These are quickly filtered using a coarse porosity Schlenk frit, washed with small quantities of cold ($-20\,°C$) diethyl ether, and vacuum dried. Yield: 0.92 g (53%) of pure product.

Anal. Calcd. for $C_{24}H_{39}P_3W$: C, 47.70; H, 6.50. Found: C, 47.50; H, 6.62.

Properties

The complex can be handled in air without decomposition, but it is best stored in an inert atmosphere. It is soluble in aromatic solvents, THF, and

---

*This complex is best prepared by the method of Chatt et al.[10] as follows. A mixture of WCl_6 (2.5 g) and amalgamated zinc (prepared by placing a single 5-g piece of Zn in 10 mL of 0.1 M mercuric acetate solution for 5 min, then removing, washing and drying the piece), in dry, degassed dichloromethane (50 mL) is stirred under an N_2 or Ar atmosphere for 3 min. A (5-g) sample of PPh_3 is then added and stirring (or shaking if necessary) is continued for a further 3 min. The resulting yellow-orange precipitate of $WCl_4(PPh_3)_2$ is filtered on a Schlenk frit under N_2 or Ar, washed with acetone (5 mL), then with ether (3×5 mL), and dried *in vacuo*. The piece of zinc can be removed with tweezers, or if its shape has been chosen with care, it will not be able to pass into the fritted Schlenk vessel, and so will not be present in the product. To the $WCl_4(PPh_3)_2$ formed above (2.1 g, 40%), in dry, degassed dichloromethane (50 mL) is added dimethylphenylphosphine (2.25 g), and the mixture is heated at reflux under N_2 or Ar for 10 min. The resulting burgundy colored solution is concentrated to one-third volume and red-brown oil is precipitated with 10 mL of dry, degassed hexanes. The supernatant liquor is then decanted under N_2 or Ar, and the volume reduced to 10 mL *in vacuo*. The purple-red product then crystallizes (1.5 g) at 0 °C over 24 h.

[†]The checkers did not observe this color change until the NaOH solution was added.

dichloromethane (with decomposition), moderately soluble in diethyl ether, and insoluble in cold aliphatic solvents. Its ^{1}H NMR spectrum in C_6D_6 shows hydride resonances at $\delta - 1.91$ ($^2J_{PH} = 36$ Hz) with satellites due to coupling to ^{183}W($^1J_{WH} = 27$ Hz). The relaxation time of 181 ms (235 K, toluene-d_8, 250 MHz) is consistent with a classical structure in solution.[8a] In the solid state a classical structure is also adopted as judged by X-ray and neutron diffraction studies of the P(i-Pr)$_2$Ph derivative.[8b]

D. HEXAHYDRIDOTRIS(TRICYCLOHEXYLPHOSPHINE)-MOLYBDENUM(VI)

$$MoCl_4(C_4H_8O)_2 + 3P(C_6H_{11})_3 + 4Na[AlH_2(OCH_2CH_2OCH_3)_2]$$
$$+ 12H_2O + 4NaOH \longrightarrow MoH_6(P(C_6H_{11})_3)_3 + 4NaCl$$
$$+ 4Na[Al(OH)_4] + 8HOCH_2CH_2OCH_3 + 3H_2$$

Procedure

To an argon-purged 100-mL Schlenk tube with a magnetic stirring bar are added $MoCl_4(thf)_2$ (0.57 g, 1.5 mmol),[6a] tricyclohexylphosphine (0.9 g, 3.3 mmol), and dry degassed THF (30 mL). The mixture is heated at reflux for 15 min, and the deep red solution is cooled first to room temperature, then with a CO_2–acetone bath. A solution of $Na[AlH_2(OCH_2CH_2OCH_3)_2]$ (3.4 M in toluene; 5 mL, 17 mmol) is added dropwise over 10 min to the stirring mixture. This mixture is allowed to warm to room temperature and stirred overnight. The volatile components are removed *in vacuo*, and diethyl ether (50 mL) is added. The mixture is hydrolyzed by careful addition of 10% aqueous NaOH over 10 min. The ether layer is decanted, dried with $MgSO_4$, and filtered. The solution is evaporated *in vacuo*, the residue is extracted with hexane (3 × 15 mL), and the combined extracts are evaporated *in vacuo* to give an orange-yellow oil (0.3 g). The solution contains the title polyhydride in excess phosphine. It has not been obtained in pure form.

Properties

The air-sensitive oil is freely soluble in all common organic solvents. Its ^{1}H NMR spectrum in C_6D_6 at 25 °C shows a quartet for the hydride resonances at $\delta - 4.42$ ($^2J_{PH} = 36.3$ Hz), indicating the presence of three phosphine ligands. The ^{1}H-decoupled ^{31}P NMR spectrum shows a singlet at 62.3 ppm (relative to external 85% H_3PO_4), which becomes a septet on selective decoupling of the phosphine ligands, proving that six hydride ligands are present.

References

1. (a) G. G. Hlatky and R. H. Crabtree, *Coord. Chem. Rev.*, **65**, 1 (1985); (b) R. H. Crabtree, *Chem. Rev.*, **85**, 245 (1985).
2. R. H. Crabtree and M. Lavin, *J. Chem. Soc. Chem. Commun.*, **1985**, 794 and 1661.
3. D. Hamilton and R. H. Crabtree, *J. Am. Chem. Soc.*, **108**, 3321 (1986).
4. H. J. Wasserman, R. R. Ryan, and G. J. Kubas, and J. J. Wasserman, *J. Am. Chem. Soc.*, **108**, 2294 (1986) and references therein.
5. (a) R. H. Crabtree and G. G. Hlatky, *Inorg. Chem.*, **21**, 1273 (1982); (b) R. H. Crabtree and G. G. Hlatky, *Inorg. Chem.*, **23**, 2388 (1984); (c) R. H. Crabtree, G. G. Hlatky, and E. M. Holt, *J. Am. Chem. Soc.*, **105**, 7302 (1983).
6. (a) J. R. Dilworth and R. L. Richards, *Inorg. Synth.*, **20**, 119 (1980); (b) F. Pennella, *J. Chem. Soc. Chem. Commun.*, 158 (1971); (c) B. Bell, J. Chatt, G. J. Leigh, and T. Ito, *J. Chem. Soc. Chem. Commun.*, **1972**, 34; (d) P. Meakin, L. J. Guggenber, W. G. Peet, E. L. Muetterties, and J. P. Jesson, *J. Am. Chem. Soc.*, **95**, 1467 (1973); (e) F. Pennella, *Inorg. Synth.*, **15**, 42 (1974).
7. L. J. Guggenberger, *Inorg. Chem.*, **12**, 2295 (1973).
8. (a) R. H. Crabtree and D. Hamilton, *J. Am. Chem. Soc.*, **110**, 4126 (1988); (b) D. Gregson, J. A. K. Howard, J. N. Nicholls, J. L. Spencer, and D. G. Turner, *J. Chem. Soc. Chem. Commun.*, **1980**, 572; (c) J. L. Spencer, personal communication, 1985.
9. E. Carmona-Guzman and G. Wilkinson, *J. Chem. Soc. Dalton Trans.*, **1977**, 1716.
10. J. Chatt, G. A. Heath, and R. L. Richards, *J. Chem. Soc. Dalton Trans.*, **1974**, 2074.

3. HEPTAHYDRIDOBIS(TRIPHENYLPHOSPHINE)-RHENIUM(VII) AND OCTAHYDRIDOTETRAKIS-(TRIPHENYLPHOSPHINE)DIRHENIUM(IV)

Submitted by C. J. CAMERON,* G. A. MOEHRING,* and R. A. WALTON*
Checked by K. G. CAULTON† and R. L. BANSEMER†

The series of complexes $ReH_7(PR_3)_2$ [where $PR_3 = PEt_2Ph$, $PEtPh_2$, PPh_3, $\frac{1}{2}Ph_2P(CH_2)_2PPh_2$, or $AsEt_2Ph$] were first prepared by the action of $Li[AlH_4]$ on various rhenium(IV) and rhenium(V) phosphine complexes.[1] This method is still the principal route to making this class of compounds, which have proven themselves to be important starting materials in the preparation of a wide variety of rhenium hydride complexes,[1-9] including the dinuclear complexes $Re_2(\mu\text{-}H)_4H_4(PR_3)_4$.[1] The thermal decomposition of $ReH_7(PR_3)_2$ into $Re_2(\mu\text{-}H)_4H_4(PR_3)_4$ is a general route to dirhenium phosphine species with this stoichiometry. An alternative preparation of the triphenylphosphine complex involves treatment of the quadruply bonded

*Department of Chemistry, Purdue University, West Lafayette, IN 47907.
†Department of Chemistry, Indiana University, Bloomington, IN 47405.

octachlorodirhenate(III) anion or $Re_2Cl_6(PPh_3)_2$ with $Na[BH_4]$ in the presence of added phosphine.[10]

Procedure

All reactions were carried out under an N_2 atmosphere and reaction solvents were deoxygenated by purging with $N_2(g)$ prior to use. Tetrahydrofuran (THF) (peroxide free) was dried over sodium benzophenone and distilled prior to use. Methanol (spectra analyzed from Fischer) and ethanol (200 proof) were used as received. These solvents were transferred to the reaction vessel with the use of a syringe. Solvents used in the work-up and purification steps were reagent grade and no special precautions were taken to purify them. Solvents used in washing the products were not deoxygenated.

A. HEPTAHYDRIDOBIS(TRIPHENYLPHOSPHINE)-RHENIUM(VII)*

$$ReOCl_3(PPh_3)_2 + Li[AlH_4] \longrightarrow ReH_7(PPh_3)_2$$

■ **Caution.** *Li[AlH$_4$] should be transferred with a glass or ceramic spatula only, and care must be taken when hydrolyzing Li[AlH$_4$] to avoid excessive frothing.*

A 300-mL three-necked round-bottomed flask, containing $ReOCl_3(PPh_3)_2$ (4.0 g, 4.8 mmol)[11,12] and a Teflon-coated stirring bar, is fitted with a rubber septum, a gas inlet tube, and a powder addition tube containing $Li[AlH_4]$ (1.4 g, 37 mmol). The flask is purged with N_2, and 30 mL of deoxygenated THF is added by syringe. The stirred suspension is cooled to 0 °C using an ice–water bath. The $Li[AlH_4]$ is added slowly over a period of 20 min. The mixture is allowed to warm to room temperature during a subsequent 2 h and 40 min period of stirring. The flask is again cooled to 0 °C, and the powder addition tube is removed. With a flush of nitrogen through the flask, a solution of 15 mL of THF and 5 mL of H_2O is added dropwise to hydrolyze any unreacted $Li[AlH_4]$. Following hydrolysis, the flask is stoppered, and the solvent is removed by distillation on a vacuum line. The dry residue to transferred to a medium porosity, 60-mL sintered glass filter. The residue is washed twice with 30-mL portions of H_2O, twice with 30-mL portions of ethanol, and five times with 30-mL portions of diethyl ether. The washings are discarded, and the

---

*In this and the following procedure, the checkers used solvents straight from reagent bottles, with no special drying or purification. All filtrations and washing of solid products were done in air. They obtained a yield of 1.85 g (54%) of $ReH_7(PPh_3)_2$, which was very pale gray, not white. Recrystallization from CH_2Cl_2–MeOH gave a snow white material.

residue is pumped dry on a vacuum line. The dry residue is extracted three times with 30-mL portions of THF into a 250-mL filter flask. Under aspirator pressure, the volume of THF is reduced to ~ 50 mL. Diethyl ether (100 mL) and pentane (100 mL) are added to induce precipitation. The white microcrystalline $ReH_7(PPh_3)_2$ is removed by filtering, washed with three 30-mL portions of diethyl ether, and dried under vacuum. Yield: 2.0 g (58%).

Anal. Calcd. for $C_{36}H_{37}P_2Re$: C, 60.23; H, 5.20. Found: C, 60.57; H, 5.20.

This product is of sufficient purity that it does not require recrystallization.

B. OCTAHYDRIDOTETRAKIS(TRIPHENYLPHOSPHINE)-DIRHENIUM(IV)

$$2ReH_7(PPh_3)_2 \xrightarrow{\Delta} Re_2(\mu\text{-H})_4H_4(PPh_3)_4 + 3H_2$$

A 25-mL three-necked round-bottomed flask containing a Teflon-coated stirring bar and $ReH_7(PPh_3)_2$ (0.30 g, 0.42 mmol) is fitted with a reflux condenser. The flask is purged with nitrogen. Deoxygenated peroxide-free THF (5 mL) and dexygenated methanol (10 mL) are added. The mixture is heated at reflux for 1 h and then is allowed to cool to room temperature. The bright red-orange insoluble $Re_2(\mu\text{-H})_4H_4(PPh_3)_4 \cdot THF$ is filtered, washed twice with 15-mL portions of ethanol, and dried under vacuum. Yield: 0.28 g (89%).

Anal. Calcd. for $C_{76}H_{76}OP_4Re$: C, 60.79; H, 5.10. Found: C, 60.83; H, 5.26.

The THF of crystallization was determined by IR and ^1H NMR spectroscopies. This product is of sufficient purity that it does not require recrystallization.

C. OCTAHYDRIDOTETRAKIS(TRIPHENYLPHOSPHINE)-DIRHENIUM(IV)

Alternative Procedure

$$Re_2Cl_6(PPh_3)_2 + 2PPh_3 + Na[BH_4] \xrightarrow{\Delta} Re_2(\mu\text{-H})_4H_4(PPh_3)_4$$

A 100-mL three-necked round-bottomed flask, containing $Re_2Cl_6(PPh_3)_2$*
(0.24 g, 0.22 mmol),[13] PPh_3 (0.14 g, 0.65 mmol), and a Teflon-coated stirring
bar, is equipped with a septum, a condenser, and a powder addition tube
containing $Na[BH_4]$ (0.14 g, 3.7 mmol). The flask is purged with nitrogen,
and 25 mL of deoxygenated ethanol is added. The suspension is stirred for
30 min before the addition of $Na[BH_4]$. Following this addition the mixture
is stirred for 15 min, and is then heated at reflux for a further 15 min. It is
filtered while still hot and the insoluble product is washed with three 15-mL
portions of methanol, one 15-mL portion of H_2O, and one additional 15-mL
portion of methanol. The bright red-orange ethanol solvate $Re_2(\mu\text{-}H)_4H_4(PPh_3)_4 \cdot C_2H_5OH$ is dried under vacuum. Yield: 0.28 g (88%).

Anal. Calcd. for $C_{74}H_{74}OP_4Re_2$: C, 60.23; H, 5.05. Found: C, 60.27; H, 5.05.
    The ethanol of crystallization is determined by IR and ^1H NMR spectroscopies. This product is of sufficient purity so as not to require
recrystallization.

Properties

The compound $ReH_7(PPh_3)_2$ in an air-stable, white, microcrystalline solid
that is soluble in dichloromethane, benzene, and THF. The ^1H NMR
spectrum provides for easy identification (triplet $J_{PH} = 18.9$ Hz at $\delta - 4.9$ in
CD_2Cl_2), and a convenient check for contamination by $ReH_5(PPh_3)_3$ (quartet
$J_{PH} = 18.6$ Hz at $\delta - 5.2$ in CD_2Cl_2) or $Re_2(\mu\text{-}H)_4H_4(PPh_3)_4$ (quintet $J_{PH} =$
9.5 Hz at $\delta - 5.6$ in CD_2Cl_2). Impurities can be removed by thorough washing
with diethyl ether or by recrystallization from dichloromethane–methanol.
    The compound $Re_2(\mu\text{-}H)_4H_4(PPh_3)_4$ forms dark red crystals and bright
red-orange powders; it is slightly soluble in diethyl ether, THF, benzene,
carbon disulfide, and dichloromethane. The compound is most easily
characterized by ^1H NMR spectroscopy (see above) or by cyclic voltammetry.
The cyclic voltammogram in dichloromethane with 0.1 M tetrabutylammonium hexafluorophosphate as supporting electrolyte consists of a reversible oxidation at -0.20 V versus an Ag–AgCl reference electrode and a
second irreversible oxidation at $+0.60$ V. (*Note*: The ferrocenium–ferrocene

---

*This compound is prepared by suspending $(Bu_4N)_2[Re_2Cl_8]$ (1.0 g, 0.88 mmol)[14] in 200 mL of
MeOH containing 5 mL of 6N HCl. Triphenylphosphine (0.54 g, 2.1 mmol) is added and the
suspension is stirred for 30 min. The olive green precipitate is filtered off, washed with 30-mL
portions of acetone, ethanol, benzene, and diethyl ether, and dried under vacuum. Yield: 0.89 g
(91%).
    ■ **Caution.** *Benzene is a human carcinogen. It should be handled only in
a well-ventilated hood, and protective gloves should be worn.*

couple occurs at + 0.47 V under identical conditions.) The complex is stable in air for several weeks and can be recrystallized by dissolving in carbon disulfide and filtering into 2-propanol.

Recently, $ReH_7(PPh_3)_2$ has been used as the starting material in a wide range of reactions with organic molecules.[2,4,15] The reaction chemistry of $Re_2(\mu\text{-}H)_4H_4(PPh_3)_4$ is quite varied. A fairly stable paramagnetic polyhydride complex, $[Re_2(\mu\text{-}H)_4H_4(PPh_3)_4][PF_6]$, can be formed by a one-electron oxidation, while it reacts with $(C_7H_7)[PF_6]$ in acetonitrile via hydride (H^-) abstraction to produce $[Re_2(\mu\text{-}H)_4H_3(PPh_3)_4(NCMe)][PF_6]$.[16]

References

1. J. Chatt and R. S. Coffey, *J. Chem. Soc. Ser. A*, **1969**, 1963.
2. W. D. Jones and J. A. Maguire, *J. Am. Chem. Soc.*, **107**, 4544 (1985).
3. J. D. Allison, G. A. Moehring, and R. A. Walton, *J. Chem. Soc. Dalton Trans.*, **1986**, 67.
4. D. Baudry, M. Ephritikhine, and H. Felkin, *J. Organomet. Chem.*, **224**, 363 (1982).
5. R. H. Crabtree, G. G. Hlatky, C. P. Parnell, B. E. Segmüller, and R. J. Uriarte, *Inorg. Chem.*, **23**, 354 (1984).
6. D. A. Roberts and G. L. Geoffroy, *J. Organomet. Chem.*, **214**, 221 (1981).
7. S. Muralidharan, G. Ferrudi, M. A. Green, and K. G. Caulton, *J. Organomet. Chem.*, **224**, 47 (1983).
8. N. G. Connelly, J. A. K. Howard, J. L. Spencer, and P. K. Woodley, *J. Chem. Soc. Dalton Trans.*, **1984**, 2003.
9. P. D. Boyle, B. J. Johnson, A. Buehler, and L. H. Pignolet, *Inorg. Chem.*, **25**, 5 (1986).
10. P. Brant and R. A. Walton, *Inorg. Chem.*, **17**, 2674 (1978).
11. N. P. Johnson, C. J. Lock, and G. Wilkinson, *Inorg. Synth.*, **9**, 145 (1967).
12. G. W. Parshall, *Inorg. Synth.*, **17**, 110 (1977).
13. F. A. Cotton, N. F. Curtis, and W. R. Robinson, *Inorg. Chem.*, **4**, 1696 (1965).
14. T. J. Barder and R. A. Walton, *Inorg. Synth.*, **23**, 116 (1985).
15. D. Baudry, P. Boydell, and M. Ephritikhine, *J. Chem. Soc. Dalton Trans.*, **1986**, 525.
16. J. D. Allison and R. A. Walton, *J. Am. Chem. Soc.*, **106**, 163 (1984).

4. TETRAHYDRIDO(η⁵-PENTAMETHYLCYCLOPENTA-DIENYL)IRIDIUM

Submitted by THOMAS M. GILBERT* and ROBERT G. BERGMAN*
Checked by JOSEPH S. MEROLA[†]

$$[\{Ir[C_5(CH_3)_5]\}_2(\mu\text{-}H)_3][PF_6] + 3Li[BEt_3H] + 2H^+$$

$$\longrightarrow 2[Ir\{C_5(CH_3)_5\}H_4] + LiPF_6 + 3BEt_3 + 2Li^+$$

Tetrahydrido(η⁵-pentamethylcyclopentadienyl)iridium was synthesized in 1982,[1] and proved to be a useful precursor for several iridium complexes of varied oxidation state.[1,2] The tetrahydride and its derivatives are among the very few iridium(V) polyhydrides that do not contain phosphine ligands.[3,4]

No alternative preparation of the title complex has been reported. The key feature of the synthesis described here is the oxidation of the iridium(III) intermediate to the iridium(V) product during work-up of the reaction mixture.[2]

Procedure

■ **Caution.** *Li[BEt₃H] (Super Hydride, 1.0 M in tetrahydrofuran (THF), Aldrich) can ignite upon contact with water, alcohols, or air. This reagent should be handled only under an inert atmosphere.*[5]

All solvents must be appropriately dried and deoxygenated: Toluene and hexane are distilled under nitrogen from sodium benzophenone; methanol is vacuum transferred from magnesium. The procedures described must be performed under an inert atmosphere, either argon or nitrogen. All vacuum evaporations should be performed with an oil pump capable of reaching 0.1 torr; the trap should be cooled with liquid nitrogen.

The Li[BEt₃H] solution (15 mmol in hydride) is prepared by introducing 15 mL of the commercial reagent into a 50-mL one-necked side-arm flask through syringe techniques, then evaporating the THF *in vacuo* at room

*Department of Chemistry, University of California, Berkeley, and Materials and Molecular Research Division, Lawrence Berkeley Laboratory, Berkeley, CA 94720. This work was carried out under the auspices of a collaborative Lawrence Berkeley Laboratory/Industrial Research project supported jointly by the Chevron Research Company, Richmond, CA, and the Director, Office of Energy Research, Office of Basic Energy Sciences, Chemical Sciences Division of the U.S. Department of Energy under Contract No. DE-AC03-76SF00098. We are grateful to Johnson Matthey, Inc., for a generous loan of iridium trichloride.

†Corporate Research Science Laboratories, Exxon Research and Engineering Company, Annandale, NJ 08801.

temperature until bubbling ceases and a viscous fluid remains (usually 2–5 mL). The concentrate is diluted with toluene (15–20 mL), and the solution is stirred for a short period until it is homogeneous. The solvent is evaporated a second time. The resulting fluid is again diluted with toluene to a volume of 20 mL, and stirred as above. The resulting solution contains a small quantity of THF, which may assist the reaction by solubilizing the starting materials and various intermediates.*

Air-stable $[\{Ir[C_5(CH_3)_5]\}_2(\mu\text{-H})_3][PF_6]$ is prepared from $\{Ir[C_5(CH_3)_5]Cl\}_2(\mu\text{-Cl})_2$ and excess hydrogen gas in 2-propanol–acetone–water by the literature method.[3c,6] This dimer (2.01 g, 2.50 mmol)[†] is slurried in 200 mL of hexane in a 500-mL one-necked side-arm flask.

At this point, if a glove box with a freezer is available, the flask is cooled in the freezer to approximately $-40\,^\circ$C and removed, and the slurry is treated dropwise with the $Li[BEt_3H]$ solution. The reaction mixture darkens considerably during addition of the hydride reagent. After complete addition, the solution is allowed to warm to ambient temperature and stirred an additional 3 h.

The deep orange reaction mixture is cooled in the glove box freezer, then filtered through a short column (2–3 cm high) of deoxygenated, hexane-wetted alumina(III) packed in a 60-mL medium porosity frit. The excess $Li[BEt_3H]$ blackens the top of the column; bubbling is often observed. The column is washed with benzene until the washings are colorless (usually ~ 100 mL), and the filtrate and washings are combined to give a pale yellow solution. Vacuum evaporation of the solvent with minimal warming (water bath held near ambient temperature) yields a yellow-white residue.

Alternatively,[‡] if no glove box is available, the 500-ml flask is capped with a 25-mL addition funnel containing the $Li[BEt_3H]$ solution, which is itself capped with a gas inlet adapter. The setup is opened to inert gas flow on a Schlenk line, after which the flask is cooled to $-40\,^\circ$C (Dry Ice–2-propanol) and the $Li[BEt_3H]$ solution added dropwise. After complete addition, the solution is allowed to warm to ambient temperature and stirred for an additional 3 h.[§] The reaction mixture is then recooled to $-40\,^\circ$C and treated

---

*The checker reports that he first attempted to use a THF-free toluene solution of $Li[BEt_3H]$ prepared from LiH and BEt_3. This failed to yield any of the desired product. He suggests that a small amount of residual THF, which remains when the solution is prepared according to the authors' directions, is essential for this reaction. The authors have noted the presence of THF coordinated to the lithium atom of a key intermediate.[2]

†Although the stoichiometry of the reaction requires three equivalents of hydride reagent per equivalent of dimer, excess reagent appears to cause more rapid reaction and give higher yields of product.

‡The checker followed this procedure.

§After adding the $Li[BEt_3H]$ solution, the checker stored the slution overnight at $-30\,^\circ$C. He did not observe darkening of the mixture until it was allowed to begin to warm.

dropwise with excess methanol (50 mL). The solution is then warmed slowly to room temperature and the solvent evaporated *in vacuo* with minimal warming (water bath held near ambient temperature) to give a dark yellow residue.

The residue from either procedure is dissolved in pentane or hexane (10–20 mL) for transfer into a sublimation apparatus of 1–5-g product capacity. Vacuum evaporation* of the solvent, again with minimal heating, followed by passive vacuum sublimation (30–40 °C, 20 mtorr) of the residue to a water-cooled cold finger gives the pure white tetrahydride. Yield: 1.23 g (74%).[†]

Properties

The compound $[Ir\{C_5(CH_3)_5\}H_4]$ is quite sensitive to oxygen in solution, affording dark solutions in minutes at room temperature in air. The solid material is less sensitive, decomposing to amorphous brown material over the course of a few hours at room temperature. The tetrahydride is also thermally sensitive both in the solid state and in solution. As a solid, it decomposes under nitrogen to a dark solid, slowly at 50 °C and rapidly at 100 °C. The effect of strong or excessively rapid heating of the complex has not been explored. Decomposition in solution appears slower. Because of the thermal instability and the high vapor pressure of the tetrahydride, excessive heating of solutions containing the material should be avoided; excessive heating should also be avoided during sublimation of the product. The authors routinely store the pure complex at −40 °C in a nitrogen-filled glove box, but noted no decomposition (as determined by development of a dark coloration) of samples after 2 weeks at ambient temperature in the glove box.

The compound $[Ir\{C_5(CH_3)_5\}H_4]$ dissolves in all the organic solvents we have tested, including methanol, diethyl ether, benzene, and pentane. It is stable to pure dichloromethane but reacts with chloroform and carbon tetrachloride.[1] The complex is photosensitive; however, we have encountered no difficulty working in ordinary room light.

^{1}H NMR (C_6D_6): $\delta = 1.99$ [s, $C_5(CH_3)_5$]; −15.44 (s, Ir—H). IR (KBr Pellet): 2150 cm⁻¹ $(v_{Ir—H})$.

Anal. Calcd. for $C_{10}H_{19}Ir$: C, 36.24; H, 5.78. Found: C, 35.98; H, 5.83.

Other spectroscopic data have been reported.[2]

---

*The tetrahydride is fairly volatile; therefore, dynamic vacuum should be minimized during evaporation of solvent and during sublimation.
[†]Occasionally two sublimations are required, depending on the available surface area of the cold finger.

References

1. T. M. Gilbert and R. G. Bergman, *Organometallics*, **2**, 1458 (1982).
2. T. M. Gilbert, F. J. Hollander, and R. G. Bergman, *J. Am. Chem. Soc.*, **107**, 3508 (1985).
3. (a) A. P. Borisov, V. D. Makhaev, and K. N. Semenenko, *Sov. J. Coord. Chem.*, **6**, 549 (1980). (b) L. F. Rhodes and K. G. Caulton, *J. Am. Chem. Soc.*, **107**, 259 (1985). (c) T. M. Gilbert and R. G. Bergman, *J. Am. Chem. Soc.*, **107**, 3502 (1985). (d) H. Werner, J. Wolf, and A. Hohn, *J. Organomet. Chem.*, **287**, 395 (1985).
4. For examples of other Ir(V) complexes, see (a) M.-J. Fernandez and P. M. Maitlis, *Organometallics*, **2**, 164 (1983). (b) M.-J. Fernandez and P. M. Maitlis, *J. Chem. Soc. Dalton Trans.*, **1984**, 2063 and references therein. (c) J. S. Ricci, Jr., T. F. Koetzle, M.-J. Fernandez, P. M. Maitlis, and J. C. Green, *J. Organomet. Chem.*, **299**, 383 (1986).
5. Proper procedures for handling air-sensitive reagents are described in: (a) H. C. Brown, *Organic Syntheses via Boranes*, Wiley, New York, 1975; (b) D. F. Shriver, *Manipulation of Air-Sensitive Compounds*, McGraw-Hill, New York, 1969; (c) *Handling Air-Sensitive Reagents*, Technical Information Bulletin AL-134, available from Aldrich Chemical Company, Milwaukee, WI.
6. C. White, A. J. Oliver, and P. M. Maitlis, *J. Chem. Soc. Dalton Trans.*, **1973**, 1901.

5.   TRIS[1, 3-BIS(DIPHENYLPHOSPHINO)PROPANE]-HEPTAHYDRIDOTRIIRIDIUM(2 +) BIS(TETRAFLUOROBORATE)* AND BIS[1, 3-BIS-(DIPHENYLPHOSPHINO)PROPANE]-PENTAHYDRIDODIIRIDIUM(1 +) TETRAFLUOROBORATE[†]

$$3[Ir(dppp)(cod)]^+ + 10H_2 \longrightarrow [Ir_3(dppp)_3H_7]^{2+} + H^+ + 3C_8H_{16}\,^{[††]}$$

$$2[Ir(dppp)(cod)]^+ + 7H_2 \longrightarrow [Ir_2(dppp)_2H_5)^+ + H^+ + 2C_8H_{16}\,^{[††]}$$

cod = 1, 5-cyclooctadiene   dppp = 1, 3-bis(diphenylphosphino)propane[‡]

Submitted by H. H. WANG, A. M. MUETING, J. A. CASALNUOVO, S. YAN, J. K.-H. BARTHELMES, and L. H. PIGNOLET[§]
Checked by MARTIN P. McGRATH and ROBERT H. CRABTREE[¶]

Two polyhydrido iridium clusters with chelating diphosphine ligands are prepared by the reaction of $[Ir(dppp)(cod)][BF_4]$ with hydrogen in a methanol solution. These compounds are formulated as $[Ir_3(dppp)_3H_7][BF_4]_2$, **1**, and $[Ir_2(dppp)_2H_5][BF_4]$, **2**. Their structures

---

* Heptahydridotris[1, 3-propanediylbis(diphenylphosphine)]triiridium(2 +) bis(tetrafluoroborate).
[†] Pentahydridobis[1, 3-propanediylbis(diphenylphosphine)]diiridium(1 +) tetrafluoroborate.
[††] The stoichiometry of these reactions has not been verified by experiment.
[‡] 1, 3-propanediylbis(diphenylphosphine).
[§] Department of Chemistry, University of Minnesota, Minneapolis, MN 55455.
[¶] Department of Chemistry, Yale University, New Haven CT 06520.

have been determined by single crystal X-ray diffraction.[1] Compound **1** contains an unusual tri-coordinate hydrido ligand within a triangle of iridium atoms. It also contains three doubly bridging and three terminal hydrido ligands. Compound **2** contains three doubly bridging and two terminal hydrido ligands. This reaction may be extended using other cationic iridium(I) cyclooctadiene diphosphine complexes. The molar distribution ratio of products **1** and **2** ($\sim$ 3:1, respectively, in the above reaction) varies depending on the steric constraints imposed by the chelate bite angles of different diphosphine ligands.[2] Formation of the trimeric species is maximized by using 1, 2-bis(diphenylphosphino)ethane, producing $[Ir_3(dppe)_3H_7][BF_4]_2$, **3**, whereas formation of the dimeric species is maximized by using 1, 4-bis(diphenylphosphino)butane, producing $[Ir_2(dppb)_2H_5][BF_4]$, **4**. These polyhydrido clusters are important because they represent a class of unusual hydrido bridged compounds and because they can serve as precursors for the synthesis of new clusters.

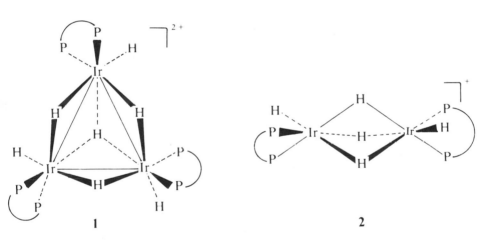

Procedure

A. (η-1, 5-CYCLOOCTADIENE) [1, 3-BIS(DIPHENYLPHOSPHINO)-PROPANE]IRIDIUM TETRAFLUOROBORATE

The starting compound $[Ir(dppp)(cod)][BF_4]$ is prepared by analogy to the synthesis of $[Rh(dppp)(nbd)][BF_4]$ (nbd = norbornadiene).[3] All manipulations are carried out under a purified N_2 atmosphere using standard Schlenk techniques. The compounds $Ag[BF_4]$ (430 mg, 2.2 mmol) and $[Ir(cod)Cl]_2$ (750 mg, 1.1 mmol)[4] are combined in a 100-mL, pear shaped, sidearm flask, to which 25 mL of acetone is added, immediately producing an orange solution and a white precipitate of AgCl. The slurry is heated at reflux for 30 min. Once cool, the slurry is transferred by means of a cannular tube to a fritted filter containing a 1-cm layer of diatomaceous earth and washed through with acetone. The ligand dppp (Strem, 920 mg, 2.2 mmol), dissolved in a minimum amount of toluene, is added slowly by means of a cannular tube to the orange filtrate, immediately turning it deep red.* A deep red precipitate is obtained upon removal of the acetone under vacuum. The toluene is removed by means of a cannular tube, and the precipitate is washed with toluene, followed by diethyl ether, and dried under vacuum. Recrystallization is accomplished in a narrow Shlenk tube by slow diffusion of a diethyl ether layer into the compound dissolved in a minimum amount of dichloromethane, producing deep red crystals in 82% yield, mp = 223 °C (dec). $^{31}P\{^{1}H\}$ NMR (CH_2Cl_2, 25 °C, relative to external H_3PO_4): δ 0.209 (s).

Anal. Calcd. for $IrP_2C_{35}H_{38}BF_4$: C, 52.57; H, 4.79. Found: C, 52.21; H, 4.72.

B. $[Ir_3(dppp)_3H_7][BF_4]_2$ AND $[Ir_2(dppp)_2H_5][BF_4]$

Compounds **1** and **2** are isolated as products of the same reaction. The manipulations are carried out under a H_2 or N_2 atmosphere. A suspension of $[Ir(dppp)(cod)][BF_4]$ (505 mg, 0.632 mmol) in 10 mL of methanol in a 100-mL, pear shaped, sidearm flask equipped with magnetic stir bar, is stirred vigorously under 1 atm of H_2. Within 5 to 10 min the solution becomes intensely yellow. The volume is reduced to < 5 mL under a slow stream of H_2 or N_2. Diethyl ether is added to precipitate a yellow-orange solid, which is then removed by filtering, washed with diethyl ether, and dried under vacuum. The product is dissolved in a minimum amount of dichloromethane in a narrow Schlenk tube and layered with about three times this volume of diethyl ether. Slow solvent diffusion yields rectangular yellow crystals of pure $[Ir_3(dppp)_3H_7][BF_4]_2 \cdot 1CH_2Cl_2$, **1**. Yield: (54%).

Anal. Calcd. for $Ir_3P_6C_{81}H_{85}B_2F_8 \cdot CH_2Cl_2$: C, 47.32; H, 4.31. Found: C, 47.17; H, 4.56.

Upon the addition of more diethyl ether to the mother liquor from which **1** is crystallized, platelike yellow crystals of $[Ir_2(dppp)_2H_5][BF_4]$, **2**, are formed. Yield: (16%).

---

*The checkers found it better to evaporate all the solvents under a vacuum at this stage. This gives a red solid, which was dissolved in CH_2Cl_2 (20 mL) and filtered through diatomaceous earth. The resulting solution is concentrated to ~ 10 mL and the crystalline product is precipitated with Et_2O (40 mL). This procedure removes some silver salts that would otherwise remain in the product, although the yield is slightly reduced (to 70%)

Anal. Calcd. for $Ir_2P_4C_{54}H_{57}BF_4$: C, 49.85; H, 4.42. Found: C, 49.66; H, 4.60.*

Properties

Pure $[Ir_3(dppp)_3H_7][BF_4]_2$, **1**, is obtained upon recrystallization from $CH_2Cl_2-Et_2O$. However, upon further recrystallization from the mother liquor, both **1** and **2** are obtained and $[Ir_2(dppp)_2H_5][BF_4]$, **2**, must be separated from **1** by physical means under the microscope. This is easily accomplished since the intensely yellow colored rectangular crystals of **1** lose solvent upon removal from the mother liquor and become opaque, whereas the pale yellow colored platelike crystals of **2** do not. Alternatively, compounds **1** and **2** may be separated by use of preparative HPLC with a Dynamax Macro C-18 column (Rainin Instrument Co., 21.4 mm id × 25 cm) and a UV detector (254 nm). Conditions: 70% ethanol, 10% acetonitrile, 20% 0.625 M Na[BF$_4$], degassed under helium, flow rate = 17.3 mL min^{-1}. Retention times: **1** = 2.9 min, **2** = 4.7 min. Purity of samples of the complexes may be checked utilizing a Dynamax C-18 Scout column (Rainin Instrument Co., 4.6 mm id × 25 cm). Conditions: 70% ethanol, 10% acetonitrile, 20% 0.125 M Na[BF$_4$], degassed under Helium, flow rate = 0.8 mL min^{-1}. Retention times: **1** = 3.0 min, **2** = 4.7 min.

Compounds **1** and **2** are readily soluble in dichloromethane, acetone, and acetonitrile. They are somewhat air sensitive in solution, but stable in the solid state. Compound **1** has a broad Ir—H stretching vibration at 2200 cm^{-1}, and **2** at 2140 cm^{-1}. ^{1}H NMR in the hydride region of **1** (acetone- d_6, 25 °C): $\delta - 7.98$ (d, $J_{transP-H} = 68$ Hz, int = 3, μ-H), $\delta - 8.83$ (q, $J_{transP-H} = 37$ Hz, int = 1, μ_3-H), $\delta - 15.0$ (br s, int = 3, terminal Ir—H). ^{1}H NMR in the hydride region of **2** (acetone-d$_6$, $- 45$ °C): $\delta - 6.89$ (d, $J_{transP-H} = 70$ Hz, int = 2, μ-H), $\delta - 7.95$ (t, $J_{transP-H} = 65$ Hz, int = 1, μ-H), $\delta - 20.56$ (m, $J \approx 19$ Hz, int = 2, terminal H). The complete analysis of these data has been published.[1]

Analogous Complexes

The compound $[Ir_3(dppe)_3H_7][BF_4]_2$, **3**, may be obtained by the same procedure using dppe as the ligand. However, upon standing overnight under a hydrogen atmosphere intensely colored yellow crystals of **3** fall out of the methanol solution. Recrystallization from $CH_2Cl_2-Et_2O$ affords pure crystals of **3**.

---

*The checkers never obtained pure **1** on crystallization, but only mixtures of **1** and **2**, and agglomeration made physical separation of the crystals difficult.

Anal. Calcd. for $Ir_3P_6C_{78}H_{79}B_2F_8 \cdot CH_3OH$: C, 47.81; H, 4.22. Found: C, 47.79; H, 4.10. IR: $v_{(Ir-H)} = 2168\,cm^{-1}$.

^1H NMR in the hydride region (CD_2Cl_2, 25 °C): $\delta - 6.26$ (d, $J = 72\,Hz$, int = 3), $\delta - 7.20$ (br, q, $J = 42\,Hz$, int = 1), $\delta - 15.66$ (m, int = 3).

The compound $[Ir_2(dppb)_2H_5][BF_4]$, **4**, is produced by following the same procedure using dppb. However, the product falls out of the methanol–diethyl ether solution after standing overnight at $- 10$ °C as yellow needlelike crystals. Yield: 78%.

Anal. Calcd. for $Ir_2P_4C_{56}H_{61}BF_4$: C, 50.60; H, 4.63. Found: C, 50.80; H, 4.75. IR: $v_{(Ir-H)} = 2164\,cm^{-1}$ (br), $2185\,cm^{-1}$ (sh).

^1H NMR in the hydride region (acetone-d_6, $- 60$ °C): $\delta - 6.92$ (d, $J = 76\,Hz$, int = 2), $\delta - 7.62$ (t, $J = 63\,Hz$, int = 1), $\delta - 21.77$ (m, $J \approx 22\,Hz$, int = 2).

References

1. H. H. Wang and L. H. Pignolet, *Inorg. Chem.*, **19**, 1470 (1980).
2. Hsien-Hau Wang, Ph.D. Thesis, University of Minnesota, MN, 1981.
3. R. R. Schrock and J. A. Osborn, *J. Am. Chem. Soc.*, **101**, 3141 (1979).
4. J. L. Herde, J. C. Lambert, and C. V. Senoff, *Inorg. Synth.*, **15**, 18 (1974).

6. HETEROBIMETALLIC HYDRIDE COMPLEXES

Submitted by E. G. LUNDQUIST and K. G. CAULTON*
Checked by J. L. SPENCER[†]

Heterobimetallic complexes have recently attracted considerable attention in light of the promise of enhanced reactivity as a result of the cooperativity between adjacent, but electronically different, metal centers.[1] A large number of these bimetallic compounds have been synthesized by the reactions of organometallic halides with anionic metal carbonyls. Here, we describe an extension of this route to the synthesis of hydride rich, Os—Zr and Os—Rh complexes by the reaction of organometallic halides with a metal polyhydride anion. These preparations demonstrate the synthetic utility of transition metal polyhydride anions.

*Department of Chemistry, Indiana University, Bloomington, IN 47405.
[†]Department of Chemistry, University of Salford, Salford, M5, 4WT, United Kingdom.

A.   mer-TRICHLOROTRIS(DIMETHYLPHENYLPHOSPHINE)-OSMIUM(III)*

$$OsO_4 + \text{excess } HCl + (\text{excess}) PMe_2Ph$$

$$\xrightarrow[\text{EtOH}]{80\,^\circ C} mer\text{-}OsCl_3(PMe_2Ph)_3 + 2OPMe_2Ph + 2H_2O$$

■   **Caution.** *Osmium tetraoxide forms highly toxic vapors that can cause severe eye burns, skin burns, acute lung congestion, and temporary blindness. A well-ventilated hood and limited handling time are recommended.*

Procedure

A 1-g vial of OsO_4 (Johnson Matthey) is scored and broken, using a glass rod, under the surface of 30 mL of ethanol and 2 mL of 12 M HCl contained in a 100-mL Erlenmeyer flask. The resulting yellow solution is filtered (to remove broken glass) into a 250-mL flask equipped with a magnetic stirring bar and reflux condenser. This solution is freeze–pump–thaw degassed twice, and against a N_2 flow, 3.5 mL of PMe_2Ph (Aldrich) is added. The resulting dark red-brown solution gradually lightens and is refluxed for 6 h at 80 °C. Refluxing for shorter periods of time, using less phosphine or less HCl will result in lower yields and the formation of $OsOCl_3(PMe_2Ph)_2$. Cooling the bright red solution and removing 10 mL of solvent under vacuum precipitate $OsCl_3(PMe_2Ph)_3$ as a crystalline, air-stable bright red material with physical properties matching those reported.[2,3] Once filtered, this material should be washed with ethanol (2 × 10 mL) to insure removal of any residual PMe_2Ph or $OPMe_2Ph$. Yields of > 90% can be consistently obtained following this procedure.

Reduction of $OsCl_3(PMe_2Ph)_3$ with $NaBH_4$ in ethanol, as described in the literature,[3] proceeds smoothly, giving good yields of $OsH_4(PMe_2Ph)_3$.

B.   BIS[1, 1(η⁵)-CYCLOPENTADIENYL]TRIS(DIMETHYL-PHENYLPHOSPHINE-2κP)-TRI-μ-HYDRIDO-1-HYDRIDOOSMIUMZIRCONIUM†

$$OsH_4(PMePh)_3 + (\text{excess})KH \xrightarrow[\text{THF}]{70\,^\circ C} K[OsH_3(PMe_2Ph)_3] + H_2$$

---

*Because of the absence of a detailed literature method for the preparation of $mer\text{-}OsCl_3(PMe_2Ph)_3$ from OsO_4, the following synthesis is included.

†Zirconium is number 1 and osmium is number 2, according to the IUPAC order of preference.

$$K[OsH_3(PMe_2Ph)_3] + Cp_2Zr(H)Cl$$

$$\longrightarrow Cp_2Zr(H)(\mu\text{-}H)_3Os(PMe_2Ph)_3 + KCl$$

■ **Caution.** *Potassium hydride is an extremely flammable, corrosive solid that will ignite violently in moist air. Since an excess of KH is used, care must be taken in the disposal* of unreacted material.*

Procedure

All operations must be performed under an N_2 atmosphere with rigorously dried solvents.

Inside a glove box,[†] a 100-mL Schlenk flask, equipped with a magnetic stirring bar and reflux condenser, is charged with 0.900 g (1.48 mmol) of $OsH_4(PMe_2Ph)_3$, 0.500 g (12.5 mmol) of KH[‡] (Aldrich) and 35 mL of tetrahydrofuran (THF). Outside the glove box, the heterogeneous mixture is refluxed (at 65 °C) for 3 h, giving a bright yellow solution. The solution is then allowed to cool and filtered to remove KH, using a medium porosity Schlenk frit. The filtrate flows directly into a 100-mL Schlenk flask containing 0.380 g (1.48 mmol) of solid $Cp_2Zr(H)Cl$ (Aldrich).[4]

The resulting solution is stirred for 2.5 h, after which time the THF solvent is evaporated under vacuum, leaving a yellow oil. This oil is dissolved in 15 mL of benzene,[§] and the solution is filtered to remove KCl. ■ **Caution.** *Benzene is a carcinogen. It should be handled with gloves in a well-ventilated hood.* Benzene is removed under vacuum, and the gummy solid is dried under vacuum at 25 °C for 2 h. Trituration with cold pentane[¶] (2×20 mL) leaves 0.86 g (70%)[⊥] of $Cp_2Zr(H)(\mu\text{-}H)_3Os(PMe_2Ph)_3$, mp under N_2 138–140 °C (dec). This product can be recrystallized, but with significant loss of material, by cooling a saturated 9:1 pentane–toluene solution for 2 days at -20 °C.

Anal. Calcd. for $C_{34}H_{47}P_3OsZr$: C, 49.20; H, 5.66. Found: C, 45.27; H, 5.64.

Spectroscopically pure material repeatedly gave low carbon analyses.

---

*Potassium hydride can be disposed of by slow addition of 2-propanol to a toluene solution containing KH. A well-ventilated area is needed since H_2 is evolved vigorously.

[†]Vacuum Atmospheres Company glove box equipped with a HE-493 purifier unit. The checker reports that he did not use a glove box to load the reaction flask, since KH is used in excess and $OsH_4(PMe_2Ph)_3$ is reasonably air stable and can be weighed and handled in the atmosphere.

[‡]Potassium hydride is obtained as a dispersion in mineral oil. The oil is removed by washing the dispersion with pentane (3×50 mL) in a dry N_2 atmosphere.

[§]The checker used toluene in place of benzene with no problem.

[¶]The checker found it convenient to leave the sample under cold (4 °C) pentane overnight at this stage.

[⊥]The checker worked at 25% of the recommended scale and obtained a yield of 76%.

Properties

The product $Cp_2Zr(H)(\mu-H)_3Os(PMe_2Ph)_3$ is extremely oxygen and moisture sensitive. Solutions of this compound decompose slowly, even with rigorously dried glassware and solvents, producing $OsH_4(PMe_2Ph)_3$ and insoluble zirconium-contaning products. The compound $Cp_2Zr(H)(\mu-H)_3Os(PMe_2Ph)_3$ is best characterized using NMR spectroscopy.* The ^1H NMR spectrum (C_6D_6) shows the bridging hydrides as a second-order pattern centered at -8.50 ppm. The terminal zirconium-bound hydride appears as a broad singlet at 4.70 ppm. The $^{31}P\{^1H\}$ NMR spectrum (C_6D_6) shows a singlet at -24.50 ppm. Other physical properties, including a crystal structure of the analogous $Cp_2Zr(Cl)(\mu-H)_3Os(PMc_2Ph)_3$ complex, are given in the literature.[5]

C. [2(η^4)-1,5-CYCLOOCTADIENE]TRIS(DIMETHYLPHENYL-PHOSPHINE-1κP)-TRI-μ-HYDRIDOOSMIUMRHODIUM[†]

$$OsH_4(PMe_2Ph)_3 + \text{excess KH} \xrightarrow[\text{THF}]{70\,°C} K[OsH_3(PMe_2Ph)_3] + H_2$$

$$K[OsH_3(PMe_2Ph)_3] + \tfrac{1}{2}[Rh(1,5\text{-cod})Cl]_2$$
$$\longrightarrow KCl + (1,5\text{-cod})Rh(\mu-H)_3Os(PMe_2Ph)_3$$

Procedure

The procedure in Section B is used to prepare a THF solution of $K[OsH_3(PMe_2Ph)_3]$ from 0.750 g (1.20 mmol) of $OsH_4(PMe_2Ph)_3$, 0.500 g (12.5 mmol) of KH and 30 mL of THF. The solution is filtered directly into a 100-mL Schlenk flask containing 0.300 g (0.600 mmol) of [RhCl(1,5-cod)]$_2$.[6]

The resulting dark red solution is stirred for 1.5 h, and the THF solvent is evaporated under vacuum. The residue is dissolved in 10 mL of benzene and the solution is filtered to remove KCl. Benzene is removed under vacuum leaving a burgundy solid. This product can be recrystallized easily by cooling a saturated 10:1 pentane–toluene solution to $-20\,°C$ for 1 day. Yield: 0.70 g (71%)[‡] of dark red crystals. mp under N_2 125–130 °C.

Anal. Calcd. for $C_{32}H_{48}P_3OsRh\cdot\tfrac{1}{2}$ toluene: C, 49.31, H, 6.00. Found: C, 49.30; H, 5.63.

*^{31}P chemical shifts are reported referenced to 85% H_3PO_4 with downfield shifts recorded as positive.
[†]Osmium is number 1 and rhodium is number 2.
[‡]The checker worked at 20% of the recommended scale and obtained a yield of 56%.

Properties

The product $(1,5\text{-cod})Rh(\mu\text{-H})_3Os(PMe_2Ph)_3$ is oxygen and moisture sensitive, although not as prone to decomposition as $Cp_2Zr(H)(\mu\text{-}H)_3Os(PMe_2Ph)_3$. The Rh—Os complex is best characterized by NMR spectroscopy. The ^{1}H NMR spectrum (C_6D_6) shows the hydrides as a second-order pattern centered at -9.20 ppm. The $^{31}P\{^1H\}$ (C_6D_6) NMR consists of a doublet at -20.6 ppm $(J_{P-Rh} = 8\,Hz)$. More detailed spectroscopic properties can be found in the literature.[7]

References

1. D. A. Roberts and G. L. Geoffroy, in *Comprehensive Organometallic Chemistry*, G. Wilkinson, F. G. A. Stone, and E. W. Abel (eds.), Pergamon, Oxford, 1982, Chapter 40.

2. J. Chatt, G. J. Leigh, D. M. P. Mingos, and R. J. Paske, *J. Chem. Soc. Ser. A*, **1968**, 2636.

3. P. G. Douglas and B. L. Shaw, *J. Chem. Soc. Ser. A*, **1970**, 334.

4. P. C. Wailes and H. Weigold, *Inorg. Synth.*, **19**, 223 (1979).

5. J. W. Bruno, J. C. Huffman, M. A. Green, and K. G. Caulton, *J. Am. Chem. Soc.*, **106**, 8310 (1984).

6. G. Giordano and R. H. Crabtree, *Inorg. Synth.*, **19**, 219 (1979).

7. E. G. Lundquist, J. C. Huffman, and K. G. Caulton, *J. Am. Chem. Soc.*, **108**, 8309 (1986).

7. μ-HYDRIDO-TETRAKIS(TERTIARY PHOSPHINE)DIPLATINUM CATIONS

Submitted by STANISLAV CHALOUPKA and LUIGI M. VENANZI*
Checked by ROBERT L. COWAN and WILLIAM C. TROGLER†

Binuclear complexes containing bridging hydride ligands are of frequent occurrence and species containing up to four such bridges have been reported.[1] The group of compounds containing a single M—H—M unit is quite extensive, and several classes of homometallic and heterometallic compounds have been obtained.[1] The M—H—M unit is conveniently described in terms of a three-center two-electron bonding scheme[2] analogous to that developed for boranes. One characteristic structural feature of the single M—H—M unit is its nonlinearity and M—H—M angles from $\sim 100°$ (see ref. 3) to $\sim 160°$ (ref. 1). Small angles have been associated with

*Laboratorium für Anorganische Chemie, ETHZ, Universitätstrasse 6, CH-8092 Zürich, Switzerland.
†Department of Chemistry, University of California, San Diego, La Jolla, CA 92093.

strong direct metal–metal interactions. An analogy can be drawn here with the molecule H_3^+, which should have a regular triangular shape in its ground state.[4] The borane nomenclature of "closed" and "open" interactions has also been carried over into transition metal chemistry.[2]

Complexes containing Pt—H—Pt bonds such as $[(Pcy_3)(SiEt_3)Pt(\mu-H)_2-Pt(SiEt_3)(Pcy_3)]$ (cy = cyclohexyl),[5] $[(Ph_2PCH_2CH_2PPh_2)HPt(\mu-H)_2Pt(Ph_2-PCH_2CH_2PPh_2)]^+$,[6] and $[HPt(\mu-H)(\mu-Ph_2PCH_2PPh_2)_2PtH]^+$,[7] were among the first published examples of doubly and singly hydride-bridged complexes.

Three types of binuclear trihydrido diplatinum cationic complexes are known: (a) the monohydrido-bridged *trans–trans* complexes (Type I),[7–9] (b) the *trans–cis* dihydrido-bridged complexes (Type II),[9,10] and the *cis–cis* dihydrido-bridged complexes (Type III).[11,12]

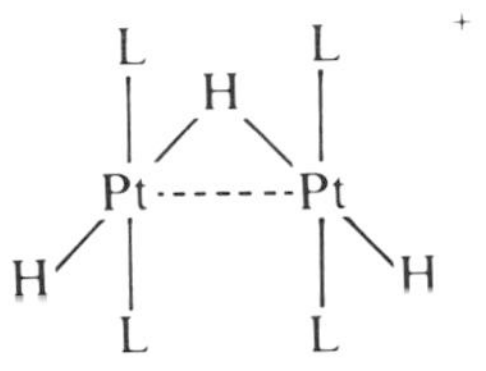

Type I 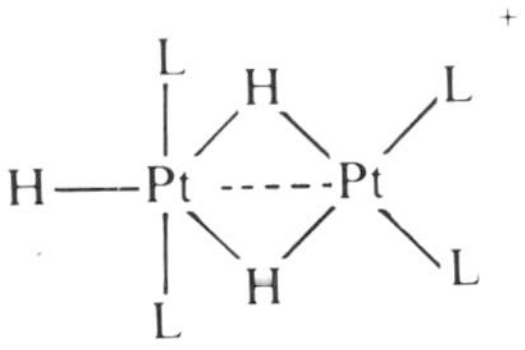   Type II

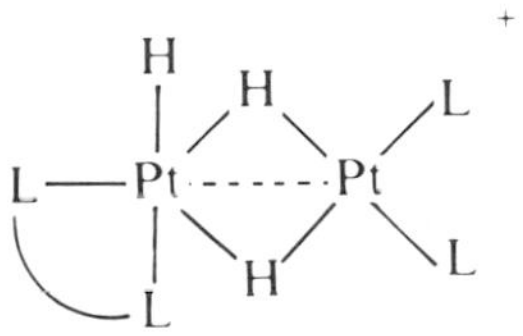

Type III   L = tertiary phosphine

Compounds of Type I in which one or both the terminal hydride ligands have been replaced by aryl groups are also known,[13,14] as are compounds of Type II, which contain a phenyl group instead of the terminal hydride ligand.[10]

Compounds of Types I–III are most easily formed when a hydridocation of the type $[PtH(solvent)L_2]^+$ (L = tertiary phosphine)[15] is allowed to react with a hydride-generating reagent such as $[BH_4]^-$ or formate. In the former case an intermediate complex containing the coordinated BH_4 ligand is formed.[16] In the latter case, the formato-complex intermediate decomposes with loss of CO_2 and formation of the dihydrido species.[14] When monodentate ligands are used, the formation of complexes of Types I or II depends on reaction conditions and on the known equilibration *trans*-$[PtH_2L_2] \rightleftharpoons cis$-

PtH_2L_2].[17] If the rate of formation of *trans*-[PtH_2L_2] is slow relative to the "coupling reaction" *trans*-[PtH(solvent)L_2]$^+$ + *trans*-[PtH_2L_2] the *trans–trans* isomer is formed. If, however, the formation of *trans*-[PtH_2L_2] is faster than the coupling reaction, the equilibrium *cis–trans*-[PtH_2L_2] is established and the *trans–cis* isomer, Type II, is preferentially formed. In any case the latter is thermodynamically more stable relative to the former.[8-10]

Some reagents, for example, pyridine or iodide, catalyze the isomerization of species of Type I to the corresponding isomers of Type II.[9] Compounds of these two types react with donor molecules giving products that can be envisaged as being produced by independent reaction of the component entities $PtHL_2$ and PtH_2L_2.[8] Compounds containing triarylphosphine ligands undergo an easy decomposition reaction in which a P—C bond is cleaved resulting in the formation of cationic species of the type [$(PAr_3)(Ar)Pt(\mu\text{-}H)(\mu\text{-}PAr_2)Pt(PAr_3)_2$]$^+$.[18]

Finally, X-ray structural studies of these compounds show that the Pt—H—Pt unit in complexes of Type I should be described as "open" since this angle, as estimated from the overall molecular geometry, is likely to be larger than $160°$.[10-12]

General Procedure

All the reactions are best carried out (a) under an atmosphere of nitrogen (it is sufficient to have a nitrogen atmosphere over the solutions and during the transfer), (b) with mechanical stirring, and (c) in the absence of light (the reaction vessels are best kept covered with aluminum foil). The solvents (puriss grade) can be used as purchased. The purity and color of the products obtained depend strongly on the quality of the silver salt used: This should be of high purity and, preferably, a freshly opened bottle should be used. All the yields are based on the amount of platinum starting material used.

A. μ-HYDRIDO-DIHYDRIDOTETRAKIS(TRIETHYLPHOSPHINE)-DIPLATINUM(II) TETRAPHENYLBORATE(1−)

trans-[$PtHCl(PEt_3)_2$] + $AgCF_3SO_3$ + solvent

$\longrightarrow$ *trans*-[$PtH(solvent)(PEt_3)_2$](CF_3SO_3) + AgCl

trans-[$PtH(solvent)(PEt_3)_2$](CF_3SO_3) + $NaHCO_2$

$\longrightarrow$ *trans*-[$PtH(HCO_2)(PEt_3)_2$] + $NaCF_3SO_3$ + solvent

trans-[$PtH(HCO_2)(PEt_3)_2$] $\longrightarrow$ *trans*-[$PtH_2(PEt_3)_2$] + CO_2

trans-[PtH$_2$(PEt$_3$)$_2$] + trans-[PtH(solvent)(PEt$_3$)$_2$](CF$_3$SO$_3$) + Na[BPh$_4$]

$\longrightarrow$ [(PEt$_3$)$_2$HPt(μ-H)PtH(PEt$_3$)$_2$] [BPh$_4$] + NaCF$_3$SO$_3$ + solvent

solvent = MeOH or H$_2$O

Procedure

A 304-mg (0.65 mmol) quantity of trans-[PtHCl(PEt$_3$)$_2$] (see ref. 19) is placed in a 50-mL wide Schlenk flask and dissolved in a mixture of 5 mL of methanol and 15 μL of water. A 167-mg (0.65 mmol) quantity of silver trifluoromethane-sulfonate is then added, and the resulting suspension is stirred for 1 h. The AgCl precipitate is removed by filtering on a glass frit, and the yellowish solution thus obtained is cooled to $-70\,°$C. Sodium formate (22 mg, 0.32 mmol) is added, the solution is then allowed to warm to $-50\,°$C, and 115 mg (0.34 mmol) of sodium tetraphenylborate(1-) is added to the solution. The flask is then loosely stoppered and allowed to warm to room temperature. The precipitate formed is collected by filtering on a glass frit and dried under high vacuum.* The crude product is recrystallized by placing it in a flask that is kept at $-10\,°$C in an ice–salt cooling bath, dissolving the solid in ~ 2 mL of acetone, which has been pre-cooled to $\sim -10\,°$C, and slowly adding ~ 20 mL of methanol in such a way that it forms a separate layer floating on top of the acetone solution. The mixture is kept at $-10\,°$C until precipitation is complete (~ 60 min), and the solid is collected by filteration at low temperature.† Yield: 308 mg (80%).

Anal. Calcd. for C$_{48}$H$_{83}$BP$_4$Pt$_2$: C, 48.65; H, 7.06; P, 10.45. Found: C, 48.48; H, 7.11; P, 10.72.

Properties

The compound [(PEt$_3$)$_2$HPt(μ-H)PtH(PEt$_3$)$_2$] [BPh$_4$) is a white crystalline, air-stable solid when pure, which melts with decomposition above 130 °C. The solid is fairly stable at room temperature, but it is light sensitive and is best stored in a refrigerator. It is soluble in acetone and dichloromethane. The solutions are relatively unstable at room temperature and thus are best prepared just before use employing pre-cooled solvents. The decomposition

---

*The checkers report that a gray oil formed, which later solidified into the product.
†The checkers allowed the mixture to crystallize overnight in a $-25\,°$C freezer. They obtained 192 mg (50%) of off-white crystals. Slightly lower yields were obtained when the crystallization was carried out at $-10\,°$C (2 h).

rate is fast when the product is impure. The decomposition product is $[(PEt_3)_2HPt(\mu\text{-}H)_2P(PEt_3)_2]$ $[BPh_4]$.

$$
\begin{array}{ccc}
PEt_3 & & PEt_3{}^+ \\
| & & | \\
H^1\!-\!Pt\!-\!H^2\!-\!Pt\!-\!H^1 \\
| & & | \\
PEt_3 & & PEt_3
\end{array}
$$

Infrared-active vibrations (Pt—H) are observed as follows:
$\nu_{(Pt-H^1)} = 2140\,cm^{-1}$ (m); $\nu_{(Pt-H^2)} = 1500 - 1800\,cm^{-1}$ (w, br).

A comprehensive set of NMR parameters and a figure of the $^1H\,NMR$ spectrum have appeared elsewhere.[9]

The easiest spectroscopic characterization of this complex is done through its $^{31}P\,NMR$ spectrum, recorded in acetone-d_6 solution. This shows one central multiplet, flanked by its ^{195}Pt satellites, with the following parameters: $\delta_{(P)}$ 20.1; $^1J_{Pt,P} = 2580\,Hz$ and $^3J_{Pt,P} = 18\,Hz$.

This spectrum[9] is similar to that of the related compound $[(PEt_3)(Ph)Pt\text{-}(\mu\text{-}H)Pt(Ph)(PEt_3)_2]$ $[BPh_4]$,[14] and the reader is referred to those references for a detailed description of the appearance and interpretation of the $^{31}P\,NMR$ spectra of compounds of this type.

The $^1H\,NMR$ spectral data, recorded in dichloromethane-d_2 solution, are $\delta_{(H^1)}$ −8.19 and $\delta_{(H^2)}$ −6.42.

An X-ray structure determination of this compound[19] shows a disordered arrangement of the cations in the crystal lattice, and allows only the precise location of the metal and donor phosphorus atoms. This, however, is indicative of a bent arrangement of the Pt—H—Pt moiety, typical of compounds containing a single M—H—M bridge.[13,14]

B. DI-μ-HYDRIDO-HYDRIDOTETRAKIS(TRIETHYL-PHOSPHINE)DIPLATINUM(II) TETRAPHENYLBORATE(1 −)

trans-$[PtHCl(PEt_3)_2]$ + $AgCF_3SO_3$ + EtOH

$\longrightarrow [PtH(EtOH)(PEt_3)_2](CF_3SO_3)$ + AgCl

trans-$[PtH(EtOH)(PEt_3)_2](CF_3SO_3)$ + NaBH$_4$

$\longrightarrow$ *trans*-$[PtH(BH_4)(PEt_3)_2]$ + NaCF$_3$SO$_3$ + EtOH

trans-$[PtH(BH_4)(PEt_3)_2]$ + EtOH $\longrightarrow$ *trans*-$[PtH_2(PEt_3)_2]$ + BH$_3$·EtOH

trans-$[PtH_2(PEt_3)_2] \rightleftharpoons cis\text{-}[PtH_2(PEt_3)_2]$

cis-[PtH$_2$(PEt$_3$)$_2$] + $trans$-[PtH(EtOH)(PEt$_3$)$_2$](CF$_3$SO$_3$) + Na[BPh$_4$]

$\longrightarrow$ [(PEt$_3$)$_2$HPt(μ-H)$_2$Pt(PEt$_3$)$_2$][BPh$_4$] + NaCF$_3$SO$_3$ + EtOH

Procedure

A 304-mg (0.65 mmol) quantity of $trans$-[PtHCl(PEt$_3$)$_2$],[19] is placed in a 25-mL Schlenk flask, fitted with a magnetic stirrer, and dissolved in 10 mL of ethanol. The solution is treated with 167 mg (0.65 mmol) of silver trifluoromethanesulfonate and the mixture is stirred for 1 h. The precipitated silver chloride is removed by filtering, the solution is cooled to $-70\,°C$, using an acetone–CO$_2$ bath, and 25 mg (0.66 mmol) of sodium tetrahydroborate(1 $-$) is added to the stirred solution. The temperature is allowed to rise to $-30\,°C$ and then 115 mg (0.336 mmol) of sodium tetraphenylborate is added. The flask is loosely stoppered and allowed to warm gradually to room temperature. The precipitate is collected on a glass frit and dried under high vacuum. The crude product is recrystallized as described for [(PEt$_3$)$_2$HPt(μ-H)PtH(PEt$_3$)$_2$] [BPh$_4$]. Yield: 311 mg (81%).*

Anal. Calcd. for C$_{48}$H$_{83}$BP$_4$Pt: C, 48.65; H, 7.06; P, 10.45. Found: C, 48.64; H. 6.97; P, 10.66.

Properties

The complex [(PEt$_3$)$_2$HPt(μ-H)$_2$Pt(PEt$_3$)$_2$] [BPh$_4$] is a white, crystalline, air-stable solid that melts with decomposition at 125 °C. It is soluble in acetone and dichloromethane and its solutions are air stable.

Infrared-active (Pt—H) vibrations are observed as follows: $v_{(Pt^1-H^1)} =$ 2150 cm^{-1} (m). Two weak, very broad absorption bands between 1550 and 2000 cm^{-1} are probably due to Pt—H^2 vibrations.

A comprehensive set of NMR parameters for this compound, as well as figures of the spectra, have appeared elsewhere.[9,10]

* The checkers obtained 230 mg (60%) of off-white crystals. Crystallization was effected overnight at $-25\,°C$.

As for the previous compound, the characterization of this product is most easily carried out through its ^{31}P NMR spectrum, which shows two central multiplets, flanked by their respective ^{195}Pt satellites, with the following parameters obtained for acetone-d_6 solutions: $\delta_{(P1)}$ 24.2, $^1J_{Pt^1,P^2} = 2540$ Hz; $\delta_{(P2)}$ 20.8, $^1J_{Pt^2,P^2} = 2741$ Hz. The main features of the ^{1}H NMR spectrum are two sets of multiplets centered as follows: $\delta_{(H1)}$ -4.93 and $\delta_{(H2)}$ -3.55.

The X-ray crystal structure of this compound is reported elsewhere.[10] The positions of the heavy atoms indicate that Pt^1 is five coordinate, whereas Pt^2 is four coordinate, and the $Pt(\mu$-$H)_2Pt$ bridge is unsymmetrical in the solid state. As the two bridging hydride ligands appear as magnetically equivalent even at low temperatures, one presumes that the bridge is dynamic on the NMR time scale. However, at temperatures of 30 °C and below no exchange between terminal (H1) and bridging (H2) hydride ligands is observed. Bridge asymmetry in the solid state and hydride equivalence in solution have been observed also for the related compound $[(PEt_3)_2(Ph)Pt(\mu$-$H)_2Pt(PEt_3)_2][BPh_4]$.[10]

C. DI-μ-HYDRIDO-HYDRIDOTETRAKIS(TRIPHENYL-PHOSPHINE)DIPLATINUM(II) TETRAPHENYLBORATE(1 −)

trans-$[PtHCl(PPh_3)_2] + AgCF_3SO_3$

$\longrightarrow$ *trans*-$[PtH(CF_3SO_3)(PPh_3)_2] + AgCl$

trans-$[PtH(CF_3SO_3)(PPh_3)_2] + MeOH$

$\longrightarrow$ *trans*-$[PtH(MeOH)(PPh_3)_2](CF_3SO_3)$

trans-$[PtH(MeOH)(PPh_3)_2](CF_3SO_3) + NaBH_4$

$\longrightarrow PtH_2(PPh_3)_2 + NaCF_3SO_3 + BH_3 \cdot MeOH$

$PtH_2(PPh_3)_2 +$ *trans*-$[PtH(MeOH)(PPh_3)_2](CF_3SO_3) + Na[BPh_4]$

$\longrightarrow [(PPh_3)_2HPt(\mu$-$H)_2Pt(PPh_3)_2][BPh_4] + NaCF_3SO_3 + MeOH$

Procedure

A 300-mg (0.40 mmol) quantity of *trans*-$[PtHCl(PPh_3)_2]$.[20,]* is placed in a 25-mL Schlenk flask and dissolved in 8 mL of dichloromethane and 2 mL of methanol. The solution is cooled to -20 °C, and 105 mg (0.40 mmol) of silver trifluoromethanesulfonate is added to the stirred solution. The silver chloride precipitate is removed by filtering, and the solution is taken to dryness under

---

*It is essential for the successful preparation of this binuclear complex that the starting material should be of high purity, otherwise colored products are obtained. An indication of this is the appearance of the solution, which should be colorless.

vacuum at $-20\,°C$. The residue is dissolved in 10 mL of methanol, which has been pre-cooled to $-70\,°C$ and 20 mg (0.53 mmol) of sodium tetrahydroborate(1 −) is then added. The solution is allowed to warm to $-60\,°C$, and 140 mg (0.41 mmol) of sodium tetraphenylborate is added. The temperature is then allowed to rise to $0\,°C$, and the precipitate is collected on a glass frit, washed three times with 5 mL of 80% aqueous methanol pre-cooled to $-20\,°C$, and dried under high vacuum. The crude product is recrystallized from dichloromethane–methanol, pre-cooled to $-10\,°C$, as described for [(PEt$_3$)HPt(μ-H)PtH(PEt$_3$)$_2$] [BPh$_4$] above. Yield: 242 mg (76%).

Anal. Calcd. for $C_{96}H_{80}BP_4Pt_2$: C, 65.57; H, 4.58; P, 7.04. Found: C, 65.73; H, 4.61; P, 7.22.

Properties

The complex [(PPh$_3$)$_2$HPt(μ-H)$_2$Pt(PPh$_3$)$_2$] [BPh$_4$] is a white, crystalline, light-sensitive, air-stable solid that melts with decomposition at $>135\,°C$. It is soluble in acetone and dichloromethane and its solutions are air stable.

The infrared-active (Pt^1—H^1) vibration is observed at 2200 cm^{-1}. The stretching vibrations associated with the Pt—H—Pt moiety, which are expected to occur between 1500 and 1800 cm^{-1},[21] cannot be identified because they overlap with bands arising from the phenyl substituents.

A comprehensive set of NMR parameters for this complex has been reported elsewhere.[10] For identification purposes the set of data given below can be used (these were obtained for dichloromethane-d_2 solutions). ^{31}P NMR spectrum: This shows two central multiplets, flanked by their respective ^{195}Pt satellites with the following values: $\delta_{(P^1)}$ 31.9, $^1J_{Pt^1,H^2}$ 2803 Hz; $\delta_{2(P^2)}$ 26.6, $^1J_{Pt^2,P^2} = 3008$ Hz. ^{1}H NMR spectrum: $\delta_{(H^1)}$ -4.75 and $\delta_{(H2)}$ -3.20.

The cationic complex easily decomposes in solution giving the cation

Its X-ray crystal structure and NMR data (^{1}H, ^{31}P, and ^{195}Pt) have been reported elsewhere.[18]

References

1. R. G. Teller and B. Bau, *Struct. Bonding*, **44**, 1 (1985).

2. R. Bau, R. G. Teller, S. W. Kirtley, and T. F. Koetzle, *Acc. Chem. Res.* **12**, 176 (1979).

3. The Pt—H—Au angle in [(PEt$_3$)$_2$(C$_6$F$_5$)Pt(μ-H)Au(PPh$_3$)] [BF$_4$] is 103.5(4)°. A. Albinati, H. Lehner, L. M. Venanzi, and M. Wolfer, unpublished observations. See also M. Wolfer, ETH Zurich, Dissertation Nr. 8151, 1986.

4. B. M. Gimark, *Molecular Structure and Bonding*, Academic Press, New York, 1979, p. 23.

5. M. Green, J. A. K. Howard, J. L. Spencer, J. Proud, F. G. A. Stone, and C. A. Tsipis, *J. Chem. Soc., Chem. Commun.*, **1976**, 671.

6. G. Minghetti, G. Banditelli, and A. L. Bandini, *J. Organomet. Chem.*, **139**, C80 (1977).

7. M. P. Brown, R. J. Puddephatt, M. Rashidi, and R. K. Seddon, *Inorg. Chim. Acta*, **23**, L27 (1977).

8. L. M. Venanzi, *Coord. Chem. Rev*, **43**, 251 (1982) and references quoted therein.

9. R. S. Paonessa and W. C. Trogler, *Inorg. Chem.*, **22**, 1038 (1983).

10. F. Bachechi, G. Bracher, D. M. Grove, B. Kellenberger, P. S. Pregosin, L. M. Venanzi, and L. Zambonelli, *Inorg. Chem.*, **22**, 1031 (1983).

11. C. B. Knobler, H. D. Kaesz, G. Minghetti, A. L. Bandini, G. Banditelli, and F. Bonati, *Inorg. Chem.*, **22**, 2324 (1983) and references quoted therein.

12. T. H. Tulip, T. Yamagata, T. Yoshida, R. D. Wilson, J. A. Ibers, and S. Otsuka, *Inorg. Chem.*, **18**, 2339 (1979).

13. G. Bracher, D. M. Grove, L. M. Venanzi, F. Bachechi, P. Mura, and L. Zambonelli, *Angew. Chem. Int. Ed. Engl.*, **17**, 778 (1978).

14. D. Carmona, R. Thouvenot, L. M. Venanzi, F. Bachechi, and L. Zambonelli, *J. Organomet. Chem.*, **250**, 589 (1983).

15. C. Sorato and L. M. Venanzi, *Inorg. Synth.*, **26**, xxx (1989).

16. B. Kellenberger and L. M. Venanzi, unpublished observations.

17. R. S. Paonessa and W. C. Trogler, *J. Am. Chem. Soc.*, **104**, 1138 (1982).

18. J. Jans, R. Naegeli, L. M. Venanzi, and A. Albinati, *J. Organomet. Chem.*, **247**, C37 (1983).

19. J. Chatt and B. L. Shaw, *J. Chem. Soc.*, 5075 (1962); G. W. Panshall, *Inorg. Synth.*, **12**, 28 (1970).

20. J. C. Bailar and H. Itatani, *Inorg. Chem.*, **4**, 1618 (1965)

21. F. Bachechi, B. Kellenberger, and L. M. Venanzi, unpublished observations.

Chapter Two

TRANSITION METAL CHALCOGENIDE COMPLEXES

8. TETRAPHENYLPHOSPHONIUM SALTS OF $[Mo_2(S)_n(S_2)_{6-n}]^{2-}$ THIOANIONS AND DERIVATIVES

Submitted by A. I. HADJIKYRIACOU* and D. COUCOUVANIS*
Checked by J. H. ENEMARK[†] and G. BACKES-DAHMANN[†]

A great variety of binary Mo–S complex anions is formed and can be isolated in reactions of the tetrathiomolybdate anion $[MoS_4]^{2-}$ with various sulfide and polysulfide anions. The nature of the anionic products that can be isolated from these reactions depends on (a) the amount of excess sulfur used (and the types of S_x^{-2} ligands present in the reaction mixtures), (b) the type of counterion used in the isolation of the complex anions, and (c) the type of solvent employed in the synthetic procedure. In a recent article,[1] we described a scheme that interrelates the various $[Mo_2(S)_n(S_2)_{6-n}]^{2-}$ anions. In this scheme (Fig. 1), any of the six homologs can hypothetically be obtained from any other by either the addition of sulfur, or the abstraction of sulfur by triphenylphosphine. Experimentally, the correctness of this scheme has been verified by the successful synthesis of most of the $[Mo_2(S)_n(S_2)_{6-n}]^{2-}$ complexes, or of their internal redox isomers. In the $[Mo_2(S)_n(S_2)_{6-n}]^{2-}$ series, the homologs with $n = 4,$[2] $5,$[1] and 6[1] have been characterized structurally. Those with $n = 2$ and 3 have been characterized structurally as the "internal-redox" isomers, $[(S_4)Mo(S)(\mu\text{-}S)_2(S)Mo(S_2)]^{2-}$ (ref. 3) and

*Department of Chemistry, University of Michigan, Ann Arbor, MI 48109.
[†]Department of Chemistry, University of Arizona, Tucson, AZ 85721.

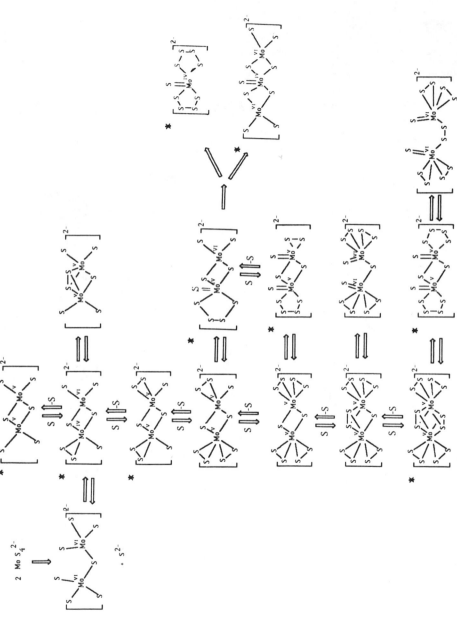

Fig. 1. The $[Mo_2(S)_n(S_2)_{6-n}]^{2-}$ anions. The starred entries denote the anions for which the structures have been

$[Mo(S)(MoS_4)(S_4)]^{2-}$ (ref. 1), respectively. The anions with $n = 0$ have been isolated and structurally characterized as both the $[Mo_2(S_2)_6]^{2-}$ (ref. 4) and $[(S_4)Mo(S)(\mu\text{-}S)_2(S)Mo(S_4)]^{2-}$ (refs. 3a and 5) isomers. The anion with $n = 1$ has not been isolated in any form.

General Procedures and Techniques

The tetraphenylphosphonium salts (Ph_4P^+) of the $[Mo_2(S)_n(S_2)_{6-n}]^{2-}$ thioanions are generally isolated as air stable crystalline materials that range in color from light red to dark (almost black) red. During the synthesis and work-up of these compounds, dry and degassed solvents must be used. In general, the best yields and highest purity of the products are obtained when the syntheses are carried out in an inert atmosphere. A Schlenk line[6] with a nitrogen atmosphere has been found satisfactory. Preferably, solutions and solvents are cannula transferred using rubber septa. The synthesis of $(Ph_4P)_2[MoS_4]$, obtained by cation exchange from $(NH_4)_2[MoS_4]$,[7] is also described. The synthesis of $(C_7H_7)SSS(C_7H_7)$, dibenzyl trisulfide (Bz_2S_3) has been reported previously.[8] This reagent is also commercially available (Aldrich).

■ **Caution.** *Stench. A well-ventilated fume hood must be used when handling* Bz_2S_3.

A. BIS(TETRAPHENYLPHOSPHONIUM) TETRATHIOMOLYBDATE(VI), $(Ph_4P)_2[MoS_4]$

$$(NH_4)_2[MoS_4] + 2(Ph_4P)Cl \longrightarrow (Ph_4P)_2[MoS_4] + 2(NH_4)Cl$$

Procedure

An amount of $(Ph_4P)Cl$ (15.00 g, 40.05 mmol) is placed in a 250-mL Erlenmeyer flask, along with a Teflon-coated stirring bar. Deionized water (100 mL) is added, and the salt is completely dissolved within a few minutes of stirring to give a clear colorless solution.* Next, a 250-mL Erlenmeyer flask containing a large Teflon-coated stirring bar is charged with fresh $(NH_4)_2[MoS_4]$ (5.00 g, 19.23 mmol). Deionized water (150 mL) is added, and the complex is dissolved within 2 to 3 min of stirring. The red solutions is quickly filtered (within 5 min) into a 500-mL filtering flask, using a medium

---

*If 15 g of $(Ph_4P)Cl$ does not dissolve in 100 mL of deionized water at room temperature, the $(Ph_4P)Cl$ reagent is not pure. In this case, the suspension should be filtered to remove the organic water insoluble impurities.

porosity fritted funnel. As soon as the filtration is completed, the product $(Ph_4P)_2[MoS_4]$ is precipitated by adding the $(Ph_4P)Cl$ solution to the vigorously stirred $(NH_4)_2[MoS_4]$ solution. The thick suspension obtained is stirred for an additional 5 min, and then it is vacuum filtered through a medium porosity fritted funnel (150-mL capacity). The crude product isolated on the fritted funnel is washed with three 50-mL portions of water, three 50-mL portions of ethanol, and three 50-mL portions of diethyl ether. The orange-red powder thus obtained is dried under vacuum for 4–5 h (ambient temperature, 10^{-1} torr). The crude product is recrystallized by placing it in a 500-mL Erlenmeyer flask along with a stirring bar, and then by dissolving it in a minimal amount of *N*, *N*-dimethylformamide (DMF) (~ 300 mL) with 3–5 min of stirring. The solution is quickly vacuum filtered (within 10 min) into a 2000-mL filtering flask, using a medium porosity fritted funnel. As soon as the filtration is completed, the red filtrate is flooded with diethyl ether (~ 800 mL) and allowed to stand for 2 h. The product is isolated by vacuum filtration through a medium porosity fritted funnel (150-mL capacity) and washed with two 50-mL portions of diethyl ether. The yield* after drying is 10.0 g (57%) of red microcrystalline $(Ph_4P)_2[MoS_4]$.

Anal. Calcd. for $C_{48}H_{40}P_2MoS_4$: C, 63.86; H, 4.43. Found: C, 63.14; H, 4.78.

Properties

Solutions of $[MoS_4]^{2-}$ slowly oxidize and hydrolyze after standing in air at room temperature for several days. FTIR (KBr pellet, Mo—S vibration): 467 cm^{-1} (s). UV/Vis (DMF solution, $\sim 10^{-4}$ *M*): 474 nm ($\varepsilon = 15{,}500$), and 322 nm ($\varepsilon = 23{,}000$).

B. $(Ph_4P)_2[(Mo_2S_{10})_{0.72}(Mo_2S_{12})_{0.28}] \cdot \frac{1}{2}HCON(CH_3)_2$

$$2(Ph_4P)_2[MoS_4] + (\text{excess})Bz_2S_3$$
$$\longrightarrow (Ph_4P)_2[Mo_2S_{10}] + (Ph_4P)_2S_y + Bz_2S_2$$
$$2(Ph_4P)_2[MoS_4] + (\text{excess})Bz_2S_3$$
$$\longrightarrow (Ph_4P)_2[Mo_2S_{12}] + (Ph_4P)_2S_y + Bz_2S_2$$

---

*The yield obtained for each member of the $[Mo_2(S)_n(S_2)_{6-n}]^{2-}$ series depends highly on the purity of the $(Ph_4P)_2[MoS_4]$. Once recrystallized, this compound should be deep red in microcrystalline form. If, following recrystallization, it is orange or pink, it should be discarded and the cation exchange repeated with *fresh* $(NH_4)_2[MoS_4]$. The cation exchange described above may be carried out under air, provided that the exposure of the $[MoS_4]^{2-}$ solution in air is short (<1 h).

Procedure

The compound $(Ph_4P)_2[MoS_4]$ (4.00 g, 4.43 mmol) is placed in a graduated 250-mL Schlenk flask along with a Teflon-coated stirring bar. The complex is dissolved with stirring in freshly distilled DMF (110 mL) to give a red solution. Solid Bz_2S_3 (10.00 g, 35.97 mmol) is added* to this solution with stirring. Within a minute, the reaction mixture turns green. As the reaction progresses, the green color decays and eventually the solution becomes dark red-brown. The reaction mixture is stirred for a total of 16 h. It is then vacuum-filtered into a graduated 500-mL Schlenk flask through a medium porosity fritted funnel. The filtrate is flooded with diethyl ether (200 mL) to induce crystallization. After standing at room temperature for 16 h[†], the product is isolated by vacuum filtration and washed with two 20-mL portions of diethyl ether. The crude product is finally recrystallized from a DMF (60 mL)–diethyl ether (120 mL) mixture. Yield: 2.1 g (76%) of dark brown microcrystalline $(Ph_4P)_2[(Mo_2S_{10})_{0.72}(Mo_2S_{12})_{0.28}]\cdot\frac{1}{2}DMF$.

Anal. Calcd. for $C_{49.5}H_{43.5}P_2Mo_2S_{10.56}N_{0.5}O_{0.5}$: C, 47.73; H, 3.52; P, 4.97; S, 27.16; N, 0.56. Found: C, 48.77; H, 3.60; P, 4.65; S, 27.14; N, 0.57.

FT–IR (KBr pellet, Mo—S vibrations): 449 cm^{-1} (w). UV–VIS (DMF solution, $\sim 10^{-3}$ M): 570 nm ($\varepsilon = 2000$), 436 nm ($\varepsilon = 4100$), and 290 nm (sh). **Observed X-Ray Powder Pattern Spacings** (Å, CuK$_\alpha$): 11.0 (vs), 10.1 (m), 9.1 (m), 8.4 (s), 7.4 (vs), 7.0 (m), 5.5 (m), 5.1 (m), 4.8 (m), 4.6 (m), 4.3 (m), 4.1 (m), 3.9–3.8 (m, diffuse), 3.5–3.35 (m, diffuse), and 3.3–3.2 (m, diffuse). A single-crystal X-ray structure determination for this compound[3] has shown that it contains both the $[Mo_2S_{10}]^{2-}$ ion, and as a minor (28%) component, the $[Mo_2S_{12}]^{2-}$ ion in the same site.

C. BIS(TETRAPHENYLPHOSPHONIUM) DI-μ-THIO-TETRATHIODIMOLYBDATE(V), $(Ph_4P)_2[Mo_2S_6]$

$$(Ph_4P)_2[Mo_2S_{10}] + 4Ph_3P \longrightarrow (Ph_4P)_2[Mo_2S_6] + 4Ph_3PS$$

$$(Ph_4P)_2[Mo_2S_{12}] + 6Ph_3P \longrightarrow (Ph_4P)_2[Mo_2S_6] + 6Ph_3PS$$

Procedure

The compound $(Ph_4P)_2[(Mo_2S_{10})_{0.72}(Mo_2S_{12})_{0.28}]\cdot\frac{1}{2}DMF$ (4.00 g, 3.21 mmol) is placed in a graduated 250-mL Schlenk flask, along with Ph_3P

---

*The stoichiometry of this reaction was investigated, and the use of eight equivalents of dibenzyl trisulfide was found necessary.
[†]After flooding with diethyl ether, the checkers placed the flask in a refrigerator for 6 h.

(5.00 g, 19.08 mmol). Dry DMF (50 mL) is added, and the mixture is stirred for 4 h at 80 °C. The red solution so obtained is allowed to cool to room temperature and then filtered. The filtrate is flooded with 150 mL of tetrahydrofuran (THF) and allowed to stand for 2 h at room temperature. A brown-red microcrystalline solid* is obtained by filtration. Following washing with toluene (three 20-mL portions) and diethyl ether (two 30-mL portions), the crude product is recrystallized by dissolving it in a minimal amount of DMF and flooding the solution with three volumes of THF. Yield: 2.0 g (58%) of dark red crystals.

Anal. Calcd. for $C_{48}H_{40}P_2Mo_2S_6$: C, 54.24; H, 3.77; P, 5.84; Mo, 18.08; S, 18.08. Found: C, 54.42; H, 3.86; P, 5.56; Mo, 17.94; S, 18.35.

FT–IR (KBr pellet, Mo—S vibrations): 452 cm^{-1} (w), 475 cm^{-1} (m), and 503 cm^{-1} (m). UV–VIS (DMF solution, 10^{-3} M): 483 nm (sh), 454 nm ($\varepsilon =$ 9900), 362 nm (sh), 310 nm ($\varepsilon = 19{,}300$), and 290 nm ($\varepsilon = 26{,}000$).

Observed X-Ray Powder Pattern Spacing (Å, CuK$_\alpha$): 12.2 (m), 9.4 (w), 8.4 (w), 7.8 (vs), 7.0 (s), 6.5 (w), 5.8 (m), 5.6 (w), 5.4 (w), 4.9 (w), 4.8 (m), 4.6 (w), and 4.4 (m). This pattern is very nearly the same as the calculated one on the basis of the single crystal X-ray structure determination[1] for the complex $(Ph_4P)_2[Mo_2S_6]$.

D. BIS(TETRAPHENYLPHOSPHONIUM) (η^2-DISULFIDO)-DI-μ-THIO-TRITHIODIMOLYBDATE(IV, VI), $(Ph_4P)_2[Mo_2S_7] \cdot HCON(CH_3)_2$

$$(Ph_4P)_2[Mo_2S_6] + Bz_2S_3 \longrightarrow (Ph_4P)_2[Mo_2S_7] + Bz_2S_2$$

Procedure

The compound $(Ph_4P)_2[Mo_2S_6]$ (2.00 g, 1.88 mmol) is placed in a graduated 250-mL Schlenk flask along with a Teflon-coated stirring bar. The complex is dissolved with stirring in freshly distilled DMF (60 mL) to give a red solution with a yellow cast. Dibenzyl trisulfide (0.53 g 1.91 mmol) is dissolved in freshly distilled DMF (20 mL) by swirling. The Bz_2S_3 solutions is added dropwise with stirring to the thiomolybdate solution over a 2 to 3-min period. The reaction mixture is stirred for 12 h at room temperature; it gradually turns dark brown-red, which is the final color. The mixture is then

---

*The checkers report that crystallization did not occur immediately after flooding the filtrate with THF, but took place after ~1 h with the flask refrigerated at 4 °C. If the DMF–THF mixture is cooled to 4 °C to accelerate crystallization, the product should be washed throughly with toluene to remove $Ph_3P{=}S$ that crystallizes upon cooling.

vacuum-filtered into a graduated 250 mL Schlenk flask through a medium porosity fritted funnel. The filtrate is flooded with dry diethyl ether (100 mL) to induce crystallization, and it is allowed to stand at room temperature for 2 h. Brown-red microcrystalline $(Ph_4P)_2[Mo_2S_7] \cdot DMF$ is collected by vacuum filtration and washed with two 20-mL portions of diethyl ether*. Yield: 1.4 g (77%) of $(Ph_4P)_2[Mo_2S_7] \cdot DMF$.

Anal. Calcd. for $C_{51}H_{47}NOP_2Mo_2S_7$: C, 52.44; H, 4.03; P, 5.31; Mo, 16.45; S, 19.19. Found: C, 52.50; H, 3.80; P, 5.35; Mo. 17.16; S, 21.02.

FT–IR (KBr pellet, Mo—S vibrations): 454 cm^{-1} (w), 480 cm^{-1} (m), and 504 cm^{-1} (m). **UV–VIS** (DMF solution, $10^{-3} M$): 560 nm ($\varepsilon = 2400$), 452 nm ($\varepsilon = 5200$), 422 nm (sh), 362 nm (sh), and 295 nm (sh).

Observed X-Ray Powder Pattern Spacings. (Å, CuK$_\alpha$): 12.5 (w), 10.7 (m), 10.0 (s), 8.4 (vs), 7.7 (s), 7.0 (m), 6.5 (m), 5.9 (m), 5.0 (m), 4.8 (m), 4.6 (m), 4.4 (m), and 4.2 (m). This pattern is very nearly the same as the one calculated on the basis of the single crystal X-ray structure determination[1] for the complex $(Ph_4P)_2Mo_2S_7 \cdot DMF$.

E. BIS(TETRAPHENYLPHOSPHONIUM) BIS(η^2-DISULFIDO)-DI-μ-THIO-DITHIODIMOLYBDATE(V), $(Ph_4P)_2[Mo_2S_8]$

$$(Ph_4P)_2[Mo_2S_6] + 2Bz_2S_3 \longrightarrow (Ph_4P)_2[Mo_2S_8] + 2Bz_2S_2$$

Procedure

The compound $(Ph_4P)_2[Mo_2S_6]$ (2.00 g, 1.88 mmol) is placed in a graduated 250-mL Schlenk flask along with a Teflon-coated stirring bar. The thiomolybdate is dissolved with stirring in freshly distilled DMF (70 mL) to give a red solution with a yellow cast. Dibenzyltrisulfide (1.04 g, 3.74 mmol) is dissolved in DMF (20 mL) by swirling. The Bz_2S_3 solution is added dropwise with stirring to the thiomolybdate solution over a 2 to 3 min period. The reaction mixture is then stirred for 12 h. As the reaction progresses, the solution turns darker. Finally, it attains a green cast. The reaction mixture is then filtered into a graduated 250-mL Schlenk flask through a medium porosity fritted funnel. The filtrate is flooded with diethyl ether (100 mL) to induce crystallization and allowed to stand for 2 h. Red-purple microcrystalline $(Ph_4P)_2[Mo_2S_8]$ is collected by vacuum filtration and washed with two 20-mL portions of diethyl ether. The yield after drying *in vacuo* is 1.8 g (85%).

---

*Throughout the experiment (precipitation of crude product and recrystallization), the diethyl ether used should be bone dry. In solution, $(Ph_4P)_2[Mo_2S_7]$ is the most moisture-sensitive member of the series.

Anal. Calcd. for $C_{48}H_{40}P_2Mo_2S_8$: C, 51.15; H, 3.55; P, 5.51; Mo, 17.05; S, 22.74. Found: C, 49.96; H, 3.61; P, 5.35; Mo, 16.81; S, 23.04.

FT–IR (KBr pellet, Mo—S vibrations: $512\,cm^{-1}$ (m), and $453\,cm^{-1}$ (w). **UV–Vis** (DMF or MeCN solution, $10^{-3}\,M$): 574 nm ($\varepsilon = 3300$), 464 nm ($\varepsilon = 2700$), and 290 nm (sh).

Observed X-Ray Powder Pattern Spacings. (Å, CuK_α): 12.0 (m), 11.0 (m), 9.4 (s), 8.4 (s), 7.5 (s), 6.4 (m), and 4.7 (s).

Properties

Analytically pure crystalline samples of these thiomolybdates undergo no measurable oxidation or hydrolysis after several weeks in air at room temperature. However, solutions of the anions $[Mo_2S_6]^{2-}$ and $[Mo_2S_7]^{2-}$ completely oxidize and hydrolyze after standing under air for several days at ambient temperature. Solutions of $[Mo_2S_8]^{2-}$, $[Mo_2S_7]^{2-}$, and $[Mo_2S_9]^{2-}$ oxidize and hydrolyze much faster (within a few hours) at elevated temperatures ($\sim 100\,°C$) under air. Therefore, solutions of these complexes should be handled under an inert atmosphere. The $(Ph_4P)^+$ salts of these thiomolybdates are very soluble in DMF ($\sim 1\,g\ 50\,mL^{-1}$) and much less soluble in MeCN.

The recommended entry to the synthesis of any of the $[Mo_2(S)_n(S_2)_{6-n}]^{2-}$ anions reported herein is the $[(Mo_2S_{10})_{0.72}(Mo_2S_{12})_{0.28}]^{2-}$ anion. From the latter, the $[Mo_2S_6]^{2-}$ anion can be obtained by the stoichiometric reaction described. In addition to the reported syntheses, the following interconversions between the $[Mo_2(S)_n(S_2)_{6-n}]^{2-}$ anions occur readily[1b] under similar reaction procedures:

$$(Ph_4P)_2[Mo_2S_8] + (\text{excess}) Bz_2S_3$$
$$\longrightarrow (Ph_4P)_2[(Mo_2S_{10})_{0.72}(Mo_2S_{12})_{0.28}] + Bz_2S_2$$

$$(Ph_4P)_2[Mo_2S_8] + 2Ph_3P \longrightarrow (Ph_4P)_2[Mo_2S_6] + 2Ph_3PS$$

$$(Ph_4P)_2[Mo_2S_6] + (\text{excess}) Bz_2S_3$$
$$\longrightarrow (Ph_4P)_2[(Mo_2S_{10})_{0.72}(Mo_2S_{12})_{0.28}] + Bz_2S_2$$

$$(Ph_4P)_2[(Mo_2S_{10})_{0.72}(Mo_2S_{12})_{0.28}] + 3Ph_3P$$
$$\longrightarrow (Ph_4P)_2[Mo_2S_9] + 3Ph_3PS$$

The complexes $(Ph_4P)_2[Mo_2S_8]$ and $(Ph_4P)_2[Mo_2S_9]$ do not interconvert by the addition or subtraction of sulfur at ambient or elevated (90 °C) temperatures.

Cleaning Glassware

Use a well-ventilated hood. Using spatula, remove as much of the solid left in the flask as possible. The remaining solid is dissolved by rinsing with DMF, and the solution is decanted. The flask is rinsed with acetone and then acetone is completely drained. Finally, concentrated nitric acid is added to the flask to destroy the smell and remaining traces of sulfur compounds.

■ **Caution.** *Do not mix acetone with concentrated nitric acid. They react violently. Some sulfur compounds, such as (NH_4)$_2$[MoS_4], react violently with concentrated nitric acid. Use nitric acid to destroy only traces of sulfur compounds left in the flasks to be cleaned. Use a well-ventilated hood, be cautious, wear rubber gloves, and use a shield.*

References

1. (a) D. Coucouvanis and A. J. Hadjikyriacou, *Inorg. Chem.*, **25**, 4317 (1986); (b) A. I. Hadjikyriacou and D. Coucouvanis, *Inorg. Chem.*, **26**, 2400 (1987).
2. W. H. Pan, M. A. Harmer, T. R. Halbert, and E. I. Stiefel, *J. Am. Chem. Soc.*, **106**, 459 (1984).
3. (a) M. Draganjac, E. Simhon, L. T. Chan, M. Kanatzidis, N. C. Baenziger, and D. Coucouvanis, *Inorg. Chem.*, **21**, 3321 (1982); (b) W. Clegg, G. Christou, C. D. Garner, and G. M. Sheldrick, *Inorg. Chem.*, **20**, 1562 (1981).
4. A. Müller, W. O. Nolte, and B. Krebs, *Angew. Chem. Int. Ed. Engl.*, **17**, 279 (1978).
5. S. A. Cohen and E. I. Stiefel, *Inorg. Chem.*, **24**, 4657 (1985).
6. D. F. Shriver, *The Manipulation of Air-Sensitive Compounds*, McGraw-Hill, New York, 1969, pp. 145–158.
7. W. H. Pan, M. E. Leonowicz, and E. I. Stiefel, *Inorg. Chem.*, **22**, 672 (1983).
8. D. Coucouvanis, M. G. Kanatzidis, E. Simhon, and N. C. Baenziger, *J. Am. Chem. Soc.*, **104**, 1874 (1982).

9. MOLYBDENUM–SULFUR CLUSTERS

Submitted by ACHIM MÜLLER* and ERICH KRICKEMEYER*
Checked by ANASTASIOS HADJIKYRIACOU[†] and DIMITRI COUCOUVANIS[†]

The compounds $(NH_4)_2[Mo_3S(S_2)_6]\cdot nH_2O$ ($n = 0$–$2)^{1-4}$ and $(NH_4)_2[Mo_2$-$(S_2)_6]\cdot 2H_2O^{1,4,5}$ were the first reported discrete pure transition metal sulfur clusters. The preparation of both compounds is very simple.

The ion $[Mo_3S(S_2)_6]^{2-}$ is the most stable molybdenum sulfur species in

*Faculty of Chemistry, University of Bielefeld, Postfach 8640, D-4800 Bielefeld, Federal Republic of Germany.
†Department of Chemistry, The University of Michigan, Ann Arbor, MI 48109-1055.

solution and has been discussed as a model for crystalline MoS_2, to which its $NH_4{}^+$ salt decomposes on heating[3,4] (MoS_2 seems to be one of the most versatile heterogeneous catalysts).

The compound $(NH_4)_2[Mo_2(S_2)_6]\cdot 2H_2O$ can be used as a generator for S_2 (ref. 6) and is an often used precursor for other molybdenum–sulfur species.

A. DIAMMONIUM TRIS(μ-DISULFIDO)TRIS(DISULFIDO)-μ₃-THIO-*TRIANGULO*-TRIMOLYBDATE(IV) HYDRATE

$$(NH_4)_6[Mo_7^{VI}O_{24}]\cdot 4H_2O + (NH_4)_2S_x \longrightarrow (NH_4)_2[Mo_3^{IV}S(S_2)_6]\cdot nH_2O$$
$$(n = 0-2)$$

Procedure

■ **Caution.** *Hydrogen sulfide and carbon disulfide are extremely poisonous and flammable. All procedures must be performed in a well-ventilated hood.*

1. An H_2S stream ($20\,L\,h^{-1}$) is bubbled into a suspension of 27.0 g of S_8 in 150 mL of 10% aqueous NH_3 solution in a 300-mL Erlenmeyer flask for 1.5 h. The sulfur dissolves to give a wine red polysulfide solution.

2. A 4.0 g (3.2 mmol) quantity of $(NH_4)_6[Mo_7O_{24}]\cdot 4H_2O$ is dissolved in 20 mL of water in a 300-mL Erlenmeyer flask. The polysulfide solution (120 mL) is added, and the flask is covered with a watch glass. The reaction mixture is then kept on an oil bath (95–98 °C) for 5 days without stirring. The red crystals of $(NH_4)_2[Mo_3S(S_2)_6]\cdot nH_2O$ that form are removed by filtering and then washed successively with 300 mL of water, 25 mL of ethanol, three times with 20-mL volumes of carbon disulfide (removal of sulfur), and finally with 20 mL of diethyl ether. The crystals are dried in air.

The yield is almost quantitative (5.5–5.8 g; depending on the crystal water content).*

Anal. Hydrogen and nitrogen are determined by elemental analysis and sulfur as $(SO_4)^{2-}$ by ion-HPLC. Calcd. for $(NH_4)_2[Mo_3S(S_2)_6]\cdot H_2O$: H, 1.33; N, 3.69; S, 54.93. Found: H, 1.10; N, 3.35; S, 55.80.

B. DIAMMONIUM BIS(μ-DISULFIDO)TETRAKIS(DISULFIDO)-DIMOLYBDATE(V) DIHYDRATE

$$(NH_4)_2[Mo^{VI}O_2S_2] + (NH_4)_2S_x \longrightarrow (NH_4)_2[Mo_2^V(S_2)_6]\cdot 2H_2O$$

*The checkers obtained 4.6 to 5.1 g of red crystals (82–91% for $n = 0$).

Procedure

■ **Caution.** *Hydrogen sulfide and carbon disulfide are extremely poisonous and flammable. All procedures must be performed in a well ventilated hood.*

1. A stream of H_2S ($20\,L\,h^{-1}$) is bubbled into a suspension of $10.0\,g$ of S_8 in $350\,mL$ of 10% aqueous NH_3 solution in a 500-mL Erlenmeyer flask for 1.5 h. The sulfur dissolves, and the resulting polysulfide solution is orange-red.

2. In a 1-L flask, $6.0\,g$ ($26.30\,mmol$) of $(NH_4)_2[MoO_2S_2]$ (ref. 7) (freshly prepared, dried over P_4O_{10} in a vacuum) in $375\,mL$ of methanol (analyzed reagent quality) is heated at reflux on an oil bath for 2 h (the color changes from orange over dark red to orange). After cooling to 20 °C the polysulfide solution is added. The reflux condenser is removed, and the flask is covered with a watch glass. The dark red reaction mixture is kept on an oil bath (55 °C) for 20 h without stirring. Precipitation of the required product starts after a short time and can be increased by cooling to 5 °C with an ice bath. Black crystals of $(NH_4)_2[Mo_2(S_2)_6]\cdot2H_2O$ are filtered off. They are washed successively with $50\,mL$ of 2-propanol, three times with 20-mL volumes of carbon disulfide (removal of sulfur), and then with $25\,mL$ of diethyl ether; they are dried in air.

Yield: $6.2\,g$ (73%).* Additional product ($0.5\,g$, 5%) precipitates from the filtrate upon cooling to 5 °C.

Anal. Hydrogen and nitrogen are determined by elemental analysis and sulfur as $(SO_4)^{2-}$ by ion HPLC. Calcd. for $(NH_4)_2[Mo_2(S_2)_6]\cdot2H_2O$: H, 1.86; N, 4.32; S, 59.31. Found: H, 1.58; N, 4.70; S, 58.90.

C. DIAMMONIUM TRIS(μ-DISULFIDO)TRIS(DISULFIDO)-μ₃-THIO-*TRIANGULO*-TRIMOLYBDATE(IV) HYDRATE AND DIAMMONIUM BIS(μ-DISULFIDO)-TETRAKIS(DISULFIDO)DIMOLYBDATE(V) DIHYDRATE

$$(NH_4)_6[Mo_7^{VI}O_{24}]\cdot4H_2O + (NH_4)_2S_x \longrightarrow (NH_4)_2[Mo_3^{IV}S(S_2)_6]\cdot nH_2O$$
$$+ (NH_4)_2[Mo_2^V(S_2)_6]\cdot2H_2O$$

Procedure

■ **Caution.** *Hydrogen sulfide and carbon sulfide are extremely poisonous and flammable. All procedures must be performed in a well-ventilated hood.*

*The checkers obtained a yield of 3.0 to $3.3\,g$ (30–35%).

1. A stream of H_2S ($20\,L\,h^{-1}$) is bubbled for 1.5 h into a suspension of $1.2\,g\,S_8$ in 100 mL of 10% aqueous NH_3 solution in a 300-mL Erlenmeyer flask. The sulfur dissolves, and the resulting polysulfide solution is orange.

2. A 4.0-g (3.2 mmol) quantity of $(NH_4)_6[Mo_7O_{24}]\cdot4H_2O$ is dissolved in 20 mL of water in an 100-mL Erlenmeyer fask (wide neck). After addition of 80 mL of the polysulfide solution the flask is covered with a watch glass. The reaction mixture is heated on an oil bath (95 °C) for 16 h without stirring. The precipitated red crystals of $(NH_4)_2[Mo_3S(S_2)_6]\cdot nH_2O$ are removed by filtering. They are washed successively with 300 mL of water, 25 mL of ethanol, three times with 10-mL volumes of carbon disulfide (removal of sulfur), and then with 10 mL of diethyl ether; they are dried in air. Yield: $(NH_4)_2[Mo_3S(S_2)_6]\cdot nH_2O$ 1.6 g (28% for $n = 1$).* The filtrate is cooled to 20 °C and the rest of the polysulfide solution is added. On standing for three days in a closed 100-mL Erlenmeyer flask under an argon atmosphere at room temperature, almost black crystals of $(NH_4)_2[Mo_2(S_2)_6]\cdot2H_2O$ precipitate from the solution. They are removed by filtering, washed twice with 10-mL portions of ice cold water, then with 20 mL of 2-propanol, three times with 15-mL portions of carbon disulfide (removal of sulfur), and finally with 20 mL of diethyl ether. The crystals are dried in air. Yield: $(NH_4)_2[Mo_2(S_2)_6]\cdot2H_2O$ 3.0 g (41%).*

Properties

The compound $(NH_4)_2[Mo_3S(S_2)_6]\cdot nH_2O$ is air stable, extraordinarily inert against hydrogen chloride solutions, and diamagnetic. Its shows characteristic IR absorptions (CsI pellet) at 544 (m; $\nu_{(S-S)_{bridge}}$), 510/504 (m; $\nu_{(S-S)_{term}}$), and 458 cm^{-1} (w; $\nu_{(Mo-\mu_3-S)}$ (ref. 8). UV–Vis (in N, N-dimethylacetamide): 540 (ν_1; $\varepsilon_i = 1.1 \times 10^3\,M^{-1}\,cm^{-1}$) and 465 nm ($\nu_2$; $\varepsilon_i = 3.6 \times 10^3\,M^{-1}\,cm^{-1}$).[9]

Freshly prepared $(NH_4)_2[Mo_3S(S_2)_6]\cdot nH_2O$ is not very soluble in water, but it shows a better solubility in N, N-dimethylformamide. On heating, the compound decomposes with the loss of nH_2O, $2NH_3$, H_2S, and 6S, to MoS_2.[3]

The black compound $(NH_4)_2[Mo_2(S_2)_6]\cdot2H_2O$ is stable in air. Freshly prepared, it is slightly soluble in water, methanol, and ethanol and has a characteristic IR absorption band (CsI pellet) at 530 cm^{-1} (m; $\nu_{(S-S)}$). The UV–Vis spectrum (in methanol) is not very characteristic (see color) and shows shoulders at ~ 560, ~ 460, ~ 375, ~ 340, and ~ 310 nm and a band at 255 nm.

---

*The checkers obtained yields of 1.3 to 1.5 g (21–26%) for $(NH_4)_2[Mo_3S(S_2)_6]\cdot nH_2O$ and 2.3 to 2.6 g (32–36%) for $(NH_4)_2[Mo_2(S_2)_6]\cdot2H_2O$.

References

1. A. Müller, R. G. Bhattacharrya, and B. Pfefferkorn, *Chem. Ber.*, **112**, 778 (1979).
2. A. Müller, S. Pohl, M. Dartmann, J. P. Cohen, J. M. Bennett, and R. M. Kirchner, Z. *Naturforsch. Teil B*, **34**, 434 (1979).
3. E. Diemann, A. Müller, and P. J. Aymonino, *Z. Anorg. Allg. Chem.*, **479**, 191 (1981); A. Müller and E. Diemann, *Chimia*, **39**, 312 (1985).
4. A. Müller, *Polyhedron*, **5**, 323 (1986).
5. A. Müller, W. O. Nolte, and B. Krebs, *Angew. Chem. Int. Ed. Engl.*, **17**, 279 (1978).
6. A. Müller, W. Jaegermann, and J. H. Enemark, *Coord. Chem. Rev.*, **46**, 245 (1982).
7. J. W. McDonald, G. D. Friesen, L. D. Rosenhein, and W. E. Newton, *Inorg. Chim. Acta*, **72**, 205 (1983).
8. Complete spectrum: A. Müller, in *Analytical Applications of FT–IR to Molecular and Biological Systems*, J. R. Durig (ed.), Reidel, Dordrecht 1980, p. 257.
9. A. Müller, R. Jostes, W. Jaegermann, and R. G. Bhattacharrya, *Inorg. Chim. Acta*, **41**, 259 (1980).

10. BIS(η⁵-METHYLCYCLOPENTADIENYL)TITANIUM PENTASULFIDE,* BIS(η⁵-METHYLCYCLOPENTADIENYL)- DIVANADIUM PENTASULFIDE,† AND BIS(η⁵-METHYLCYCLOPENTADIENYL)DIVANADIUM TETRASULFIDE‡

**Submitted by JAMES DARKWA,§ DEAN M. GIOLANDO,§
CATHERINE JONES MURPHY,§ and THOMAS B. RAUCHFUSS§
Checked by A. MÜLLER¶**

The cyclopentadienyl metal sulfides constitute a class of compounds that display varied reactivity toward small molecules, unsaturated organic compounds, and other metal complexes.

Bis(η⁵-cyclopentadienyl)titanium pentasulfide was prepared first by Samuel and Schmidt from $TiCp_2Cl_2$ (Cp = cyclopentadienyl) and polysulfide salts.[1] It can also be prepared from the reaction of $TiCp_2(CO)_2$ or $TiCp_2(CH_2R)_2$ with elemental sulfur. The method presented here is a modification of Köpf's procedure[2] and is equally applicable to the Cp, MeCp, and *i*-PrCp (*i*-Pr = isopropyl) derivatives. This reaction has been successfully conducted at 40 times

---

* Bis(η⁵-methylcyclopentadienyl) (pentasulfido-S^1, S^5)titanium.
† Bis(η⁵-methylcyclopentadienyl) (μ-disulfido-*S:S′*) (μ-disulfido-η^2:η^2) (μ-sulfido)divanadium.
‡ Bis(η⁵-methylcyclopentadienyl) (μ-disulfido-*S:S′*) (di-μ-sulfido)divanadium.
§ School of Chemical Sciences, University of Illinois, Urbana, IL 61801.
¶ Lehrstuhl für Anorganische Chemie I, Universität Bielefeld, Bielefeld, Federal Republic of Germany

the scale described here. Bis(η^5-cyclopentadienyl)divanadium pentasulfide was first prepared by the reaction of $VCp(CO)_4$ and sulfur[3] and by the thermolysis of VCp_2S_5.[4] The structure and even the molecularity of $V_2Cp_2S_5$ remained a mystery until our work on the MeCp derivatives.[5] The preparation of $V_2(MeCp)_2S_5$ is an adaptation of Dahl and Petersen's method and entails the *in situ* generation of $V(MeCp)_2S_5$. This preparation is streamlined in the sense that it proceeds directly from an inexpensive starting material, VCl_3, to the product. Previous synthesis involved the use of $V(MeCp)_2Cl_2$ which is tedious to prepare and usually forms in $< 50\%$ yield.

The compound $V_2(MeCp)_2S_5$ is a useful precursor to $V_2(MeCp)_2S_4$, which in turn is a versatile synthetic intermediate.

A.   METHYLCYCLOPENTADIENE

Procedure

■   **Caution.**   *This procedure should be carried out in well-ventilated fume hood. Cyclopentadienes are toxic by inhalation and are highly flammable.*

Commercial methylcyclopentadiene dimer (Aldrich) is cracked using a 40-cm Vigreux column with a pot temperature of 210 to 225 °C. The first $\sim 10\%$ of the distillate is discarded and $\sim 70\%$ of the remainder is collected at 63 to 65 °C at a drip rate of $\sim 100\,mL\,h^{-1}$. The crude methylcyclopentadiene is redistilled using the same column to give a product containing $\sim 2\%$ of C_5H_6 impurity.

Properties

Methylcyclopentadiene obtained by cracking the commercially available dimer contains up to $\sim 10\%$ cyclopentadiene. This can lead to major purity problems with derived organometallic complexes, particularly for those compounds that contain several $CH_3C_5H_4$ ligands. The diene can be purified simply by redistillation of the crude methylcyclopentadiene monomer. In our procedure the same fractional distillation apparatus is used for both the initial cracking and the redistillations. The purity of the methylcyclopentadiene can be assayed by ^1H NMR spectroscopy.[7] Methylcyclopentadiene can be stored at -30 °C for a few days.

B.   BIS(η^5-METHYLCYCLOPENTADIENYL)TITANIUM PENTASULFIDE

$$Ti(MeCp)_2Cl_2 + S_x^{2-} \longrightarrow Ti(MeCp)_2S_5 + 2Cl^-$$

Procedure

■ **Caution.** *This synthesis should be carried out in a well-ventilated fume hood. Carbon disulfide is extremely flammable and very toxic by inhalation.*

A slurry of 0.50 g (1.80 mmol) of $Ti(MeCp)_2Cl_2$ in 10 mL of acetone* is prepared in an open 25-mL Erlenmeyer flask with magnetic stirring. In one portion, 1 mL of freshly filtered ammonium polysulfide solution† is added to the $Ti(MeCp)_2Cl_2$ solution. After stirring for 15 min, the solution is filtered in air, and the red crystals are washed with water, methanol, and then two 5-mL portions of CS_2. Yield: 0.510 g (77%).

Anal. Calcd. for $C_{12}H_{14}S_5Ti$: C, 39.37; H, 3.85; S, 43.69; Ti, 13.08. Found: C, 39.59; H, 3.88; S, 43.58; Ti, 12.96.

Properties

The synthesis is also applicable to $TiCp_2S_5$. Red crystalline $Ti(RCp)_2S_5$ (R = Me and H) is air stable in solution and in the solid state. It is moderately soluble in halogenated hydrocarbon solvents and practically insoluble in alcohols and hydrocarbons. The ^1H NMR spectrum $(CDCl_3)$ consists of multiplets at $\delta = 6.18, 6.13, 6.09,$ and 5.92 ppm and singlets at $\delta = 2.26$ and 1.99 ppm. These data are consistent with nonequivalent MeCp groups arising from the chair conformation of the TiS_5 ring.

The pentasulfido complex reacts with tertiary phosphines to give dimers $[Ti(MeCp)_2]_2S_x$ $(x = 4$ and 6).[8,9] The pentasulfido complex also gives heterocycles of the type $Ti(MeCp)_2S_4CR_2$ when treated with ammonium sulfide in the presence of certain alkylating agents.[9,10] In fact, the Me_2CS_4 chelate is a minor side product in the synthesis of $Ti(MeCp)_2S_5$. The compound $Ti(MeCp)_2S_5$ reacts with electrophilic acetylenes to give dithiolenes

---

*$Ti(MeCp)_2Cl_2$ is prepared by the following procedure. Freshly cracked and distilled MeCpH (excess, 30 mL) is added dropwise to 1.38 g (60 mmol) of freshly pressed Na wire in 250 mL of THF. After all of the Na metal has been consumed the mixture is stirred for an additional 1 h. The colorless solution of MeCpNa is added dropwise to a solution of 3.29 mL (30 mmol) of $TiCl_4$ in 200 mL of toluene. The resultant deep red slurry is stirred for 10 h. A 6 M HCl solution (200 mL) is cautiously added to the reaction mixture, and the aqueous phase is extracted with 100-mL portions of CH_2Cl_2. The combined organic layers are concentrated to 300 mL and stored at $-25\,°C$ for 12 h. A further recrystallization affords 6.8 g (24.5 mmol, 82%) of bronze crystals of $Ti(MeCp)_2Cl_2$.

†The ammonium polysulfide solution is prepared by adding 12 g of elemental sulfur to 40 g of 20% amonium sulfide (Alfa). The deep red solution is stirred for 10 h. Undissolved sulfur is not filtered, so that the polysulfide solution remains saturated. The solution is filtered prior to use.

$Ti(MeCp)_2S_2C_2R_2$.[9,11] Thermolysis of $TiCp_2S_5$ at 140 °C induces a rearrangement to $TiCp(S_5C_5H_5)$.[12]

A prominent application of $TiCp_2S_5$ involves its reactions with main group halides according to the following stoichiometry

$$TiCp_2S_5 + ECl_2 \longrightarrow TiCp_2Cl_2 + ES_5$$

In this way S_6, S_7, S_9, S_{10}, S_{11}, S_{13}, S_{15}, S_{20}, $S_{10}(CO)_4$, and S_5Se_2 have been prepared.[1,13] In this application $TiCp_2S_5$ is preferred relative to $Ti(MeCp)_2S_5$ because the poorly soluble $TiCp_2Cl_2$ is easily separated from the products.

C. BIS(η^5-METHYLCYCLOPENTADIENYL)DIVANADIUM PENTASULFIDE

$$2NaMeCp + VCl_2(thf)_x \longrightarrow V(MeCp)_2 + 2NaCl$$

$$V(MeCp)_2 + \tfrac{5}{8}S_8 \longrightarrow V(MeCp)_2S_5$$

$$V(MeCp)_2S_5 \xrightarrow[\Delta]{} V_2(MeCp)_2S_5 + CH_3C_5H_4S_x$$

tetrahydrofuran = thf (ligand) and THF (solvent)

Procedure

■ **Caution.** *This synthesis should be conducted in a well-ventilated fume hood. Tetrahydrofuran, toluene, and dichloromethane are all harmful if inhaled and can cause skin and eye irritation. Tetrahydrofuran forms explosive peroxides; only fresh, peroxide-free material should be used.*

All operations are performed under a nitrogen atmosphere unless otherwise stated. In a 500-mL flask with a side arm, a slurry of VCl_3 (Cerac) (16.0 g, 101.7 mmol) in 100 mL of dry THF* is heated at reflux for 12 h to give a pink slurry of $VCl_3(thf)_3$.[†] To the $VCl_3(thf)_3$ thus obtained is added 3.3 g of zinc dust while the slurry is warm, and the resulting mixture is stirred for 15 min to give a purple solution of "$VCl_2(thf)_x$".[14]

A separate 500-mL round-bottomed flask with a side arm is charged with 12.0 g of a 40% sodium dispersion in mineral oil (Aldrich) and 100 mL of dry THF. Doubly distilled methylcyclopentadiene (20 mL, 234.8 mmol) is then added dropwise to the sodium using a pressure equalizing dropping funnel. When all the methylcyclopentadiene is added, the mixture is heated at reflux

---

*Aldrich Gold Label was redistilled from Sodium benzophenone.
[†]A more detailed description of $VCl_3(thf)_3$ can be found in *Inorg. Synth.* **21**, 138 (1982).

for 1 h to ensure complete conversion to the sodium methylcyclopentadienide.

After the sodium methylcyclopentadienide solution has cooled to room temperature, it is added via a cannula to the slurry of "$VCl_2(thf)_x$", and this mixture is heated at reflux for 1 h. The resulting dark violet mixture is allowed to cool, and the solvent is removed *in vacuo*.

Dry toluene, 120 mL, is added to the oily residue in order to extract the $V(MeCp)_2$ formed.[15] The mixture is Schlenk filtered into a 500-mL two-necked flask containing 10.1 g of elemental sulfur. The solution assumes the red-brown color characteristic of $V(MeCp)_2S_5$.[3] The mixture is heated at reflux for 5 h to induce the formation of the divanadium complex. The hot reaction mixture is then filtered in air through a bed comprised of 1 in. each of Celite® and silica gel, washing with CH_2Cl_2 (~1 L) until the eluate is colorless. The filtrate is concentrated on a rotary evaporator to 50 mL (at which stage black crystals form) and then stored at $-25\,°C$ for 16 h. The black crystals are removed by filtering and washed with hexanes (10 mL). Yield: 10.73 g, 50% based on VCl_3.

Anal. Calcd. for $C_{12}H_{14}S_5V_2$: C, 34.28; H, 3.36; S, 38.13; V, 24.23. Found: C, 34.28; H, 3.46; S, 38.03; V, 23.89.

Properties

Black crystalline $V_2(MeCp)_2S_5$ is air stable both in solution and in the solid state. It sublimes at $150\,°C$ (0.05 torr) with decomposition. The purity of $V_2(MeCp)_2S_5$ is easily checked by thin layer chromatography (TLC) on silica gel, eluting with toluene–hexane. It is moderately soluble in CH_2Cl_2 and toluene. The ^1H NMR spectrum ($CDCl_3$), consisting of multiplets at $\delta = 6.64$ (2H), 6.57 (2H), and 6.42 (4H) and a singlet at $\delta = 2.38$ ppm (6H), indicates that the unsymmetrical V_2S_5 score is retained in solution. IR (Nujol mull): 1067 (m), 1052 (m), 1034 (m), 931 (w), 917 (w), 889 (w), 815 (s), 602 (w), 566 (m), 534 (m), 469 (w), and 430 (w) cm^{-1}. The dark green selenium analog, $V_2(MeCp)_2Se_5$, can be prepared in a similar manner using red selenium.

D. BIS(η⁵-METHYLCYCLOPENTADIENYL)DIVANADIUM TETRASULFIDE

$$V_2(CH_3C_5H_4)_2S_5 + PBu_3 \longrightarrow V_2(CH_3C_5H_4)_2S_4 + SPBu_3$$

Procedure

All operations are performed under an inert atmosphere. A solution of tributylphosphine (Aldrich) (1.50 mL, 6.02 mmol) in CH_2Cl_2 (10 mL) is added dropwise via a dropping funnel to a solution of $V_2(MeCp)_2S_5$ (2.01 g, 4.80 mmol) in 20 mL of CH_2Cl_2. The addition takes ~15 min during which time the purple solution of $V_2(MeCp)_2S_5$ turns red. The reaction is complete immediately after the addition of Bu_3P. The resulting solution is concentrated to 10 mL, diluted with 10 mL of hexanes, and further concentrated to 5 mL to give purple crystals of $V_2(MeCp)_2S_4$. The crystals are isolated by filtration under nitrogen and washed with hexanes. Yield: 1.58 g, 84%.

Anal. Calcd. for $C_{12}H_{14}S_4V_2$: C, 37.11; H, 3.63; S, 33.02; V, 26.23. Found: C, 37.11; H, 3.59; S, 33.25; V, 25.98.

Properties

The purple $V_2(MeCp)_2S_4$ is very soluble in CH_2Cl_2 and moderately soluble in aromatic hydrocarbons. Its solutions are air sensitive, and the solid is moderately air sensitive. It reacts with Bu_3P, further desulfurizing to $V_4(MeCp)_4S_4$ and $V_5(MeCp)_5S_6$.[6,16] When heated, $V_2(MeCp)_2S_4$ decomposess to give $V_2(MeCp)_2S_5$. It can, however, be converted back to the pentasulfide more cleanly by treatment with elemental sulfur. It also reacts with electrophilic acetylenes[17] and diazenes.[18] It is a versatile ligand and forms cluster derivatives with many organometallic compounds.[19] Its ^1H NMR spectrum ($CDCl_3$), consisting of multiplets at $\delta = 6.67$ (4H), and 6.54 (4H) and a singlet at $\delta = 2.44$ ppm (6H), indicates a symmetrical V_2S_4 core structure.

Comments on Methylcyclopentadienyl Compounds

The MeCp ligand offers three advantages relative to Cp: stereochemical information by ^1H (and ^{13}C) NMR, improved solubility, and superior crystals for X-ray crystallography. The ^1H NMR spectra shown in Fig. 1 demonstrate the sensitivity of the ring proton chemical shifts to their environment. This method works particularly well for proton NMR frequencies $\geqslant 200$ MHz and relies on the fact that $J_{(H,H')} \ll |\delta_H - \delta_{H'}|$. The solubilities of MeCp complexes are significantly improved relative to Cp complexes; they are more soluble in chlorocarbons and aromatic solvents, yet they still can be precipitated using alkanes or alcohols. The change from MeCp to Cp does not lead to dramatically altered structural or reactivity characteristics (unlike the case with C_5Me_5). With regard to crystallographic considerations, we and others have been unable to characterize $V_2Cp_2S_5$, $V_2Cp_2S_4$, and $V_4Cp_4S_4$ structurally, but the crystallography on the MeCp analogs is straightforward.

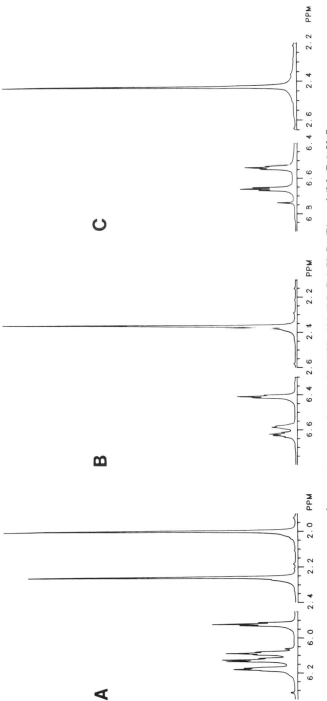

Fig. 1. 300 MHz ^{1}H NMR spectra for $(MeCp)_2TiS_5$ (A), $(MeCp)_2V_2S_5$ (B), and $(MeCp)_2V_2S_4$ (C). Spectrum A is consistent with nonequivalent MeCp groups (two Me signals), each of which is bisected by the same plane of symmetry. Spectrum B shows that the MeCp groups are equivalent (one Me signal), but that these groups do not lie on a plane of symmetry since there are clearly more than two ring proton multiplets. In contrast, spectrum C is consistent with a compound featuring equivalent MeCp groups which are bisected by a symmetry plane. The peak at $\delta 6.75$ is assigned to the C_5H_5 signal in $Cp(MeCp)V_2S_4$.

References

1. M. Draganjac and T. B. Rauchfuss, *Angew. Chem. Int. Ed. Engl.*, **24**, 742 (1985).
2. H. Köpf and B. Block, *Chem. Ber.*, **102**, 1504 (1969).
3. R. A. Schunn, C. J. Fritchie, and E. T. Prewitt, *Inorg. Chem.*, **5**, 892 (1966).
4. E. G. Muller, J. L. Petersen, and L. F. Dahl, *J. Organomet. Chem.*, **111**, 91 (1976).
5. C. M. Bolinger, T. B. Rauchfuss, and A. L. Rheingold, *Organometallics*, **1**, 1551 (1982).
6. C. M. Bolinger, Ph.D. Thesis, University of Illinois, Urbana-Champaign, IL, 1984.
7. N. M. Sergeyev, in *Progress in Nuclear Magnetic Resonance Spectroscopy*, J. W. Emsley, J. Feeney, and L. H. Sutcliff (eds), Pergamon, Oxford, 1973, pp. 71–144.
8. C. M. Bolinger, T. B. Rauchfuss, and S. R. Wilson, *J. Am. Chem. Soc.*, **103**, 5620 (1981).
9. D.M. Giolando, T. B. Rauchfuss, A. L. Rheingold, and S. R. Wilson, *Organometallics*, **6**, 667 (1987).
10. D. M. Giolando, T. B. Rauchfuss, *Organometallics*, **3**, 1551 (1984).
11. C. M. Bolinger and T. B. Rauchfuss, *Inorg. Chem.*, **21**, 3947 (1982).
12. D. M. Giolando and T. B. Rauchfuss, *J. Am. Chem. Soc.*, **106**, 6455 (1984).
13. R. Steudel, *Top. Curr. Chem*, **102**, 149 (1982).
14. R. J. Bouma, J. H. Teuben, W. R. Buekema, R. K. Bansemer, J. C. Huffman, and K. G. Caulton, *Inorg. Chem.*, **23**, 2715 (1984).
15. F. H. Köhler and W. Prössdorf, *Z. Naturforsch. Tiel B* **32**, 1026 (1977).
16. J. Darkwa, J. R. Lockemeyer, P. D. W. Boyd, T. B. Rauchfuss, and A. L. Rheingold, *J. Am. Chem. Soc.*, **110**, 141 (1988).
17. C. M. Bolinger, T. B. Rauchfuss, and A. L. Rheingold, *J. Am. Chem. Soc.*, **105**, 6321 (1983).
18. C. M. Bolinger, T. B. Rauchfuss, and S. R. Wilson, *J. Am. Chem. Soc.*, **106**, 7800 (1984).
19. C. M. Bolinger, T. D. Weatherill, T. B. Rauchfuss, A. L. Rheingold, C. S. Day, and S. R. Wilson, *Inorg. Chem.*, **25**, 634 (1986).

11. CYCLOMETALLAPOLYSULFANES (AND SELANES) OF BIS(η^5-CYCLOPENTADIENYL) TITANIUM(IV), ZIRCONIUM(IV), MOLYBDENUM(IV), AND TUNGSTEN(IV)

Submitted by ALAN SHAVER,* JAMES M. MCCALL,† and
GABRIELA MARMOLEJO‡
Checked by F. BOTTOMLEY,§ E. C. FERRIS,§ and J. A. GLADYSZ¶

Since the mid 1950s the development of the chemistry of the catenating element carbon with transition metals has been on the frontier of research for organometallic chemists. However, even by the late 1960s, the number of transition metal complexes containing simple polysulfido ligands, S_x^{2-}, where $x > 1$, was relatively small, given the catenating propensity of sulfur. Since then the number of such complexes has increased greatly, and their syntheses, structures, and reactivities were recently reviewed.[1] The complexes MCp_2S_5, where $Cp = \eta^5\text{-}C_5H_5$ and $M = $ Ti, Zr, and Hf; MCp_2S_4, where $M = $ Mo, and W; $MCp_2^*S_3$, where $Cp^* = \eta^5\text{-}C_5(CH_3)_5$ and $M = $ Ti, Zr, and Hf present a fascinating series of structures and reactivities that provide insight into aspects of metal–sulfur chemistry. The compound $TiCp_2S_5$ was first prepared in 1966[2]; it has been isolated from several different reactions involving $TiCp_2$ and sulfur species.[3-6] It possesses a classic cyclohexane-like TiS_5 ring[7,8] with axial and equatorial Cp rings, the barrier to interconversion being $\sim 76\,kJ\,mol^{-1}$.[9] Treatment of $TiCp_2S_5$ with S_xCl_2 is a general route to sulfur allotropes such as S_7.[3] One method of preparing $TiCp_2S_5$ is via treatment of $TiCp_2Cl_2$ with aqueous ammonium polysulfide.[5] More recently, it has been prepared[6] using tetrahydrofuran (THF) solutions of lithium polysulfides prepared[10] from $Li[BEt_3H]$ (Super Hydride®) and sulfur. The latter route, described here, has the advantage of being generally applicable to the preparation of MCp_2S_5 and $MCp_2^*S_3$, where $M = $ Ti, Zr, and Hf,[6] and $ThCp_2^*S_5$.[11] It avoids the use of H_2S required to prepare aqueous ammonium polysulfide. Substituted cyclopentadienyl rings can also be present, and the complexes $Ti(\eta^5\text{-}C_5H_4(CH_3)_2)S_5$, $Ti(\eta^5\text{-}C_5H_4Si(CH_3)_3)_2S_5$, Ti-$(CH_2(\eta^5\text{-}C_5H_4)_2)S_5$, and $Ti(\eta^5\text{-}C_5H_5)(\eta^5\text{-}C_5(CH_3)_5)S_5$ have been prepared via this route. The complexes MCp_2S_5, where $M = $ Zr, and Hf,

*Department of Chemistry, McGill University, 801 Sherbrooke St. W., Montreal, Quebec, Canada, H3A 2K6.
†Domtar Inc., Senneville, Quebec, Canada, H9X 3L7.
‡Flamencos 63-1 Mexico D.F., 03900, Mexico.
§Department of Chemistry, University of New Brunswick, Fredericton, New Brunswick, Canada E3B 5A3.
¶Department of Chemistry, The University of Utah, Salt Lake City, UT 84112.

could not be prepared under aqueous conditions. They too have cyclohexane-like rings, but the barriers to interconversion are ~ 49 and $58 \, \text{kJ mol}^{-1}$, respectively.[6] The polyselanes, MCp_2Se_5 can be prepared for $M = Ti$, Zr, and Hf via an analogous route using THF solutions of lithium polyselenides.[12] The complexes $MCp_2^*S_3$, where $M = Ti$, Zr, and Hf, prepared via treatment of $MCp_2^*Cl_2$ with the lithium polysulfide, were the first complexes possessing a bidentate S_3^{2-} ligand.[6,13] The MS_3 ring is markedly nonplanar, with pseudoaxial and equatorial sites for the Cp* ligand; the barrier to inter-conversion is $\sim 40 \, \text{kJ mol}^{-1}$.

The complex $MoCp_2S_4$ has been prepared via treatment of $MoCp_2Cl_2$ with Na_2S_2 (ref. 14) in ethanol. Lithium polysulfide solutions in THF can also be used, but the yields are lower.[15] The two complexes $MoCp_2S_2$ and $MoCp_2S_4$ can be interconverted by treating the former with sulfur[14] or by treating the latter with two equivalents of triphenylphosphine.[15] Compounds MCp_2S_4, where $M = Mo$ and W, possess nonplanar MS_4 rings[16] while the S_2^{2-} ligand is bonded side-on in $MoCp_2S_2$.[15] The complex WCp_2S_2 has not been isolated. Preparations reported here are modifications of the Na_2S_2 route.

A. BIS(η^5-CYCLOPENTADIENYL)[PENTASULFIDO(2−)]-TITANIUM(IV), $TiCp_2S_5$

$$2Li[BEt_3H] + \tfrac{5}{8}S_8 \longrightarrow Li_2S_5 + 2Et_3B + H_2$$

$$TiCp_2Cl_2 + Li_2S_5 \longrightarrow TiCp_2S_5 + 2LiCl$$

Procedure

■ **Caution.** *All reactions involving solutions of lithium polysulfides or polyselenides generate compounds that are malodorous and toxic to varying degrees and consequently must be performed in a well-ventilated fume hood. Li[BEt₃H] can ignite on contact with water, alcohols, or air. This reagent should be handled only under an inert atmosphere.*

The preparation of the THF solutions of lithium polysulfides or poly-selenides is general and is described in detail here. The reaction apparatus consists of a 300-mL three-necked round-bottomed flask containing a stirring bar and fitted with a nitrogen inlet, a pressure equalizing dropping funnel (100 mL), and a serum stopper. The flask is charged with sulfur powder (0.48 g, 1.9 mmol) (Aldrich, sublimed, 100 mesh) and the apparatus is evacuated (oil pump, 0.01 torr) through the nitrogen inlet and refilled with N_2 (Linde, prepurified) three times. Super Hydride (6.0 mL of a $1 \, M$ solution of Li[BEt₃H] in THF, 6.0 mmol) (Aldrich), is slowly added by means of syringe.

to the stirred sulfur powder. The resulting exothermic and effervescent reaction mixture is stirred for 20 min.

A red solution of $TiCp_2Cl_2$ (0.75 g, 3.0 mmol) (Alfa) in THF (75 mL, heated to reflux over sodium benzophenone and distilled under N_2 just before use) is prepared under N_2 in a 300-mL three-necked round-bottomed flask containing a stirring bar and fitted with a nitrogen inlet, and two stoppers. One of the stoppers is removed, and the solution is further degassed by a rapid purge of N_2 for 3 min and transferred to the dropping funnel via a 50-mL syringe. The $TiCp_2Cl_2$ is added dropwise over 20 min to the lithium polysulfide, to give a deep red reaction solution, which is stirred under N_2 at room temperature for 14 h. The solution is then stripped to dryness at room temperature (oil pump, 0.01 torr) and the residue is extracted with CH_2Cl_2 (100 mL, spectrograde). The solution is filtered in air through Celite® (5 mL) and the solvent is removed at room temperature by means of a rotary evaporator (15 torr) to give the product as dark red crystals (0.99 g, 98%). The product is spectroscopically (NMR) pure, mp 196–198 °C, lit.[5] 201–202 °C.*

Anal. Calcd. for $C_{10}H_{10}S_5Ti$: C, 35.49; H, 2.98; S, 47.37. Found: C, 35.22; H, 3.34; S, 47.27.

B. BIS(η^5-CYCLOPENTADIENYL)[PENTASELENIDO(2 −)]-TITANIUM(IV), $TiCp_2Se_5$

$$2Li[BEt_3H] + 5Se \longrightarrow Li_2Se_5 + 2Et_3B + H_2$$

$$TiCp_2Cl_2 + Li_2Se_5 \longrightarrow TiCp_2Se_5 + 2LiCl$$

Procedure

As in the procedure in Section A, powdered gray selenium (1.38 g, 15.0 mmol) is treated with Super Hydride (6.0 mL of a 1 M solution of $Li[BEt_3H]$ in THF, 6.0 mmol) to give a brown solution of lithium polyselenides. A suspension of $TiCp_2Cl_2$ (0.75 g, 3.0 mmol) in dry THF (50 mL) is purged with N_2 and added dropwise to the lithium polyselenides. The reaction mixture becomes purple; it is stirred under N_2 at ambient temperature for 4.75 h. The volatile components are removed *in vacuo* (0.01 torr). The black residue is extracted under N_2 with CH_2Cl_2 (6 × 50 mL), and the solution is filtered under N_2 through Celite (5 mL). The purple filtrate is concentrated *in vacuo* to ~ 50 mL

*The checker (JAG) obtained a 92% yield of microcrystalline product with mp 190 to 192 °C. He reports the presence of a small amount of semisolid residue between the crystals.

and cooled at $-20\,°C$ overnight. Crops of the product, air-stable purple crystals, are collected by filtration from successive concentrations of the mother liquors (1.23 g, 72%). The product foams without melting at 209–211 °C (lit.[5] mp 211 °C).*

Anal. Calcd. for $C_{10}H_{10}Se_5Ti$: C, 20.97; H, 1.76. Found: C, 21.12; H, 1.77.

C. BIS(η^5-PENTAMETHYLCYCLOPENTADIENYL)-[TRISULFIDO(2−)] TITANIUM(IV), $TiCp_2^*S_3$

$$2Li[BEt_3H] + \tfrac{5}{8}S_8 \longrightarrow Li_2S_5 + 2Et_3B + H_2$$

$$TiCp_2^*Cl_2 + Li_2S_5 \longrightarrow TiCp_2^*S_3 + \tfrac{2}{8}S_8$$

$$Cp^* = \eta^5\text{-}C_5(CH_3)_5$$

Procedure

Powdered sulfur (0.16 g, 0.63 mmol) is treated with Super Hydride (2.0 mmol) as in the procedure in Section A to give the lithium polysulfide solution. A suspension of $TiCp_2^*Cl_2$ (ref. 17) (available from Strem) (0.39 g, 1.0 mmol) in dry THF (30 mL) is added dropwise under N_2 and the reaction mixture is heated at reflux for 10.5 h. The volatile components are removed under vacuum (0.01 torr, room temperature) and the residue is extracted under N_2 with CH_2Cl_2 (4 × 25 mL). The solution is filtered under N_2 through Celite (5 mL) and the solvent is removed under vacuum as above to give the product as a black microcrystalline solid which is spectroscopically (NMR) pure.[†] The product is dissolved in dry THF (5 mL), pentane (10 mL) is added, and the solution is stored at $-20\,°C$ for 5 to 10 days to give black crystals. The mother liquors are concentrated twice to give two more crops of crystals (0.23 g combined, (55%); mp 149–152 °C). The moderate yield is a result in part of the high solubility of the complex.

Anal. Calcd. for $C_{20}H_{30}S_3Ti$: C, 57.95; H, 7.29, S, 23.20. Found: C, 57.43; H, 7.40, S, 22.83.

---

*The checker (JAG) obtained a 59% yield from four successive concentrations of the mother liquors. He reports the presence of a small amount of semisolid residue in the product, which foamed at 207 °C.

†The checker (JAG) obtained a black (but not microcrystalline) solid in 87% yield. The product was pure by ^1H NMR, but attempts to recrystallize it were unsuccessful.

D. BIS(η^5-CYCLOPENTADIENYL)[TETRASULFIDO(2−)]-MOLYBDENUM(IV), MoCp$_2$S$_4$

$$Na + C_2H_5OH \longrightarrow NaOC_2H_5 + \tfrac{1}{2}H_2$$

$$NaOC_2H_5 + H_2S \longrightarrow NaSH + C_2H_5OH$$

$$2NaSH + \tfrac{1}{8}S_8 \longrightarrow Na_2S_2 + H_2S$$

$$MoCp_2Cl_2 + 3Na_2S_2 \longrightarrow MoCp_2S_4 + 2Na_2S + 2NaCl$$

■ **Caution.** *The preparation involves the use of H_2S, which is very toxic and has a bad odor. Extreme care must be exercised. The reactions must be conducted in a well-ventilated fume hood and every effort must be made to ensure containment of the gas. The use of an H_2S manifold such as the one illustrated in Fig. 1 is recommended.*

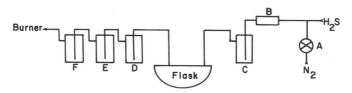

Fig. 1. H$_2$S manifold (schematic).

The gas is dried by means of a calcium chloride column (B), and is passed through a Nujol bubbler (C) before entering the reaction vessel. Unreacted gas is passed through three washing towers containing, successively, 5 *M* NaOH (D), saturated aqueous lead acetate (E), and 5 *M* NaOH (F) and finally into the air intake of a Bunsen burner that is burning natural gas. Connecting the H$_2$S line to a source of N$_2$ via stopcock A permits the system to be purged before and, more importantly, after the reaction, which significantly reduces residual contamination.

Procedure

A 100-mL three-necked round-bottomed flask, is fitted with a stirring bar, an inlet bubbling tube connected to the H$_2$S manifold, a stopper, and an exit tube connected to the washing towers. The system is purged with N$_2$ (via A), and the flask is charged with absolute ethanol (25 mL). Sodium (0.57 g, 25 mmol, fresh pea sized fragments rapidly cut from a large lump in air) is carefully added in portions to the ethanol with stirring.

■ **Caution.** *Sodium metal reacts violently with water liberating highly flammable H_2 gas. The metal causes burns on contact with the skin.*

After all the sodium has reacted, the solution is treated with H_2S (A closed) at a moderate rate for 15 to 30 min until the solution is light lemon yellow in color. The solution is then purged with N_2 for 30 min. The solution is warmed to 60 °C, sulfur (0.46 g, 14 mmol) (Aldrich, sublimed 100 mesh) is added, and the mixture is stirred under N_2 for 1 h. The compound $MoCp_2Cl_2$ (1.19 g, 4 mmol)[18] (available from Strem) is added, and the mixture is stirred for another 1 h under N_2 and then cooled. Water (100 mL) is added, and the reaction mixture is transferred in air to a separatory funnel (500 mL) where it is extracted with $CHCl_3$ (5 × 20 mL).

■ **Caution.** *Chloroform is harmful if inhaled. It is a suspected carcinogen and may cause adverse reproductive effects. Handle it only in a well-ventilated fume hood.*

The combined $CHCl_3$ extracts are stripped to dryness at room temperature (oil pump, 0.01 torr), and the residue is pumped overnight. Two consecutive recrystallizations from boiling N,N-dimethylformamide (DMF) (10 mL) gives the product as red crystals (1.17 g, 70%), dec ~ 175 °C, lit.[14] dec ~ 180 °C.*

Anal. Calcd. for $C_{10}H_{10}MoS_4$: C, 33.89; H, 2.85; S, 36.18. Found: C, 33.69; H, 2.38; S, 36.07.

Properties

The polysulfane complexes $TiCp_2S_5$, $MoCp_2S_4$, and $TiCp_2^*S_3$ are red to dark red materials, the titanium complexes being so dark as to appear black. All are thermally stable and surprisingly air stable in the solid state and in solution. They are fairly soluble in most organic solvents except simple hydrocarbons. The NMR data for the complexes[6] (ppm relative to TMS, $CDCl_3$, room temperature) are $TiCp_2S_5$, 6.03, 6.32; $MoCp_2S_4$, 5.23; and $TiCp_2^*S_3$, 1.87. The compound $TiCp_2Se_5$ is purple and its properties are similar to those of its sulfur analog; NMR: 5.92, and 6.34. All the complexes give parent ion peaks in their mass spectra.

References

1. M. Draganjac and T. B. Rauchfuss, *Angew. Chem. Int. Ed. Engl.*, **24**, 742 (1985).
2. E. Samuel, *Bull. Soc. Chim. Fr.*, **1966**, 2548.
3. M. Schmidt, B. Block, H. D. Block, H. Köpf, and E. Wilhelm, *Angew. Chem. Int. Ed. Engl.*, **7**, 642 (1968).
4. E. Samuel and G. Giannotti, *J. Organomet. Chem.*, **113**, C17 (1976).

*The checkers (FB and ECF) recommend recrystallization from hot chloroform (yield 74%) instead of hot DMF.

5. (a) H. Köpf, B. Block, and M. Schmidt, *Chem. Ber.*, **101**, 272 (1968). (b) J. Darkwa, D. M. Giolando, C. A. Jones, and T. B. Rauchfuss, *Inorg. Synth.*, **27**, 51 (1990).

6. A. Shaver and J. M. McCall, *Organometallics*, **3**, 1823 (1984).

7. E. F. Epstein, I. Bernal, and H. Köpf, *J. Organomet. Chem.*, **26**, 229 (1971).

8. E. G. Müller, J. L. Petersen, and L. Dahl, *J. Organomet. Chem.*, **111**, 91 (1976).

9. E. W. Abel, M. Booth, and K. G. Orrell, *J. Organomet. Chem.*, **160**, 75 (1978).

10. J. A. Gladysz, V. K. Wong, and B. G. Jick, *Tetrahedron*, **35**, 2329 (1979).

11. D. A. Wrobleski, D. T. Cromer, J. V. Ortig, T. B. Rauchfuss, R. R. Ryan, and A. P. Sattelberger, *J. Am. Chem. Soc.*, **108**, 174 (1986).

12. J. A. Gladysz, J. L. Hornby, and J. E. Garbe, *J. Org. Chem.*, **43**, 1204 (1978).

13. P. H. Bird, J. M. McCall, A. Shaver, and U. Siriwardane, *Angew. Chem. Int. Ed. Engl.*, **21**, 384 (1982).

14. H. Köpf, S. K. S. Hazani, and M. Leitner, *Z. Naturforsch.*, **33b**, 1398 (1978).

15. A. Shaver and G. Marmolejo, unpublished results.

16. (a) H. D. Block and R. Allmann, *Cryst. Struct. Commun.*, **4**, 53 (1975). (b) B. R. Danis and I. Bernal, *J. Cryst. Struct.*, **2**, 135 (1972).

17. (a) J. E. Bercaw, R. H. Marvich, L. G. Bell, and H. H. Brintzinger, *J. Am. Chem. Soc.*, **94**, 1219 (1972). (b) J. E. Bercaw, *J. Am. Chem. Soc.*, **96**, 5087 (1974).

18. (a) M. L. H. Green and P. J. Knowles, *J. Chem. Soc. Perkin 1*, 989 (1973). (b) R. B. King, *Organometallic Synthesis*, Vol. 1, Academic Press, New York, 1965, p. 79.

12. BIS(HYDROGENSULFIDO) COMPLEXES OF BIS(η5-CYCLOPENTADIENYL) TITANIUM(IV) AND TUNGSTEN(IV)

Submitted by ALAN SHAVER,* GABRIELA MARMOLEJO,† and
JAMES M. MCCALL‡
Checked by THOMAS B. RAUCHFUSS,§ LENORE KOCZON,§ and
SCOTT SIMERLY§

Metal hydrogensulfido complexes (LMSH) are interesting species, the chemistry of which is beginning to attract attention. The reactivity of such complexes with organic reagents has been investigated,[1] and have been isolated from model systems.[2] The complexes $MCp_2(SH)_2$ (M = Ti, Mo, and W) have been known for some time, and have recently been shown to be

*Department of Chemistry, McGill University, 801 Sherbrooke St. W., Montreal, Quebec, Canada, H3A 2K6.
†Flamencos 63-1 Mexico D.F., 03900, Mexico.
‡Domtar Inc., Senneville, Quebec, Canada H9X 3L7.
§School of Chemical Sciences, University of Illinois, Urbana, IL 61801.

precursors to heterometallic dimers[3] and to complexes containing novel catenated polysulfido ligands.[4] The literature method for the preparation of $TiCp_2(SH)_2$ involves treating $TiCp_2Cl_2$ with H_2S in the presence of Et_3N in anhydrous ether, followed by extraction with water to remove the by-product Et_3NHCl.[5] This procedure is difficult to reproduce,[3b] and a report has appeared cautioning against the isolation of $TiCp_2(SH)_2$.[6] The anhydrous conditions described here are a significant improvement, giving the product in nearly quantitative yield. The complex $WCp_2(SH)_2$ was prepared in 1967 via treatment of the dichloride with $NaSH$ in ethanol,[7] and the method described here is a modification of this procedure.

■ **Caution.** *The preparations involve the use of H_2S, which is very toxic and has a bad odor. Extreme care must be exercised. The reactions must be conducted in a well-ventilated fume hood and every effort must be made to ensure containment of the gas. The recommended procedure for drying and handling H_2S is described in the preceding synthesis.*[8]

A. BIS(η^5-CYCLOPENTADIENYL)BIS(HYDROGENSULFIDO) TITANIUM(IV), $TiCp_2(SH)_2$

$$TiCp_2Cl_2 + 2H_2S + 2Et_3N \longrightarrow TiCp_2(SH)_2 + 2Et_3NHCl$$

Procedure

The reaction apparatus consists of a 1-L three-necked round-bottomed flask equipped with a stirring bar, an inlet bubbling tube connected to the H_2S manifold, a small (50 mL) pressure equalizing dropping funnel containing 7 mL of Et_3N (50 mmol), and an exit tube connected to the washing towers (see Fig. 1 in Section 11). The system is purged with N_2 (via A), and the flask is charged with 500 mL of tetrahydrofuran (THF) distilled from sodium benzophenone. The compound $TiCp_2Cl_2$ (6.0 g, 24 mmol) (Alfa) is dissolved with stirring to give a clear red solution (incomplete dissolution prior to addition of the H_2S leads to product contaminated by impurities that are difficult to remove). The solution is treated with H_2S (A closed) at a moderate rate (~ 2 bubbles/s) for 10 min; then the Et_3N is slowly added dropwise (5 min). The H_2S flow is maintained for a further 90 min and is then replaced (via A) with a N_2 purge for 30 min. The dark suspension is rapidly filtered (fume hood) in air through a pad of Celite® on a large sintered glass filter, followed by washing the filter cake with THF (4 × 10 mL) to give a malodorous red-black solution and an off-white residue (Et_3NHCl) on the frit. The combined filtrate and washings in a 1-L one-necked flask are stripped to dryness at room temperature (oil pump, 0.01 torr), and the brown residual powder is dried by pumping overnight (5.6 g, 93%, dec $\sim$ 130 °C lit.[5] dec 150–160 °C). Recrystallization from $CHCl_3$ (ref. 5) is accompanied by severe losses[3]; however, the powder is spectroscopically pure (NMR) and can be used as is.

Anal. Calcd. for $C_{10}H_{12}S_2Ti$: C, 49.18; H, 4.95; S, 26.26. Found: C, 49.97; H, 5.18; S, 23.18, 23.41.*

The compound $Ti(MeCp)_2(SH)_2$ prepared by the same method is easier to isolate in crystalline form.[9]

B. BIS(η^5-CYCLOPENTADIENYL)BIS(HYDROGENSULFIDO) TUNGSTEN(IV), $WCp_2(SH)_2$

$$Na + CH_3CH_2OH \longrightarrow NaOCH_2CH_3 + \tfrac{1}{2}H_2$$

$$NaOCH_2CH_3 + H_2S \longrightarrow NaSH + CH_3CH_2OH$$

$$WCp_2Cl_2 + 2NaSH \longrightarrow WCp_2(SH)_2 + 2NaCl$$

Procedure

Ethanol (70 mL) is placed in a 300-mL flask equipped as in Section A. Pieces of sodium (0.5 g, 22 mmol) are added under N_2, and the ethanol is stirred until the reaction is complete. The solution is treated with H_2S (ref. 10) for 15 min and then purged with N_2 for 15 min. The compound WCp_2Cl_2 (ref. 11) (1 g, 3.4 mmol) is added in one portion, and the mixture is stirred at 60 °C (oil bath) for 1 h. The brick red slurry is cooled to room temperature and H_2O (100 mL) is added. The mixture is filtered under N_2, and the brick red precipitate is discarded. The filtrate, in a 300-mL three-necked flask, is extracted under N_2 with 5 × 20 mL of $CHCl_3$; the chloroform layers are removed each time by means of a syringe and collected under N_2.

■ **Caution.** *Chloroform is harmful if inhaled. It is a suspected carcinogen and may cause adverse reproductive effects. Handle it only in a well-ventilated fume hood.*

The combined extracts are stripped to dryness at room temperature (oil pump, 0.01 torr) and pumped on overnight. The residue is dissolved in the minimum volume of $CHCl_3$ and transferred to the top of a chromatography column (20 × 2.5 cm) of deactivated alumina (10% H_2O) in hexanes. The column is eluted with $CHCl_3$, and one band containing the product is stripped to dryness (rotary evaporator) to give the analytical sample (0.78 g, 67%, dec $\sim$ 150 °C).

*The checkers worked on one half the recommended scale. Yield: 80 to 86%. They report the following analysis: C, 48.96; H, 5.10; S, 26.08.

Anal. Calcd. for $C_{10}H_{12}WS_2$: C, 31.59; H, 3.19; S, 16.87. Found: C, 31.56; H, 3.06; S, 16.46.

Properties

The complexes $MCp_2(SH)_2$, where M = Ti and W, possess unpleasant odors, especially for M = Ti, and are air sensitive. They should be stored in Schlenk tubes in a freezer; a sample of $TiCp_2(SH)_2$ showed no evidence of decomposition after 1 year at $-20\,°C$. They are soluble in $CHCl_3$, CH_2Cl_2, and THF. The compound $TiCp_2(SH)_2$ decomposes to polycrystalline oligomers[5] upon being heated in boiling benzene and also, after several months, on standing in the solid state at room temperature. Both complexes show peaks due to molecular ions in their mass spectra. Their ^1H NMR spectra in $CDCl_3$ display two peaks in the appropriate ratio, with the peaks resulting from the thiol protons appearing above TMS for $WCp_2(SH)_2$ (δ at ambient temperature):

Complex	(C_5H_5)	(SH)
$TiCp_2(SH)_2$	6.28	3.38
$WCp_2(SH)_2$	5.20	-1.49

References

1. R. J. Angelici and R. G. W. Gingerich, *Organometallics*, **2**, 89 (1983).
2. M. R. DuBois, M. C. VanDerweer, D. L. DuBois, R. C. Haltiwanger, and W. R. Miller, *J. Am. Chem. Soc.*, **102**, 7456 (1980).
3. (a) A. Shaver, J. M. McCall, P. H. Bird, and N. Ansari, *Organometallics*, **2**, 1894 (1983); (b) A. Shaver and J. M. McCall, *Organometallics*, **3**, 1823 (1984).
4. C. J. Ruffing and T. B. Rauchfuss, *Organometallics*, **4**, 524 (1985).
5. H. Köpf and M. Schmidt, *Angew. Chem. Int. Ed. Engl.*, **4**, 953 (1965).
6. R. Ralea, C. Ungurenasu, and S. Cihodaru, *Rev. Roum. Chim.*, **12**, 861 (1967).
7. M. L. H. Green and W. E. Lindsell, *J. Chem. Soc. Ser. A*, 1455 (1967).
8. A. Shaver, J. M. McCall, and G. Marmolejo, *Inorg. Synth.*, **27**, 59 (1989).
9. C. M. Bolinger, J. E. Hoots, and T. B. Rauchfuss, *Organometallics*, **1**, 223 (1982).
10. R. E. Eibeck, *Inorg. Synth.*, **7**, 128 (1963).
11. (a) M. L. H. Green and P. J. Knowles, *J. Chem. Soc. Perkin*, 1, 989 (1973); (b) R. B. King, *Organometallic Synthesis* Vol. 1, Academic Press, New York, 1965, p. 79.

13. (μ-DISULFIDO-S)(μ-DISULFIDO-η^2:η^2)BIS-(η^5-PENTAMETHYLCYCLOPENTADIENYL)(μ-THIO)-DICHROMIUM(Cr—Cr)

$$Cr_2(C_5Me_5)_2(CO)_4 + \tfrac{5}{8}S_8 \longrightarrow Cr_2(C_5Me_5)_2S_5 + 4CO$$

Submitted by JOACHIM WACHTER,* HENRI BRUNNER,* and WALTER MEIER*
Checked by UMAR RIAZ[†] and M. DAVID CURTIS[†]

The compound $Cr_2(C_5Me_5)_2S_5$ is the first chromium representative of sulfur rich dinuclear cyclopentadienyl transition metal complexes of general formula $M_2Cp_2S_x$ $(x \geqslant 4)$.[1] It is obtained from the reaction of $Cr_2(C_5Me_5)_2(CO)_4$ with sulfur in toluene solution.[2] Alternatively, it can be prepared from $Cr(C_5Me_5)_2$ and sulfur in about the same yield.

The title complex incorporates electron-rich disulfur ligands and may therefore serve as a useful substrate for the addition of electrophiles or in the synthesis of transition metal sulfide clusters, in particular those containing the M_4S_4 core.

Procedure

The procedure must be carried out under nitrogen with solvents freshly distilled under N_2 from Na–K alloy employing the Schlenk tube technique.[3] In a 250-mL flask, which is equipped with a reflux condenser and a mercury over-pressure valve, a solution of 1.46 g (3 mmol) of $Cr_2(C_5Me_5)_2(CO)_4$ (ref. 4)[‡] and 0.38 g (1.48 mmol) of S_8 in 100 mL of toluene is stirred at 45 °C for 66 h.[§] The reaction mixture is filtered, concentrated to 15 mL, and chromatographed on silica gel (0.063–0.200 mm, activity II–III, column 30 × 3 cm). Unreacted starting material, along with sulfur, is eluted with toluene as a green band, followed by a dark green band, which is eluted with 3:1 toluene–diethyl ether,[¶] while a blue green residue remains at the top of the column. After evaporation of the solvent from the second green band, 0.22 to 0.27 g

---

*Institut für Anorganische Chemie der Universität Regensburg, Universitätsstr. 31, D-8400 Regensburg, Federal Republic of Germany.
[†]Department of Chemistry, The University of Michigan, Ann Arbor, MI 48109.
[‡]The checkers obtained a 45% yield of $Cr_2(C_5Me_5)_2(CO)_4$ by following the procedure in ref. 4.
[§]Increased temperature and reaction time results in a considerable decrease of the yield.
[¶]The checkers observed a blue band between the two green bands. The blue band remained on the column after the tluene–ether elution.

(14–17%) of $Cr_2(C_5Me_5)_2S_5$ in the form of a green powder is obtained. The crude product is still contaminated by trace amounts of $Cr_2(C_5Me_5)_2S_4$.[5] An analytically pure sample can be obtained by the following procedure: A 0.34-g sample of the crude $Cr_2(C_5Me_5)_2S_5$ is dissolved in 17 mL of diethyl ether. After filtration, 5 mL of pentane is added, and the solution is stored for several days at $-25\,°C$ to give ~ 65 mg of dark green prisms. mp > 250 °C.

Anal. Calcd. for $C_{20}H_{30}Cr_2S_5$: C, 44.92; H, 5.65; S, 29.98; Cr, 19.45. Found: C, 44.88; H, 5.78; S, 29.76; Cr, 18.32.

Properties

The crystals can be stored under nitrogen for several weeks without decomposition. Solutions of $Cr_2(C_5Me_5)_2S_5$ are air sensitive. The compound is soluble in CH_2Cl_2, toluene, and acetone, but is insoluble in alcohols and pentane. IR spectrum (KBr pellet): 598 (w), 495 (m), 445 (w) cm^{-1} (v_{S-S} and v_{Cr-S} absorptions). ^1H NMR spectrum (CDCl$_3$): $\delta_{CH_3} = 2.13$. X-ray crystal structure: d(Cr—Cr) 2.489(2) Å, d(S—S) 2.101(5) Å (μ, η^1-S, S), d(S—S) 2.149(5) Å (μ, η^2-S$_2$). Reaction with PPh$_3$ gives $Cr_2(C_5Me_5)_2(\mu, \eta^2\text{-}S_2)(\mu\text{-}S)_2$,[5] reaction with H$_2$ (150 bar, 80 °C) leads to $Cr_4(C_5Me_5)_4S_4$.[5] $Cr_2(C_5Me_5)_2S_5$ serves as starting material for the formation of the following clusters: $Cr_4(C_5Me_5)_2(C_5H_5)_2S_4$,[6] $Cr_2Co_2(C_5Me_5)_2(CO)_2S_4$,[7] and $Cr_2Fe_2(C_5Me_5)_2(NO)_2S_4$.[8]

References

1. J. Wachter, *J. Coord. Chem.*, **15**, 219 (1987); L. Y. Goh and T. C. W. Mak, *J. Chem. Soc. Chem. Commun.*, 1474 (1986).

2. H. Brunner, J. Wachter, E. Guggolz, and M. L. Ziegler, *J. Am. Chem. Soc.*, **104**, 1765 (1982).

3. P. Fehlhammer, W. A. Herrmann, and K. Öfele, in *Handbuch der Präparativen Anorganischen Chemie*, G. Brauer (ed.), Vol. 3 Auflage, Enke Verlag, Stuttgart 1981.

4. R. B. King, M. Z. Iqbal, and A. D. King, jr., *J. Organomet. Chem.*, **171**, 53 (1979).

5. H. Brunner, H. Kauermann, W. Meier, and J. Wachter, *J. Organomet. Chem.*, **263**, 183 (1984); H. Brunner, J. Pfauntsch, J. Wachter, B. Nuber, and M. L. Ziegler, *J. Organomet. Chem.*, **359**, 179 (1989).

6. H. Brunner, H. Kauermann, and J. Wachter, *J. Organomet. Chem.*, **265**, 189 (1984).

7. H. Brunner and J. Wachter, *J. Organomet. Chem.*, **240**, C41 (1982); H. Brunner, W. Meier, J. Wachter, H. Pfisterer, and M. L. Ziegler, *Z. Naturforsch. Teil B*, **40**, 923 (1985).

8. H. Brunner, H. Kauermann, and J. Wachter, *Angew. Chem. Int. Ed. Engl.*, **22**, 549 (1983).

Chapter Three

EARLY TRANSITION METAL POLYOXOANIONS

14. INTRODUCTION TO EARLY TRANSITION METAL POLYOXOANIONS

WALTER G. KLEMPERER*

Early transition metal polyoxoanions have traditionally posed serious challenges to the synthetic chemist because of their stoichiometric and structural complexity. The past four decades have seen the application of modern physical chemical techniques of polyoxoanion chemistry, however, allowing an increasingly systematic approach to their synthesis.[1] This approach has in turn led to further integration of polyoxoanion chemistry into the broader areas of inorganic, organic, and organometallic chemistry, both in terms of synthetic techniques and applications.[2] The set of syntheses selected for this volume does not reflect a comprehensive attempt to represent the most important classes of complexes that have been prepared in recent years. Instead, an attempt has been made to focus selectively on systematic syntheses of well-characterized families of related species as well as syntheses of compounds that illustrate how the domain of polyoxoanion chemistry has been expanded to include organic and organometallic derivatives.

The six isopolyanions whose syntheses are described in Section 15 are relatively small and structurally simple species that are soluble in organic

*University of Illinois, Department of Chemistry, 505 South Mathews Avenue, Urbana, IL 61801.

solvents as tetrabutylammonium salts. They are thus representative of a fairly large class of polyoxoanions that have served as useful reagents for the syntheses of new derivatives in aprotic media. They also illustrate in a simple way some of the basic structure types observed in polyoxoanion chemistry. One structure type of particular relevance to this chapter consists of a neutral W_nO_{3n} cage encapsulating one or more formally anionic subunits. The hexatungstate ion shown in Section 15 Structure **b**, has such a structure where a W_6O_{18} cage encapsulates an O^{2-} ion. The W_6O_{18} cage consists of six $[WO]^{4+}$ units occupying the vertices of an octahedron plus twelve O^{2-} units bridging the edges of that octahedron so that each hexavalent tungsten center forms single bonds to four adjacent doubly bridging oxygen atoms plus a double bond to its terminal oxo ligand. The $[W_{10}O_{32}]^{4-}$ ion (Section 15 Structure **d**) has a similar structure where each tungsten center has the local environment just described for tungsten atoms in $[W_6O_{19}]^{2-}$, but is part of a larger $W_{10}O_{30}$ cage that encapsulates two O^{2-} ions.

The tungstosilicates and vanadotungstophosphates treated in Sections 16 and 17 are all derived from the Keggin anions $[SiW_{12}O_{40}]^{4-}$ and $[PW_{12}O_{40}]^{3-}$. Here, a $W_{12}O_{36}$ cage of the type W_nO_{3n}, just described for $[W_6O_{19}]^{2-}$ and $[W_{10}O_{32}]^{4-}$, encapsulates a tetrahedral oxoanion, either $[SiO_4]^{4-}$ or $[PO_4]^{3-}$. A family of closely related species has been generated through modification of this basic structural framework. One modification involves replacement of W^{VI} centers in the $W_{12}O_{36}$ cage with V^V centers to generate mixed-metal species. A second type of modification involves removal of W^{VI} centers from the $W_{12}O_{36}$ cage, along with their terminal oxo groups ($[WO]^{4+}$ units) plus bridging O^{2-} groups linking these centers, to generate defect structures. Examples of $W_{12}O_{36}$ cage defect structures include the $[W_{11}O_{35}]^{4-}$ cage of $[SiW_{11}O_{39}]^{8-}$ obtained by removal of a single $[WO]^{4+}$ unit, the $[W_{10}O_{32}]^{4-}$ cage of $[SiW_{10}O_{36}]^{8-}$ obtained by removal of two $[WO]^{4+}$ units plus two bridging O^{2-} units, and the $[W_9O_{30}]^{6-}$ cage of $[SiW_9O_{34}]^{10-}$ obtained by removal of three $[WO]^{4+}$ units plus three bridging O^{2-} units. The relationships between parent Keggin anions, their mixed-metal derivatives, and their defect structure are quite subtle since various forms of isomerism are possible. First, several topologically inequivalent geometries are possible for the parent $W_{12}O_{36}$ cage structure. These are distinguished below by α, β, and δ prefixes. Second, replacement or removal of a given number of tungsten centers to generate mixed-metal species or defect structures can generate isomeric structures defined by the locations of the centers removed from the $W_{12}O_{36}$ cage. The structures of these isomers are indicated below by numerical prefixes or subscripts as in $[\alpha\text{-}1,2,3\text{-}PV_3W_9O_{40}]^{6-}$ or $[\beta_2\text{-}SiW_{11}O_{39}]^{8-}$. Finally, alternative orientations of the tetrahedral $[SiO_4]^{4-}$ or $[PO_4]^{3-}$ subunit relative to the surrounding metal oxide cage is sometimes possible. In the case of the

$[PW_9O_{34}]^{9-}$ defect structure, for example, the $[PO_4]^{3-}$ can adopt two different orientations relative to the $[W_9O_{30}]^{6-}$ cage fragment, leading to two stable, isomeric configurations identified by the prefixes A and B.

Sections 18, 19, and 20 are concerned with compounds based upon or derived from W_nO_{3n} cages plus encapsulated anions and thus are related to the $[W_6O_{19}]^{2-}$, $[W_{10}O_{32}]^{4-}$, $[SiW_{12}O_{40}]^{4-}$, and $[PW_{12}O_{40}]^{3-}$ complexes and derivatives just described. In the α- and $[\beta-P_2W_{18}O_{62}]^{6-}$ structures, two $[PO_4]^{3-}$ units are encapsulated by a $W_{18}O_{54}$ cage. The $[As_2W_{21}O_{69}(H_2O)]^{6-}$ structure is based on a $W_{21}O_{63}$ cage that encapsulates two $[AsO_3]^{3-}$ anions, and the $[P_5W_{30}O_{110}]^{15-}$ structure is based on a $W_{30}O_{90}$ cage that encapsulates five $[PO_4]^{3-}$ anions. Defect structures of the type described previously for Keggin anions are also observed, the $[P_8W_{48}O_{184}]^{40-}$ ion providing the most impressive example. Here, eight $[PO_4]^{3-}$ anions are bound to the interior of a toroidal $[W_{48}O_{152}]^{16-}$ framework. The massive, symmetric nature of this species is perhaps best appreciated from a space-filling model:

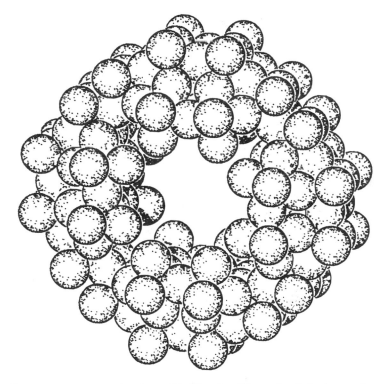

This D_{4h} structure contains four $[P_2W_{12}O_{44}]^{6-}$ units linked together symmetrically into a ring by eight doubly bridging O^{2-} units, and it can thus be viewed as the condensation product of four $[P_2W_{12}O_{48}H_4]^{10-}$ ions:

$$4[P_2W_{12}O_{48}H_4]^{10-} \longrightarrow [P_8W_{48}O_{184}]^{40-} + 8H_2O \tag{1}$$

From a synthetic point of view, the synthesis of the $[P_8W_{48}O_{184}]^{40-}$ anion represents a landmark achievement, since the structural building blocks just delineated are in fact also synthetic building blocks that themselves have a $[P_2W_{18}O_{62}]^{6-}$ defect structure. The preparation thus proceeds in three steps: (1) preparation of the parent $[\alpha\text{-}P_2W_{18}O_{62}]^{6-}$ or $(\alpha\text{-}W_{18}O_{54})(PO_4{}^{3-})_2$ ion, (2) conversion of the parent ion to the defect complex $[\alpha\text{-}P_2W_{12}O_{48}]^{14-}$ or $(\alpha\text{-}W_{12}O_{40}{}^{8-})(PO_4{}^{3-})_2$, in its diprotonated form, and finally (3) protonation of this defect complex and condensation according to eq. (1).

Syntheses of two further macroions are provided in Sections 21 and 22, the spectacular $[As_4W_{40}O_{140}]^{28-}$ and $[Sb_9W_{21}O_{86}]^{19-}$ anions. Attention is then shifted to areas that are far less developed, namely, organic and organo-metallic derivative chemistry. Organophosphorus derivatives of the type described in Section 23 have been known for over 50 years and are extremely well characterized both in terms of structure and solution dynamics. Organo-metallic derivatives are, however, of more recent origin and many of them are still incompletely characterized from a structural point of view. The organotitanium derivative whose synthesis is described in Section 24 falls into this category. Its multistep preparation represents, however, an exemplary synthesis of techniques described in earlier sections, involving the preparation of a defect Keggin anion, a mixed-metal Keggin anion, plus a tetrabutylammonium polyoxoanion salt soluble in aprotic media suitable for reaction with organometallic reagents.

References

1. M. T. Pope, *Heteropoly and Isopolymetalates*, Springer-Verlag, New York, 1983.
2. V. W. Day and W. G. Klemperer, *Science*, **228**, 533 (1985).

15. TETRABUTYLAMMONIUM ISOPOLYOXOMETALATES

WALTER G. KLEMPERER*

By virtue of their solubility in aprotic organic solvents and their availability as pure crystalline materials, tetrabutylammonium isopolyoxometalates have found widespread application as starting materials for the synthesis of

*University of Illinois, Department of Chemistry, 505 South Mathews Avenue, Urbana, IL 61801.

numerous derivatives including organometallic complexes,[1] organic reagents,[2] polyoxothioanions,[3] organic derivatives,[4] and mixed-metal polyoxoanions.[5] They have also been useful for physical studies where a polar aprotic solvent like acetonitrile, dichloroethane, methylene chloride, or nitromethane is desirable, as in dynamic NMR[6] and photochemical[7] studies.

The synthesis of alkylammonium isopolyoxometalates was pioneered by Jahr, Fuchs, and Oberhauser who first prepared salts of $[Mo_6O_{19}]^{2-}$, $[W_6O_{19}]^{2-}$, and $[W_{10}O_{32}]^{4-}$ by controlled hydrolysis of molybdic and tungstic acid esters.[8,9] More convenient preparations of these and other salts that utilize readily available starting materials were developed later.[10-14] The preparations reported here are modifications of published procedures. They have been selected on the basis of convenience and ease of reproducibility. All of the anion structures have been determined by X-ray crystallography; they are shown below for $[Mo_2O_7]^{2-}$ (**a**),[12] $[Mo_6O_{19}]^{2-}$ and $[W_6O_{19}]^{2-}$ (**b**),[15,16] $[\alpha\text{-}Mo_8O_{26}]^{4-}$ (**c**),[11] $[W_{10}O_{32}]^{4-}$ (**d**),[17] and $[V_{10}O_{28}H_3]^{3-}$ in $(V_{10}O_{28}H_3{}^{3-})_2$ (**e**).[18]

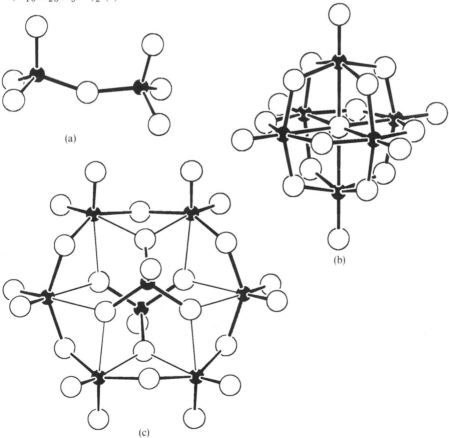

(a)

(b)

(c)

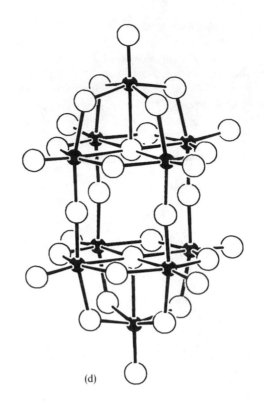

(d)

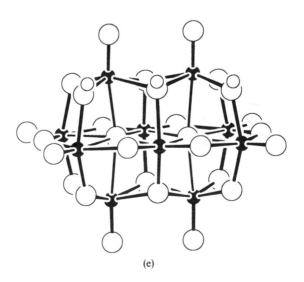

(e)

A. TETRABUTYLAMMONIUM HEXAMOLYBDATE(VI)

$$6Na_2MoO_4 + 10HCl + 2(n\text{-}C_4H_9)_4NBr$$

$$\longrightarrow [(n\text{-}C_4H_9)_4N]_2[Mo_6O_{19}] + 10NaCl + 2NaBr + 5H_2O$$

Submitted by N. H. HUR,* W. G. KLEMPERER,* and R.-C. WANG*
Checked by M. FOURNIER[†]

Procedure

A solution of 2.50 g (10.3 mmol) of commercial, ACS reagent grade, sodium molybdate dihydrate (Na$_2$MoO$_4 \cdot 2H_2O$) in 10 mL of water is acidified with 2.9 mL of 6 N aqueous HCl (17.4 mmol) in a 50-mL Erlenmeyer flask with vigorous stirring over a 1-min period at ambient temperature. A solution of 1.21 g (3.75 mmol) of commercial, 99% pure tetrabutylammonium bromide in 2 mL of water is then added with vigorous stirring to cause immediate formation of a white precipitate. (Use of impure tetrabutylammonium bromide may result in blue discoloration at this stage. Tetrabutylammonium bromide can be recrystallized by cooling a saturated, 100 °C, water solution to ambient temperature.) The resulting slurry is then heated to 75 to 80 °C with stirring for ~ 45 min.[††] During this period, the white solid slowly changes to yellow. This crude product is collected on a medium-porosity filter with suction and washed with three 20-mL portions of water. Crystallization is accomplished by dissolving the air-dried crude product (2.17 g) in 80 mL of hot acetone (60 °C) and cooling the solution to − 20 °C. After 24 h, the yellow crystalline product is collected on a filter with suction, washed twice with 20-mL portions of diethyl ether, and dried for 12 h *in vacuo* (0.1 torr). Yield: 1.98 g (1.45 mmol), 84% of theory based on Mo.

Anal. Calcd. for C$_{32}$H$_{72}$N$_2$Mo$_6$O$_{19}$: C, 28.17; H, 5.32; N, 2.05; Mo, 42.18. Found: C, 28.16; H, 5.27; N, 2.07; Mo, 42.01.

Properties

Yellow, crystalline [(*n*-C$_4$H$_9$)$_4$N]$_2$[Mo$_6$O$_{19}$] appears to be indefinitely stable when stored in a desiccator. The compound has a characteristic IR

*University of Illinois, Department of Chemistry, 505 South Mathews Avenue, Urabana, IL 61801.
†Université Pierre et Marie Curie, Laboratoire de Physicochimie Inorganique, 4 place Jussieu, 75252 Paris Cedex 05 France.
††The checker stirred the slurry for 12 h at 60 °C.

spectrum (mineral oil, 650–1000 cm^{-1}): 742 (m), 800 (s), 880 (w), 890 (w, sh), 956 (s), and 988 (w).

B. TETRABUTYLAMMONIUM OCTAMOLYBDATE(VI)

$$8Na_2MoO_4 + 12HCl + 4(n\text{-}C_4H_9)_4NBr$$

$$\longrightarrow [(n\text{-}C_4H_9)_4N]_4[Mo_8O_{26}] + 12NaCl + 4NaBr + 6H_2O$$

Submitted by N. H. HUR,* W. G. KLEMPERER,* and R.-C. WANG*
Checked by M. FOURNIER[†]

Procedure

A solution of 5.00 g (20.7 mmol) of commercial, ACS reagent grade, sodium molybdate dihydrate ($Na_2MoO_4 \cdot 2H_2O$) in 12 mL of water is acidified with 5.17 mL of 6.0 N aqueous HCl (31.0 mmol) in a 50-mL Erlenmeyer flask with vigorous stirring over a period of 1 to 2 min at room temperature. A solution of 3.34 g of commercial 99% pure tetrabutylammonium bromide (10.4 mmol) in 10 mL of water is then added with vigorous stirring to cause immediate formation of a white precipitate. After stirring the mixture for 10 min, the precipitate is collected on a medium-porosity filter with suction and washed successively with 20 mL of water, 20 mL of ethanol, 20 mL of acetone, and 20 mL of diethyl ether. This crude product (4.78 g) is dissolved in 35 mL of acetonitrile and stored for 24 h at − 10 °C. The clear, colorless, block-shaped crystals that form are collected by suction filtration and dried for 12 h *in vacuo* (0.1 torr). The transparency of the crystals is lost upon drying. Yield: 3.58 g (1.66 mmol), 64% of theory based on molybdenum.[‡]

Anal. Calcd. for $C_{64}H_{144}N_4Mo_8O_{26}$: C, 35.70; H, 6.74; N, 2.60; Mo, 35.64. Found: C, 35.64; H, 6.79; N, 2.59; Mo, 35.56.

Properties

This compound appears to be stable indefinitely when stored in a desiccator. It has a characteristic IR spectrum (mineral oil, 650–1000 cm^{-1}): 656 (s),

---

*University of Illinois, Department of Chemistry, 505 South Mathews Avenue, Urbana, IL 61801.
†Université Pierre et Marie Cure, Laboratoire de Physicochimie Inorganique, 4 place Jussieu, 75252 Paris Cedex 05, France.
‡The checker worked at double the recommended scale and obtained 9.7 g of crude product. After crystallization, however, his yield was 45%.

722 (m), 736 (m), 804 (s), 852 (m), 885 (w, sh), 904 (s), 920 (s), and 950 (m). Exposure to atmospheric moisture causes slow transformation to a hydrated anion whose IR spectrum differs markedly from the spectrum of the pure anion.

C. TETRABUTYLAMMONIUM DIMOLYBDATE(VI)

$$[(n\text{-}C_4H_9)_4N]_4[Mo_8O_{26}] + 4(n\text{-}C_4H_9)_4NOH$$
$$\longrightarrow 4[(n\text{-}C_4H_9)_4N]_2[Mo_2O_7] + 2H_2O$$

Submitted by N. H. HUR,* W. G. KLEMPERER,* and R.-C. WANG*
Checked by M. FOURNIER[†]

Procedure

A solution of 2.58 g (1.20 mmol) tetrabutylammonium octamolybdate (see Section B) in 20 mL of acetonitrile is combined with 4.8 mL of 1.0 M methanolic tetrabutylammonium hydroxide (4.8 mmol) (Eastman Kodok) in a 50-mL Erlenmeyer flask. The resulting cloudy solution is stirred for 20 min at ambient temperature, gravity filtered through medium porosity filter paper, and reduced to a viscous oil by removing the solvent under vacuum. This material is then diluted with 15 mL of acetonitrile,[‡] and 20 mL of diethyl ether is added with stirring to obtain a white precipitate. After 20 min of further stirring, the white precipitate is collected on a medium porosity filter with suction and washed with 20 mL of diethyl ether. This crude product (2.98 g) is dissolved in 15 mL of acetonitrile, and the resulting solution is gravity filtered through medium porosity filter paper to remove any insoluble material. Diethyl ether is then carefully added without stirring just until the solution becomes saturated (~ 45 mL), and the solution is stored for 24 h at $-10\,°C$. The clear, colorless, block-shaped crystals that form are collected by suction filtration and dried for 12 h *in vacuo* (0.1 torr). The crystals lose their transparency upon drying. Yield: 2.61 g (3.31 mmol), 69% of theory based on molybdenum.[#]

*University of Illinois, Department of Chemistry, 505 South Mathews Avenue, Urbana, IL 61801.
[†]Université Pierre et Marie Curie, Laboratorie de Physicochimie Inorganique, 4 Place Jussieu, 75252 Paris Cedex 05, France.
[‡]The checker reports that a solid residue was present after dilution with acetonitrile.
[#]The checker worked at double the recommended scale and obtained 4.2 g of crude product. After recrystallization his yield was 2.8 g (37%).

Anal. Calcd. for $C_{32}H_{72}N_2Mo_2O_7$: C, 48.73; H, 9.20; N, 3.55; Mo, 24.32. Found: C, 48.84; H, 9.32; N, 3.58; Mo, 24.21.

Properties

This compound appears to be stable indefinitely when stored in a desiccator. It has a characteristic IR spectrum (mineral oil, 650–1000 cm^{-1}): 735 (m, sh), 780 (s, br), 880 (s, br), 928 (m), and 975 (w).

D. TETRABUTYLAMMONIUM HEXATUNGSTATE(VI)

$$6Na_2WO_4 + 10HCl + 2(n\text{-}C_4H_9)_4NBr$$

$$\longrightarrow [(n\text{-}C_4H_9)_4N]_2[W_6O_{19}] + 10NaCl + 2NaBr + 5H_2O$$

Submitted by M. FOURNIER*
Checked by W. G. KLEMPERER[†] and N. SILAVWE[†]

This polyoxoanion is prepared according to the procedure previously described for the hexamolybdate.[19]

Procedure

All reagents are commercial analyzed reagents, used without further purification.

A mixture of 33.0 g (100 mmol) of sodium tungstate dihydrate ($Na_2WO_4 \cdot 2H_2O$, 99%), 40 mL of acetic anhydride, and 30 mL of *N,N*-dimethylformamide (DMF) is stirred, in a 250-mL Erlenmeyer flask, at 100 °C for 3 h. A white fluid cream is obtained.

A solution of 20 mL of acetic anhydride and 18 mL of 12 *N* HCl in 50 mL of DMF is added with stirring, and the resulting mixture is gravity filtered through a medium porosity filter paper in order to eliminate the undissolved white solid. After washing the solid with 50 mL of methanol, the clear filtrate is allowed to cool to room temperature. A solution of 15 g (47 mmol) of tetra-butylammonium bromide in 50 mL of methanol is added with rapid stirring, to give a white precipitate. This suspension is stirred for 5 min and the product

*Université Pierre et Marie Curie, Laboratoire de Physicochimie Inorganique, 4 place Jussieu, 75252 Paris Cedex 05, France.
[†]University of Illinois, Department of Chemistry, 505 South Mathews Avenue, Urabana, IL 61801.

is isolated on a Pyrex Büchner funnel with a fritted disk (porosity 5–15 μ) using suction. After washing with 20 mL of methanol and 50 mL of diethyl ether, this air dried crude product (22.5 g) is sufficiently pure for most purposes.

Recrystallization from a minimum amount of hot dimethyl sulfoxide (DMSO) (8 mL, 80 °C) gives, after 2 days at room temperature, 18 g of clear colorless diamond-shaped crystals, which are collected on a Pyrex Büchner funnel with fritted disk (porosity 5–15 μ), using suction. Yield: 18 g (10 mmol), 60% of theory based on W.

Anal. Calcd. for $C_{32}H_{72}N_2W_6O_{19}$: C, 20.31; H, 3.84; N, 1.48; W, 58.30. Found: C, 20.46; H, 3.83; N, 1.52; W, 58.39.

Properties

The block-shaped colorless crystals of this compound appear to be indefinitely stable to air and daylight. The material is easily characterized by its IR spectrum (KBr pellet 1000–200 cm^{-1}, *cation bands): 975 (vs), 888* (vw), 873* (vw), 812 (vs), 752* (vw), 736* (vw), 716* (vw), 664* (vw), 588 (m), 445 (vs), 402* (vw), and 368 (m).

Electrochemical properties (CH_3CN, 0.2 M [NBu$_4$][BF$_4$], 5 × 10^{-2} M^{-1} compound). One reversible monoelectronic wave at − 0.940 V *versus* SCE.

Electronic spectrum in CH_3CN: One strong absorption band at λ_{max}, 268 nm; $\varepsilon_{max} = 1.22 \times 10^4 M^{-1} cm^{-1}$; λ_{min}, 244 nm; $\varepsilon_{min} = 6.2 \times 10^3 M^{-1} cm^{-1}$. The compound slowly turns into $[W_{10}O_{32}]^{4-}$ if it is recrystallized from DMF.

E. TETRABUTYLAMMONIUM DECATUNGSTATE(VI)

$$10Na_2WO_4 + 16HCl + 4(n\text{-}C_4H_9)_4NBr$$

$$\longrightarrow [(n\text{-}C_4H_9)_4N]_4[W_{10}O_{32}] + 16NaCl + 4NaBr + 8H_2O$$

Submitted by M. FOURNIER*
Checked by W. G. KLEMPERER[†] and N. SILAVWE[†]

Procedure

All reagents are commercial analyzed reagents and are used without further purification.

*Université Pierre et Marie Curie, Laboratoire de Physicochimie Inorganique, 4 place Jussieu, 75252 Paris Cedex 05, France.
[†]University of Illinois, Department of Chemistry, 505 South Mathews Avenue, Urbana, IL 61801.

A solution of 16.0 g (48.5 mmol) of sodium tungstate dihydrate ($Na_2WO_4 \cdot 2H_2O$, 99%) in 100 mL of boiling water is acidified very quickly (10 s) with 33.5 mL of boiling 3 M HCl solution (100.5 mmol) with rapid stirring. The local white gelatinous precipitate disappears immediately. After boiling (1 or 2 min) the clear, yellow, hot solution is precipitated by addition of 6.4 g (26.4 mmol) of tetrabutylammonium bromide in 10 mL of water. The voluminous white precipitate is separated from the hot solution on a Büchner funnel with fritted disk (5–15 μ) using suction. This white solid is washed three times with 40-mL portions of boiling water, then twice with 60-mL portions of ethanol and finally twice with 100-mL portions of diethyl ether; it is then air dried (1 h). The crude product (15 g) is sufficiently pure for most purposes.

Recrystallization from hot N,N-dimethylformamide (DMF, 99%) (10 mL at 80 °C) gives, after 1 day, yellowish prismatic crystals (12.0 g), which are collected on a Büchner funnel with fritted disk (5–15 μ) using suction. The crystals contain small amounts of crystallization solvent, which can be eliminated by prolonged high vacuum drying and grinding. Alternatively, the material may be recrystallized from CH_3CN: 9.0 g of the compound is dissolved in 6 mL of acetonitrile (99%) at 80 °C, and the solution is cooled to room temperature. The crystals that form are washed with diethyl ether and air dried (1 h). Yield: 8.2 g of a high purity compound. Yield: crude product, 15 g (4.5 mmol), 93% of theory based on W. DMF crystallized product, 12 g (3.6 mmol), 74.2% of theory based on W.* CH_3CN recrystallized product, 10.9 g (3.3 mmol), 67.7% of theory based on W.

Anal. Calcd. for $C_{64}H_{144}N_4W_{10}O_{32}$: C, 23.1; H, 4.37; N, 1.69; W, 55.37. Calcd. for ($C_{64}H_{144}N_4W_{10}O_{32} \cdot 0.5\ C_3H_7ON$): C, 23.4; H, 4.4; N, 1.88; W, 54.80. Found: DMF crystals C, 22.9; H, 4.4; N, 2.0; W, 54.92. CH_3CN crystals C, 23.12; H, 4.42; N, 1.69; W, 55.29.

Properties

The pale yellow crystals of this compound are stable in air. They are daylight sensitive when traces of DMF remain in the material.

The solid has a characteristic IR spectrum (KBr pellet 1000–200 cm^{-1}, *cation band): 991 (vw), 958 (vs), 942 (s), 888 (vs), 802 (vs), 740* (sh), 582 (w), 434 (m), 425 (sh), 405 (s), 345 (w), and 331 (m).

Electrochemical: (DMF, 0.2 M [NBu$_4$][BF$_4$]), two reversible mono-electronic waves (-1.02 and -1.60 V versus saturated calomel electrode).

Electronic spectrum in CH_3CN: strong charge-transfer bands at 322 nm

---

*The checkers yield of DMF crystallized product was 8.0 g (50%).

$(\varepsilon_{max} = 1.34 \times 10^4 \ M^{-1} \ cm^{-1})$ and 262 nm $(\varepsilon_{max} = 1.84 \times 10^4 \ M^{-1} \ cm^{-1})$; $\lambda_{min} = 286$ nm $(\varepsilon_{min} = 9.2 \times 10^3 \ M^{-1} \ cm^{-1})$.

The compound slowly turns into $[W_6O_{19}]^{2-}$ if recrystallized from DMSO.

F. TETRABUTYLAMMONIUM TRIHYDROGEN DECAVANADATE(V)

$$10Na_3VO_4 + 27HCl + 3(n\text{-}C_4H_9)_4NBr$$

$$\longrightarrow [(n\text{-}C_4H_9)_4N]_3[V_{10}O_{28}H_3] + 27NaCl + 3NaBr + 12H_2O$$

Submitted by W. G. KLEMPERER* and O. M. YAGHI*
Checked by M. FOURNIER†

Procedure

A solution of 15.00 g (81.6 mmol) of commercial 99% pure sodium orthovanadate $(Na_3VO_4)^‡$ in 110.0 mL of water contained in a 250-mL Erlenmeyer flask is acidified at ambient temperature with 71.0 mL of 3 N aqueous HCl (213 mmol), added with rapid stirring from a burrette at a rate of 2 drops s^{-1}. The initially colorless solution becomes orange upon addition of the acid. The acidified solution is added in 2-mL portions over a 15-min period to a rapidly stirred solution of 60 g (186 mmol) of commercial 99% pure tetrabutylammonium bromide in 60 mL of water. After stirring for another 15 min. The orange-yellow precipitate is collected on a medium porosity filter using suction and is washed successively with 60 mL of water, 60 mL of ethanol, and 300 mL of diethyl ether. It is finally dried under vacuum (0.1 torr) for 12 h. This material (17 g) is added to 150 mL of acetonitrile, and after stirring for 10 min the solution is gravity filtered through fine porosity paper to remove insoluble material. Addition of 300 mL of anhydrous diethyl ether to the clear, dark orange filtrate yields a yellow-orange precipitate that is collected on a medium porosity filter using suction, washed with 100 mL of anhydrous diethyl ether, and dried under vacuum (0.1 torr) for 1 h to give 12.3 g of a yellow-orange solid.

*University of Illinois, Department of Chemistry, 505 South Mathews Avenue, Urbana, IL 61801.
†Université Pierre et Marie Curie, Laboratoire de Physicochimie Inorganique, 4 place Jussieu, 75252 Paris Cedex 05, France.
‡The checker started with 6.64 g (54.4 mmol) of sodium metavanadate (NaVO₃), which was converted to the orthovanadate by dissolving in 10.9 mL (109 mmol) of 10 N NaOH and then diluting with 70 mL of water. Appropriately scaled quantities were used in the subsequent steps.

Crystalline material is obtained by dissolving the crude product in 50 mL of acetonitrile and allowing diethyl ether vapor to diffuse into the solution. This is achieved by preparing the acetonitrile solution in a 250 mL beaker and placing the beaker inside a larger, closed, screwcap bottle containing ~ 200 mL of diethyl ether. The yellow-orange crystals that form after 3 days are collected on a medium porosity filter using suction, washed with 100 mL of diethyl ether with loss of transparency, and dried for 12 h under vacuum.* Yield: 1.8 g (7.00 mmol), 86% of theory based on vanadium.

Anal. Calcd. for $C_{48}H_{111}N_3V_{10}O_{28}$: C, 34.16; H, 6.63; N, 2.49; V, 30.18. Found: C, 34.16; H, 6.65; N, 2.58; V, 30.14.

Properties

The tetrabutylammonium salt of $[V_{10}O_{28}H_3]^{3-}$ is hygroscopic, but it is stable for at least 1 year when stored in a desiccator in darkness. It is most easily identified by its characteristic IR spectrum (mineral oil, $700-1000\,cm^{-1}$): 739 (m, sh), 770 (m), 803 (m), 840 (m), 880 (w), 940 (m), 968 (s), and 985 (sh).

References

1. T. M. Che, V. W. Day, L. C. Francesconi, M. F. Fredrich, W. G. Klemperer, and W. Shum, *Inorg. Chem.*, **24**, 4055 (1985).
2. W. A. Nugent, *Tetrahedron Lett.*, **37**, 3427 (1978).
3. Y. Do, E. D. Simhon, and R. H. Holm, *Inorg. Chem.*, **24**, 1831 (1985).
4. V. W. Day, M. F. Fredrich, W. G. Klemperer, and R.-S. Liu, *J. Am. Chem. Soc.*, **101**, 491 (1979).
5. M. Filowitz, R. K. C. Ho, W. G. Klemperer, and W. Shum, *Inorg. Chem.*, **18**, 93 (1979).
6. V. W. Day, W. G. Klemperer, and C. Schwartz, *J. Am. Chem. Soc.*, **109**, 6030 (1987).
7. T. Yamase, N. Takabayashi, and M. Kaji, *J. Chem. Soc. Dalton Trans.*, **1984**, 793.
8. K. F. Jahr, J. Fuchs, and R. Oberhauser, *Chem. Ber.*, **101**, 477 (1968).
9. J. Fuchs and K. F. Jahr, *Z. Naturforsch. Teil B*, **23**, 1380 (1968).
10. $Mo_6O_{19}^{2-}$: K. H. Tytko and B. Schoenfeld, *Z. Naturforsch. Teil B*, **30**, 471 (1975).
11. α-$Mo_8O_{26}^{4-}$: J. Fuchs and H. Hartl, *Angew. Chem., Int. Ed. Engl.*, **15**, 375 (1976); W. G. Klemperer and W. Shum, *J. Am. Chem. Soc.*, **98**, 8291 (1976).
12. $Mo_2O_7^{2-}$: V. W. Day, M. F. Fredrich, W. G. Klemperer, and W. Shum, *J. Am. Chem. Soc.*, **99**, 6146 (1977).
13. $W_6O_{19}^{2-}$: M. Boyer and B. LeMeur, *C. R. Acad. Sci. Ser. C*, **281**, 59 (1975); C. Sanchez, J. Livage, J. P. Launay, and M. Fournier, *J. Am. Chem. Soc.*, **105**, 6817 (1983).

*The checker crystallized the compound by dissolving 7.0 g of the crude material in 12 mL of warm (50 °C) acetonitrile. The solution deposits large, well-shaped, orange crystals after standing for 1 h at room temperature (25 °C).

14. $W_{10}O_{32}^{4-}$: A. Chemseddine, C. Sanchez, J. Livage, J. P. Launay, and M. Fournier, *Inorg. Chem.*, **23**, 2609 (1984).
15. H. R. Allcock, E. C. Bissell, and E. T. Shawl, *Inorg. Chem.*, **12**, 2963 (1973).
16. J. Fuchs, W. Freiwald, and H. Hartl, *Acta Cryst.*, **B34**, 1764 (1978).
17. J. Fuchs, H. Hartl, W. Schiller, and U. Gerlach, *Acta Cryst.*, **B32**, 740 (1976).
18. V. W. Day, W. G. Klemperer, and D. J. Maltbie, *J. Am. Chem. Soc.*, **109**, 2991 (1987).
19. M. Che, M. Fournier, and J. P. Launay, *J. Chem. Phys.*, **71**, 1954 (1974).

16. α-, β-, and γ-DODECATUNGSTOSILICIC ACIDS: ISOMERS AND RELATED LACUNARY COMPOUNDS

Submitted by ANDRÉ TÉZÉ* and GILBERT HERVÉ*
Checked by RICHARD G. FINKE[†] and DAVID K. LYON[†]

The most studied tungstic heteropolyanions have the Keggin structure characteristic of $[XW_{12}O_{40}]^{n-}$, where X is a tetrahedral heteroatom ($X = Si^{IV}$, Ge^{IV}, P^V, and As^V). The Keggin structure (α-isomer) is based on the arrangement of four groups of three edge shared W_3O_{13} octahedra around the XO_4 central tetrahedron. The overall symmetry is T_d. The β isomer is obtained by rotation of one W_3O_{13} by 60° (overall C_{3v} symmetry), and the γ isomer by rotations of two groups (overall C_{2v} symmetry).[1]

The dodecatungstosilicates are stable in acid solution. When the pH increases, hydrolytic cleavages of W—O bonds occur, leading to well-defined lacunary polyanions, with 11, 10, or 9 tungsten atoms.

The most stable compounds, or least labile when they are metastable, are obtained with Si^{IV} as the heteroelement. Three dodecatungstosilicates, four undecatungstosilicates,[2] one decatungstosilicate,[3] and two nonatungstosilicates[4] are known. They are prepared following the general route in Fig. 1.

The lacunary polyanions can act as ligands with numerous metal cations,[5] leading to mono-, di-, or trinuclear complexes according to the number of vacant sites. These complexes are the earliest model materials of mixed oxides, and they are important for catalytic purposes, particularly for the oxidation of organic substrates.

These polyanions differ only slightly in their compositions; hence, elemental analysis is not a reliable criterion of purity. It is more convenient to characterize the heteropolyanions by their physicochemical properties: UV–Vis spectrum and polarogram in solution and IR spectrum in the solid

*Laboratoire de Physicochimie Inorganique, U.A. CNRS 419, Université Pierre et Marie Curie, 75252 Paris Cedex 05, France.
†Department of Chemistry, University of Oregon, Eugene, OR 97403.

$$SiO_3^{2-} + WO_4^{2-}$$

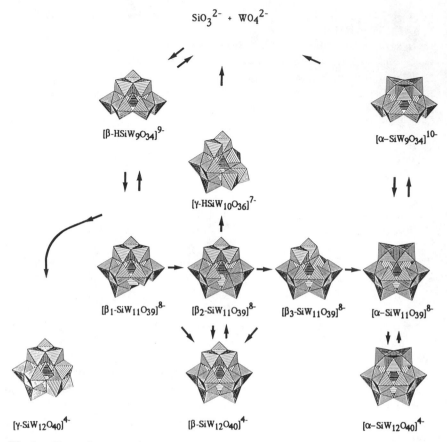

$[\beta\text{-HSiW}_9O_{34}]^{9-}$

$[\alpha\text{-SiW}_9O_{34}]^{10-}$

$[\gamma\text{-HSiW}_{10}O_{36}]^{7-}$

$[\beta_1\text{-SiW}_{11}O_{39}]^{8-}$ $[\beta_2\text{-SiW}_{11}O_{39}]^{8-}$ $[\beta_3\text{-SiW}_{11}O_{39}]^{8-}$ $[\alpha\text{-SiW}_{11}O_{39}]^{8-}$

$[\gamma\text{-SiW}_{12}O_{40}]^{4-}$ $[\beta\text{-SiW}_{12}O_{40}]^{4-}$ $[\alpha\text{-SiW}_{12}O_{40}]^{4-}$

Fig. 1. General route for synthesis of dodecatungstosilicates and related lacunary polyanions.

state. An especially attractive and powerful technique is ^{183}W NMR, but it is not as commonly available as the other methods mentioned.

The crystallized compounds have well-defined numbers of water molecules, but the water content of products obtained as powders is dependent on the drying and storage conditions. The values given are only indicative.

Materials

Starting materials for the following preparations are sodium tungstate ($Na_2WO_4 \cdot 2H_2O$) and sodium metasilicate ($Na_2SiO_3 \cdot 5H_2O$), which are obtained commercially. An appropriate amount of $Na_2SiO_3 \cdot 9H_2O$ can be substituted for the pentahydrate.

Some preparations (Procedures in Sections C, E, G, and J) require the synthesis of intermediates. Generally, pH must be controlled with a precision of about 0.3 units, and a good indicator paper can be used for this purpose. Some preparations (Sections C and G) require accurate pH readings on a calibrated pH meter.

A.   SODIUM α-NONATUNGSTOSILICATE, Na_{10} [α-SiW_9O_{34}]·SOLVENT

$$9[WO_4]^{2-} + [SiO_3]^{2-} + 10H^+ + 10Na^+ \longrightarrow Na_{10}[\alpha\text{-}SiW_9O_{34}] + 5H_2O$$

Procedure

Sodium tungstate (182 g, 0.55 mol) and sodium metasilicate (11 g, 50 mmol) are dissolved in 200 mL of hot water (80–100 °C) in a 1-L beaker containing a magnetic stirring bar. To this solution is added dropwise 130 mL of 6 M HCl in ∼ 30 min with vigorous stirring. The solution is boiled until the volume is ∼ 300 mL. Unreacted silica is removed by filtration over a fine frit or over Celite® or by centrifugation.

Anhydrous sodium carbonate (50 g) is dissolved in 150 mL of water in a separate beaker. This solution is slowly added to the first solution with gentle stirring. A precipitate forms slowly. It is removed by filtering, using a sintered glass filter, after 1 h. The solid is stirred with 1 L of 4 M NaCl solution and filtered again. It is then washed successively with two 100-mL portions of ethanol and 100 mL of diethyl ether and dried under vacuum. Yield: ∼ 110 g (85%).

Properties

Sodium nonatungstosilicate is a white solid, which is slightly soluble in water. The solution is metastable at all pH values, but it changes slowly. The best characterization is polarography in 1 M sodium acetate–1 M acetic acid buffer: The polarogram shows only one four-electron wave with $E_{1/2} = -0.80$ V *versus* SCE. The polyanion can be characterized in the solid state by its IR spectrum (KBr pellet, cm^{-1}):[6] 985, 940 (sh), 930, 865, 848 (sh), 808, 712, 552, 530, 490(sh), 435, 373, and 335.

The [α-SiW_9O_{34}]$^{10-}$ polyanion is a useful species for synthesizing the adduct complexes [$SiW_9M_3O_{40}$]$^{n-}$, especially with M = Mo^{VI}, V^V, and di-or trivalent cations of the first transition series.

B. SODIUM β-NONATUNGSTOSILICATE, $Na_9[\beta\text{-}SiW_9O_{34}H]\cdot23H_2O$

$$9[WO_4]^{2-} + [SiO_3]^{2-} + 11H^+ + 9Na^+ + 18H_2O$$
$$\longrightarrow Na_9[\beta\text{-}SiW_9O_{34}H]\cdot23H_2O$$

Procedure

Sodium metasilicate (12 g, 57 mmol) is dissolved in 250 mL of water in a 1-L beaker containing a magnetic stirring bar, and sodium tungstate (150 g, 455 mmol) is added. To this solution, 95 mL of 6 M HCl is added very slowly with vigorous stirring. Unreacted silica is removed by filtration over a fine frit or over Celite® or by centrifugation. The desired salt crystallizes from this solution at 5 °C for 2 to 3 days. The crystals are collected on a sintered glass filter, washed with two portions of 20 mL of cold water (4 °C), and air dried. Yield: ~ 50 g (35%).

Properties

Sodium β-nonatungstosilicate is a white solid (triclinic, space group $P\bar{1}$, with $a = 131.66, b = 126.09, c = 183.50$ pm, $\alpha = 69.53, \beta = 73.51, \gamma = 63.16°, Z = 2$),[7] slightly soluble in water. The polyanion is unstable in aqueous solution at any pH, but the change is slow. It can be characterized by polarography in 1 M sodium acetate–1 M acetic acid buffer: The polarogram shows two waves at − 0.82 and − 0.93 V *versus* SCE. The solid is well characterized by its IR spectrum (KBr pellet, cm^{-1}): 990 (m), 935 (m), 865 (vs), 800 (s), 745 (s), 535 (m), and 315 (s). The chemical reactivity of the $[\beta\text{-}SiW_9O_{34}H]^{9-}$ polyanion is similar to that of the α isomer.

C. POTASIUM γ-DECATUNGSTOSILICATE, $K_8[\gamma\text{-}SiW_{10}O_{36}]\cdot12H_2O$

$$[\beta_2\text{-}SiW_{11}O_{39}]^{8-} + 2CO_3^{2-} + 8K^+ + 13H_2O$$
$$\longrightarrow K_8[\gamma\text{-}SiW_{10}O_{36}]\cdot12H_2O + 2HCO_3^- + [WO_4]^{2-}$$

Procedure

This synthesis requires accurate pH readings on a calibrated pH meter.

The potassium salt of the β_2 isomer of undecatungstosilicate (15 g, 5 mmol), synthesized as described in the procedure in section F, is dissolved in 150 mL of water maintained at 25 °C. Impurities in the $K_8[\beta_2\text{-}SiW_{11}O_{39}]$ salt (mainly paratungstate) give insoluble materials, which have to be removed rapidly by filtration on a fine frit or through Celite®. The pH of the solution is quickly adjusted to 9.1 by addition of a $2\,M$ aqueous solution of K_2CO_3. The pH of the solution is kept at this value by addition of the K_2CO_3 solution for exactly 16 min. The potassium salt of the γ-decatungstosilicate is then precipitated by addition of solid potassium chloride (40 g). During the precipitation (10 min), the pH must be maintained at 9.1 by addition of small amounts of the K_2CO_3 solution. The solid is removed by filtering, washed with 1 M KCl solution, and air dried. Yield: ∼ 10 g (70%).*

Properties

The potassium salt of $[γ\text{-}SiW_{10}O_{36}]^{8-}$ is soluble in water and stable below pH 8 (in strongly acidic solution, pH < 1, it converts very slowly into $[β\text{-}SiW_{12}O_{40}]^{4-}$). A polarogram of the solution exhibits two reversible two-electron waves, with half-wave potentials -0.75 and -0.84 V *versus* SCE in 1 M acetic acid $-$ 1 M sodium acetate buffer, pH 4.7. The ^{183}W NMR spectrum of the solution in H_2O–D_2O (90/10) mixtures shows three lines with relative intensities 2:2:1, in agreement with the X-ray diffraction determination of the structure of the polyanion in the rubidium salt.[3] The chemical shifts are, respectively, -96.4, -137.2, and -158.2 ppm (external reference $2\,M$ Na_2WO_4 in alkaline D_2O).

The compound can be characterized in the solid state by its IR spectrum (KBr pellet, cm^{-1}): 989 (m), 941 (s), 905 (s), 865 (vs), 818 (vs), 740 (vs), 655 (sh), 553 (w), 528 (m), 478 (sh), 390 (sh), 360 (s), 328 (sh), 318 (m), and 303 (sh).

This lacunary polyanion is used to synthetize adduct complexes $[γ\text{-}SiW_{10}M_2O_{40}]^{n-}$, especially with V^V, Mo^{VI}, W^{VI}, and di- or trivalent cations of the first transition series.

D. POTASSIUM α-UNDECATUNGSTOSILICATE, $K_8[α\text{-}SiW_{11}O_{39}]\cdot 13H_2O$

$$11[WO_4]^{2-} + [SiO_3]^{2-} + 16H^+ \longrightarrow [α\text{-}SiW_{11}O_{39}]^{8-} + 8H_2O$$

Procedure

Sodium metasilicate (11 g, 50 mmol) is dissolved with magnetic stirring at room temperature in 100 mL of distilled water (if the solution is not completely clear, it is filtered) (Solution A). In a 1-L beaker, containing a magnetic

*The checkers started with 9 g of $K_8[β_2\text{-}SiW_{11}O_{39}]$ and obtained a yield of 2.8 g.

stirring bar, sodium tungstate (182 g, 0.55 mol) is dissolved in 300 mL of boiling distilled water (Solution B).

To the boiling Solution B, a solution of 4 M HCl (165 mL) is added dropwise in ~ 30 min, with vigorous stirring to dissolve the local precipitate of tungstic acid. Solution A is then added and, quickly, 50 mL of 4 M HCl is also added. The pH is ~ 5 to 6. The solution is kept boiling for 1 h. After cooling to room temperature, the solution is filtered if it is not completely clear. Potassium chloride (150 g) is added to the solution, which is stirred magnetically. The white solid product is collected on a sintered glass funnel (medium porosity), washed with two 50-mL portions of a KCl solution (~ 1 M), then washed with 50 mL of cold water, and finally dried in air. Yield: ~ 145 g (90%).

Properties

The potassium salt of the undecatungstosilicate is soluble in water (~ 20 mmol L^{-1}). It is stable in aqueous solution between pH 4.5 and 7. The compound is characterized in solution by polarography: The polarogram shows two two-electron waves at -0.76 and -0.93 V *versus* SCE in 1 M NaCH$_3$COO 1 M CH$_3$COOH buffer.

The $[SiW_{11}O_{39}]^{8-}$ ion reacts in solution with most of the transition metal cations to form 1:1 complexes, for example $[SiW_{11}O_{39}Co(H_2O)]^{6-}$. The electronic spectrum of the complex with vanadium(IV)[8] provides a good method of identifying the undecatungstosilicate isomer. The $[\alpha\text{-}SiW_{11}O_{39}VO]^{6-}$ complex is formed in solution by dissolution of equimolar amounts of the heteropolyanion and vanadyl sulfate. The visible spectrum of the complex shows: $\varepsilon^{sh}_{855} = 100$, $\varepsilon^{sh}_{675} = 280$, $\varepsilon^{max}_{490} = 600$, $\varepsilon^{min}_{395} = 400$.

The potassium salt is characterized in the solid state by its IR spectrum[6] (K Br pellet, cm^{-1}): 1000, 952, 885, 870 (sh), 797, 725, 625 (sh), 540, 520, 472 (sh), 430 (sh), 368, and 332.

E.   SODIUM β_1-UNDECATUNGSTOSILICATE,
Na$_8[\beta_1\text{-}SiW_{11}O_{39}]$·SOLVENT

$$Na_9[\beta\text{-}SiW_9O_{34}H]\cdot 23H_2O + 2[WO_4]^{2-} + 4H^+$$
$$\longrightarrow Na_8[\beta_1\text{-}SiW_{11}O_{39}](\text{solvated}) + 25H_2O + NaOH$$

Procedure

Sodium β-nonatungstosilicate (5.8 g, 2 mmol, from the procedure in Section B) and sodium tungstate solution (1.42 g, 4 mmol in 40 mL of water) are placed in a 200-mL beaker containing a magnetic stirring bar. To this mixture 5 mL of 1 M HCl is added dropwise over 5 min. The pH is ~ 6.0. Upon addition of absolute ethanol (50 mL) an oil is formed. The supernatant is decanted off. The oil is washed several times with absolute ethanol (5 × 25 mL) to obtain a powder that is filtered on a fine frit and air dried. Yield: 5 g (85%).*

Properties

Sodium β_1-undecatungstosilicate is a white water-soluble solid which, in solution, slowly changes into the β_2 isomer (Compound F). It is characterized by polarography in 1 M sodium acetate–1 M acetic acid: The polarogram shows two two-electron waves at − 0.63 and − 0.83 V *versus* SCE.

The visible spectrum of the $[\beta_1\text{-SiW}_{11}O_{39}VO]^{6-}$ complex (see properties of Compound D) shows: $\varepsilon^{sh}_{870} = 170$, $\varepsilon^{sh}_{670} = 420$, $\varepsilon^{max}_{590} = 520$, $\varepsilon^{min}_{540} = 475$, $\varepsilon^{max}_{450} = 580$, $\varepsilon^{min}_{390} = 380$. The polyanion is characterized in the solid state by its IR spectrum (KBr pellet, cm^{-1}): 990, 950, 900 (sh), 865, 780, 725, 620, 535, 460, 360, and 320.

F. POTASSIUM β₂-UNDECATUNGSTOSILICATE, K₈[β₂-SiW₁₁O₃₉]·14H₂O

$$11[WO_4]^{2-} + [SiO_3]^{2-} + 16H^+ + 8K^+ + 6H_2O \longrightarrow K_8[\beta_2\text{-SiW}_{11}O_{39}] \\ \cdot 14H_2O$$

Procedure

Sodium metasilicate (11 g, 50 mmol) is dissolved in 100 mL of water (Solution A). Sodium tungstate (182 g, 0.55 mol) is dissolved in 300 mL of water in a separate 1-L beaker containing a magnetic stirring bar. To this solution, 165 mL of 4 M HCl is added in 1-mL portions over 10 min, with vigorous stirring (there is a local formation of hydrated tungstic acid that slowly disappears). Then, Solution A is poured into the tungstate solution, and the pH is adjusted to between 5 and 6 by addition of the 4 M HCl solution (~ 40 mL). This pH is maintained by addition of small amounts of 4 M HCl for 100 min. Solid potassium chloride (90 g) is then added to the solution with gentle stirring. After 15 min, the precipitate is collected by filtering through a sintered glass filter. Purification is achieved by dissolving the product in 850 mL of water. The insoluble material is rapidly removed by filtration on a fine frit, and the salt is precipitated again by addition of solid KCl (80 g). The

---

*Checkers yield: 3.8 g.

precipitate is separated by filtration, washed with $2\,M$ potassium chloride solution (2 portions of 50 mL), and air dried. Yield: ~ 60 to $80\,g$ (37–50%).

Properties

Potassium β_2-undecatungstosilicate is a white solid, which is soluble in water. In solution it slowly converts into the β_3 isomer (Compound G). It is characterized by polarography in $1\,M$ sodium acetate–$1\,M$ acetic acid buffer: The polarogram shows two two-electron waves at -0.63 and -0.77 V *versus* SCE. The visible spectrum of the $[\beta_2\text{-SiW}_{11}\text{O}_{39}\text{VO}]^{6-}$ complex in aqueous solution (see properties of Compound D) shows: $\varepsilon^{sh}{}_{855} = 130$, $\varepsilon^{sh}{}_{670} = 420$, $\varepsilon^{max}{}_{510} = 780$, $\varepsilon^{min}{}_{370} = 290$.

The polyanion is characterized in the solid state by its IR spectrum (KBr pellet, cm^{-1}): 988, 945, 875, 855, 805, 730, 610 (sh), 530, 460 (sh), 395 (sh), 360, and 325.

G. POTASSIUM β_3-UNDECATUNGSTOSILICATE $K_8[\beta_3\text{-SiW}_{11}O_{39}]\cdot 14H_2O$

$$K_8[\beta_2\text{-SiW}_{11}O_{39}]\cdot 14H_2O \longrightarrow K_8[\beta_3\text{-SiW}_{11}O_{39}]\cdot 14H_2O$$

Procedure

Potassium β_2-undecatungstosilicate (30 g, 9.4 mmol, from the procedure in Section F) is dissolved in 300 mL of water. If some residue remains, it is removed by filtration. The pH is adjusted to 5.0 by addition of $4\,M$ HCl solution (~ 5 mL) and maintained at this value for $4\frac{1}{2}$ h at room temperature. Solid potassium chloride (60 g) is added to yield a precipitate. The mixture is stirred for ~ 2 h. The solid (mixture of the β_3 and α isomer) is filtered on a fine frit and the solid is stirred with water (200 mL). The insoluble part (mainly α isomer) is remove by filtration, and solid potassium chloride (40 g) is added to the filtrate. The solid product is collected by filtration, washed with $1\,M$ KCl (two portions of 20 mL), and air dried. Yield: 13.5 g (45%).*

Properties

The potassium β_3-undecatungstosilicate is a white solid, which is soluble in water. In solution it slowly converts into the α isomer. It can be characterized by polarography in $1\,M$ sodium acetate–$1\,M$ acetic acid buffer: The

---

*Checkers report a detectable (IR spectroscopy) level of α-isomer in the product. Yield: 10.6 g.

polarogram shows two two-electron waves at -0.69 and -0.89 V *versus* SCE. The visible spectrum of the $[\beta_3\text{-SiW}_{11}\text{O}_{39}\text{VO}]^{6-}$ complex in aqueous solution (see properties of Compound D) shows: $\varepsilon^{sh}_{870} = 130$, $\varepsilon^{sh}_{680} = 360$, $\varepsilon^{max}_{590} = 470$, $\varepsilon^{min}_{530} = 395$, $\varepsilon^{max}_{440} = 480$, $\varepsilon^{min}_{395} = 335$.

The polyanion is characterized in the solid state by its IR spectrum (KBr pellet, cm^{-1}): 985, 945, 878, 790, 742, 665, 615(sh), 540, 510, 450, 390, 365, and 330.

H. α-DODECATUNGSTOSILICIC ACID AND ITS POTASSIUM SALT, $H_4[\alpha\text{-SiW}_{12}O_{40}]\cdot n H_2O$ and $K_4[\alpha\text{-SiW}_{12}O_{40}]\cdot 17H_2O$

$$12[\text{WO}_4]^{2-} + [\text{SiO}_3]^{2-} + 22\text{H}^+ \longrightarrow [\alpha\text{-SiW}_{12}\text{O}_{40}]^{4-} + 11\text{H}_2\text{O}$$

The preparation of the acid has been described in *Inorganic Syntheses.*[9] The following procedure uses hydrochloric acid instead of sulfuric acid for the extraction and is safer.

Procedure

Sodium metasilicate (11 g, 50 mmol) is dissolved with magnetic stirring at room temperature in 100 mL of distilled water; if the solution is not completely clear, it is filtered (Solution A). In a 1-L beaker containing a magnetic stirring bar, sodium tungstate (182 g, 0.55 mol) is dissolved in 300 mL of boiling distilled water (Solution B).

To the boiling Solution B, a solution of 4 M HCl (165 mL) is added drop by drop over 5 min with vigorous stirring in order to dissolve the local precipitate of tungstic acid. Then solution A is added, followed quickly by addition of 50 mL of 4 M HCl. The pH is $\sim$ 5 to 6. The solution is kept at $\sim$ 100 °C for 1 h. A solution of 1 M sodium tungstate (50 mL, 50 mmol) and, immediately thereafter, 80 mL of 4 M HCl are added. After cooling to room temperature, the solution is filtered through a fine frit if it is not completely clear. Its volume is $\sim$ 450 to 500 mL. This solution is used to obtain either the acid or the potassium salt.

To obtain the acid, the solution is transferred to a 1-L separatory funnel. The acid is extracted by addition of 80 mL of diethyl ether and 120 mL of a mixture of equal volumes of diethyl ether and concentrated hydrochloric acid chilled to -20 °C. The heaviest layer is collected, and diethyl ether is removed under vacuum. The acid is crystallized by slow evaporation of its aqueous solution.

To obtain the potassium salt, the pH is adjusted to $\sim$ 2 with aqueous 1 M KOH. Solid KCl (50 g) is added. The potassium salt is removed by filtering and dried in air. Recrystallization is effected at room temperature from an aqueous solution saturated at 50 °C (pH adjusted to 2 by HCl). Yield: 130 g (75%).

Properties

The acid and its potassium salt are white solids, which are soluble in water. The polyanion is stable below pH 4.5. It is characterized by polarography of the fresh solution in 1 M sodium acetate–1 M acetic acid buffer: The polarogram shows two one-electron and one two-electron waves at $- 0.24$, $- 0.48$, and $- 0.95$ V *versus* SCE, respectively. The UV spectrum in aqueous solution shows: $\varepsilon^{max}_{258} = 47, 000$, and $\varepsilon^{min}_{232} = 19, 200$.

The polyanion is characterized in the solid state by its IR spectrum (KBr pellet, cm^{-1});[10] for the acid: 1020, 981, 928, 880, 785, 552 (sh), 540, 475, 415, 373, and 332; for the potassium salt: 1020, 999, 980, 940 (sh), 925, 894, 878, 780, 550 (sh), 530, 474, 413, 373, and 333.

I. β-DODECATUNGSTOSILICIC ACID AND ITS POTASSIUM SALT, H₄[β-SiW₁₂O₄₀]·nH₂O and K₄[β-SiW₁₂O₄₀]·9H₂O

$$12[WO_4]^{2-} + [SiO_3]^{2-} + 22H^+ \longrightarrow [\beta\text{-}SiW_{12}O_{40}]^{4-} + 11H_2O$$

Procedure

A 198-g quantity of sodium tungstate (0.6 mol) is dissolved in 250 mL of distilled water at 50 °C in a 1-L beaker, and 240 mL of 3 M aqueous HCl is added by fractions, with vigorous magnetic stirring in order to rapidly dissolve the local precipitate of tungstic acid. A solution of 11 g of sodium metasilicate (0.05 mol) in 100 mL of water is quickly added, followed by 56 mL of 6 M HCl (final pH $\sim$ 1). After reducing the volume to 400 mL by heating at 80 °C, the solution is filtered on a fine frit or through Celite® in order to remove unreacted silica.

This solution is used to obtain either the acid or the potassium salt. To obtain the acid, the solution is transferred to a 750-mL separatory funnel. The acid is extracted by addition of 80 mL of diethyl ether and 120 mL of a mixture of equal volumes of diethyl ether and concentrated HCl chilled to $- 20$ °C. The heaviest layer is collected, and diethyl ether is removed under vacuum. The acid is crystallized by slow evaporation of its water solution.

To obtain the potassium salt, the pH is adjusted to $\sim$ 2 with aqueous 1 M KOH. Solid KCl (50 g) is added. The potassium salt is removed by filtering and dried in air. Recrystallization is effected at room temperature from an aqueous solution saturated at 50 °C (pH adjusted to 2 by HCl). Yield: 110 g (65%).

Properties

The acid and the potassium salt are pale yellow solids soluble in water. The polyanion is stable below pH 4.5. It is characterized by polarography of the fresh solution in 1 M sodium acetate–1 M acetic acid buffer: The polarogram shows two one-electron and one two-electron waves at − 0.14, − 0.42, and − 0.78 V *versus* SCE, respectively. The UV spectrum in aqueous solution shows: $\varepsilon^{max}_{264} = 30,000$, and $\varepsilon^{min}_{244} = 25,500$.

The polyanion is characterized in the solid state by its IR spectrum (KBr pellet, cm^{-1});[10] for the acid: 1018, 980, 920, 790, 550 (sh), 530, 510 (sh), 427, 395 (sh), 372 (sh), 360, and 335; for the potassium salt: 1018, 984, 917, 865, 791, 550 (sh), 530, 510 (sh), 427, 395 (sh), 372 (sh), 360, and 335.

J.   TETRABUTYLAMMONIUM γ-DODECATUNGSTOSILICATE, [(*n*-C$_4$H$_9$)$_4$N]$_4$ [γ-SiW$_{12}$O$_{40}$]

$$[\gamma\text{-SiW}_{10}O_{36}]^{8-} + 2[WO_4]^{2-} + 8H^+ + 4[N(n\text{-}C_4H_9)_4]^+$$

$$\longrightarrow [N(n\text{-}C_4H_9)_4]_4 \, [\gamma\text{-SiW}_{12}O_{40}] + 4H_2O$$

Procedure

The potassium salt of the γ-decatungstosilicate (7.5 g, 2.5 mmol, from the procedure in Section C) is added to 50 mL of water; 1 M perchloric acid (20 mL, 20 mmol) is added. The mixture is gently stirred for 20 min and then the precipitate of potassium perchlorate is removed by filtering on a medium frit. Hydrochloric acid solution (1 M, 5 mL) and absolute ethanol (75 mL) are added to the filtrate, and then, drop by drop over 2 min and with vigorous stirring, a solution of 1 M aqueous sodium tungstate (5 mL, 5 mmol) is added. After 30 min the insoluble material is removed by filtration on a fine frit, and 4 g of tetrabutylammonium bromide is added to the solution. The precipitate is collected on a fine frit,* washed with ethanol (two portions of 25 mL), and diethyl ether (20 mL), and then it is dissolved in acetonitrile (25 mL). This solution is filtered on a low speed paper in order to eliminate traces of insoluble material, and the filtrate is poured into diethyl ether (250 mL) with vigorous stirring. The precipitate is collected by filtration on a fine frit and washed with diethyl ether (two portions of 25 mL). Yield: ~ 5 g (50%).

---

*The checkers found it is necessary to centrifuge the mixture in order to isolate the product.

Properties

The tetrabutylammonium salt of $[\gamma\text{-SiW}_{12}O_{40}]^{4-}$ is soluble and stable in numerous organic solvents, for example, N,N-dimethylformamide (DMF), dimethyl sulfoxide, and acetonitrile. The 250-MHz ^{183}W NMR spectrum of the DMF solution exhibits four lines, with a 2:1:2:1 intensity ratio at $\delta = -102, -115, -125,$ and -157 ppm, respectively (external reference Na_2WO_4 in 1 M NaOH). The electronic spectrum of the acetonitrile solution in the UV region exhibits several broad shoulders between 220 and 300 nm $(\varepsilon^{sh}_{274} = 23,500$ and $\varepsilon^{sh}_{234} = 32,500)$.

The compound is characterized in the solid state by its IR spectrum (KBr pellet, cm^{-1}): 1010, 973, 912, 882 (sh), 873, 840, 791, 775 (sh), 700 (sh), 552, 538, 520, 488, 460, 445, 411, 390, 351, 330, and 310.

References

1. M. T. Pope, *Heteropoly and Isopoly Oxometalates*, Springer-Verlag, New York, 1983, p. 58.
2. A. Tézé and G. Hervé, *J. Inorg. Nucl. Chem.*, **39**, 999 (1977).
3. J. Canny, A. Tézé, R. Thouvenot, and G. Hervé, *Inorg. Chem.*, **25**, 2114 (1986).
4. G. Hervé and A. Tézé, *Inorg. Chem.*, **16**, 2115 (1977).
5. M. T. Pope, *Heteropoly and Isopoly Oxometalates*, Springer-Verlag, New York, 1983, p. 93.
6. C. Rocchiccioli-Deltcheff and R. Thouvenot, *C. R. Acad. Sci. Ser. C*, **278**, 857 (1974).
7. F. Robert and A. Tézé, *Acta Cryst.*, **B37**, 318 (1981).
8. G. Hervé, A. Tézé, and M. Leyrie, *J. Coord. Chem.*, **9**, 245 (1979).
9. E. O. North, *Inorg. Synth.*, **1**, 129 (1939).
10. C. Rocchiccioli-Deltcheff, M. Fournier, R. Franck, and R. Thouvenot, *Inorg. Chem.*, **22**, 207 (1983).

17. VANADIUM(V) SUBSTITUTED DODECATUNGSTOPHOSPHATES

$$[PW_{12}O_{40}]^{3-} + n[VO_2]^+ + 2nOH^- \longrightarrow [PV_nW_{12-n}O_{40}]^{(3+n)-} + nH_2WO_4$$

Submitted by PETER J. DOMAILLE*
Checked by GILBERT HERVÉ and ANDRÉ TÉZÉ[†]

Polyoxoanions have a demonstrated utility in a diverse range of applications spanning electronics, catalysis, and biology. The authoritative monograph

*Central Research and Development Department, E.I. du Pont de Nemours and Company, Inc., Experimental Station, Wilmington, DE 19898.
[†]Laboratoire de Physicochimie Inorganique, U.A. CNRS 419, Université Pierre et Marie Curie, 75252 Paris Cedex 05, France.

by Pope[1] describes the most important literature up until 1983 along with important generalizations of the chemistry.

Tungstophosphates probably constitute the largest number of different heteropolyanions, yet the synthesis of specific (transition metal)-substituted positional isomers of these species has been hampered because of doubt in the formulation of precursor lacunary (defect) polyanions and uncertainties in product identification. This work describes the preparation of several vanadium(V) substituted dodecatungstophosphates. Replacement of tungsten(VI) by vanadium(V) produces significantly stronger oxidants.

A necessary requirement of these syntheses is isolation, or the in-place generation, of an appropriate defect polytungstate anion that is subsequently resubstituted with vanadium(V). The ion $[PVW_{11}O_{40}]^{4-}$ is derived from in-place generation of $[PW_{11}O_{39}]^{7-}$; $[\alpha-1,2,3-PV_3W_9O_{40}]^{6-}$ is prepared from preformed $[A-PW_9O_{34}]^{9-}$; β- and $[\gamma-PV_2W_{10}O_{40}]^{5-}$ are isolated from $[PW_{10}O_{36}]^{7-}$.

The reader is urged to consult Pope's monograph[1] and the published literature[2-5] for a complete description of the various isomers and numbering schemes. While a systematic IUPAC nomenclature is developing, the necessary complexity obscures the relatively straightforward differences between compounds. Fully descriptive systematic nomenclature is thus avoided here.[6]

Two important preparative details common to all syntheses are

1. Reproducible syntheses require accurate pH readings on a calibrated pH meter.
2. Vanadium(V) species are strong oxidants, particularly at low pH, and metallic spatulas are attacked, leaving an intense heteropoly "blue" vanadium(IV) product. Ceramic spatulas are recommended.

Microanalytical results in polyoxoanion chemistry are notoriously bad and, given the high molecular weights, are not particularly definitive. On the other hand, NMR is an accurate measure of compound purity and identification. Microanalytical data are considered as qualitative rather than quantitative guides.

Product identification relies heavily on solution phase ^{31}P NMR. Although differences in chemical shifts are small and are both pH- and temperature dependent, careful adherence to a systematic approach to measurements gives chemical shifts which are reproducible to better than 0.02 ppm. Because of the importance of correct technique for measuring chemical shifts, the experimental approach to recording the ^{31}P NMR spectra is described.

A. ^{31}P CHEMICAL SHIFT MEASUREMENTS OF TUNGSTOPHOSPHATES

Chemical shifts of tungstophosphates lie in the range of $+10$ to -20 ppm with respect to 85% H_3PO_4; so moderate sweep widths and digital resolution necessary for 0.5–1.0 Hz per point are selected. A spectrum of a thermostatically controlled 30 °C sample of 85% H_3PO_4 with a concentric D_2O capillary for field frequency lock is acquired, and the chemical shift is set to 0.00 ppm. The spectrum is remeasured at 5-min intervals for at least 15 min to ensure that thermal equilibrium has been reached, and that the shift remains at 0.00 ppm. Using the same type of tube and concentric capillary of D_2O, the spectrum of a solution of polyoxoanion at a known temperature and pH is then recorded with the same spectrometer conditions. The absolute measured chemical shift is subject to bulk susceptibility effects, primarily due to H_3PO_4, and is thus dependent on the NMR magnetic field direction relative to the tube axis.[3] In a standard superconducting magnet configuration, where the tube axis is parallel to the B_0 field, a value of $\delta_\parallel$ is measured, whereas with an electromagnet or permanent magnet, $\delta_\perp$ is obtained. Standard values of the susceptibility of H_3PO_4 and H_2O give a chemical shift difference $\delta_\parallel - \delta_\perp \simeq 0.70$ ppm. Table I lists the ^{31}P chemical shifts obtained with a parallel geometry for all soluble compounds described below.

TABLE I

Compound	pH	$\delta_\parallel$(ppm) $\pm$ 0.02 ppm, 30 °C[a]
$H_3[PW_{12}O_{40}]$	Independent	-14.55
$K_4[PVW_{11}O_{40}]$	Independent	-14.19
$K_5[1,2\text{-}PV_2W_{10}O_{40}]$[b]	2.0	-13.61
$K_5[1,4\text{-}PV_2W_{10}O_{40}]$[b]	2.0	-13.46
$K_6[1,2,3\text{-}PV_3W_9O_{40}]$	1.8	-13.41
	5.0	-12.82
$K_6[1,4,9\text{-}PV_3W_9O_{40}]$[b]	2.0	-11.38
	5.5	-12.16
$Cs_6[P_2W_5O_{23}]$	6.0	-2.13
$Cs_5[\gamma\text{-}PV_2W_{10}O_{40}]$	4.0	-14.55
$Cs_5[\gamma\text{-}PV_2W_{10}O_{40}]$	4.0	-12.85

[a]Negative values are to low frequency of the 85% H_3PO_4 reference.
[b]Synthetic details for these compounds are not described since the checkers were unable to repeat the preparations.

B. TETRAPOTASSIUM α-VANADOUNDECATUNGSTO-PHOSPHATE, $K_4[\alpha\text{-PV}^V W_{11} O_{40}] \cdot x H_2 O$

Procedure[2]

A mixture of 59 g (0.02 mol) of commercial tungstophosphoric acid $(H_3 PW_{12} O_{40} \cdot n H_2 O)$ is dissolved in 50 mL of water in a 300-mL beaker fitted with a magnetic stirrer at ~25 °C. Approximately 5 g (0.07 mol) of solid $Li_2 CO_3$ is added in small portions with vigorous stirring to bring the pH to 4.8. At this stage the only solution polyanion present is $[PW_{11} O_{39}]^{7-}$. In a separate 200-mL beaker, 3.1 g (0.025 mol) of sodium metavanadate $(NaVO_3)$ is dissolved in 100 mL of water by heating at 80 °C. The cooled, stirred solution is adjusted to pH 4.8 by dropwise addition of ~2 mL of 6 M HCl. This decavanadate solution is filtered through a fine frit and then added to the tungstophosphate solution with stirring. The pH of the mixture is adjusted to 2.0 with an additional 7 mL of 6 M HCl. The solution is heated to 60 °C for 10 min and cooled back to 25 °C, whereupon the pH is readjusted back to 2.0 with 2 mL of 6 M HCl, and the mixture is reheated to 60 °C. This procedure is repeated (usually two more times with a total 1 mL of HCl) until the cooled solution remains at pH 2.0 ± 0.1. The solution is reheated to 60 °C, and 20 g (0.27 mol) of solid KCl is added. The temperature is maintained at 60 °C for 15 min. Cooling the solution to 25 °C while stirring affords a canary yellow solid that is collected by suction filtration on a medium frit. The solid is washed with two 50-mL portions of water at pH 2 and 25 °C. It is then dried under suction on a coarse frit to yield 28 g (48%) of product. *240703*

Anal. Calcd. for $K_4[PVW_{11} O_{40}] \cdot 2 H_2 O$: K, 5.33; P, 1.05; V, 1.73; W, 68.9; H_2O, 1.2. Found: K, 5.06; P, 1.03; V, 1.52; W, 70.4; H_2O, 1.5.

Properties

A bright canary yellow solid characterizes this polyanion as devoid of higher vanadium content species. Slight discoloration of the material occurs with prolonged drying, perhaps as a result of photoreduction. The product is the least soluble of all α compounds described, and it is isolable in pure form by simple washing. ^{31}P and [51]V NMR spectra are single lines at -14.19 and -557.3 ppm (linewidth 25 Hz), respectively, at pH 2 and 30 °C. The IR spectra show lines characteristic of the polytungstate framework and phosphate, but with the threefold degeneracy of the phosphate removed because of the C_s symmetry of the product. The IR (medium = mineral oil, range = $1600–650 \text{ cm}^{-1}$): 1101 (m), 1077 (s), 1065 (sh), 982 (s), and 881 (s).

C. NONASODIUM NONATUNGSTOPHOSPHATE, $Na_9[A\text{-}PW_9O_{34}]^{3,7,8}$

Procedure

A mixture of 120 g (0.36 mol) of sodium tungstate dihydrate $Na_2WO_4 \cdot 2H_2O$ and 150 g of water is stirred in a 300-mL beaker with a magnetic stirring bar until the solid is completely dissolved. Phosphoric acid (85%) is added dropwise with stirring (4.0 mL, 0.06 mol). After addition of the acid is complete, the measured pH is 8.9 to 9.0. Glacial acetic acid (22.5 mL, 0.40 mol) is added dropwise with vigorous stirring. Large quantities of white precipitate form during the addition. The final pH of the solution is 7.5 ± 0.3. The solution is stirred for 1 h, and the precipitate is collected and dried by suction filtration on a medium frit. A typical yield is 90 g (85–90%). [12 M HCl (35 mL, 0.42 mol) is an alternative to acetic acid as the condensing acid, stirring is extended to 4 h before product collection.]* The crude product $Na_{9-x}H_x[A\text{-}PW_9O_{34}] \cdot xH_2O$ is used without further purification in subsequent reactions.

Anal. Calcd. for 98.8% $Na_9[PW_9O_{34}] \cdot 7H_2O + 1.2\%NaCl$: Na, 8.44; P, 1.20; W, 63.8; O, 25.3; H, 0.54; Cl, 0.71; H_2O, 4.86. Found: Na, 8.10; P, 1.20; W, 62.9; O, 25.9; H, 0.50; Cl, 0.71; H_2O, 4.86.

Properties

Heating at 120 °C induces a solid state isomerization from $[A\text{-}PW_9O_{34}]^{9-}$ to $[B\text{-}PW_9O_{34}]^{9-}$, which is characterized by a marked change in the IR phosphate stretching region (1020–1200 cm^{-1}). The extent of isomerization is somewhat erratic.† IR (medium = mineral oil, range = 1600–650 cm^{-1}): $Na_{9-x}H_x[A\text{-}PW_9O_{34}]$: 1052 (s), 1017 (m), 941 (s), 886 (m), 836 (s), 740–760 (s, broad). $Na_{9-x}H_x[B\text{-}PW_9O_{34}]$: 1174 (m), 1062 (m), 1018 (w), 995 (m), 897 (s), 823 (s), 739 (s).

D. HEXACESIUM α-1,2,3-TRIVANADONONATUNGSTO-PHOSPHATE, $Cs_6[\alpha\text{-}1,2,3\text{-}PV_3W_9O_{40}]$

Procedure[2]

A 250-mL beaker fitted with a magnetic stirrer is charged with 8.2 g (100 mmol) of sodium acetate and 100 g of water, and the mixture is stirred until the solid is dissolved. Approximately 6 mL of acetic acid is added until pH 4.8

---

*The checkers note that high yield depends largely on very slow addition of HCl.
†The checkers never obtained > 30% of isomerization.

is measured. To this solution are added 3.05 g (25 mmol) of sodium metavanadate (NaVO$_3$) and 20 g (8.2 mmol) of unheated Na$_{9-x}$H$_x$[A-PW$_9$O$_{34}$]. The solution is then stirred at 25 °C for 48 h. The wine red solution is filtered by suction through a fine frit to remove any solid. A solution of 8 g of cesium chloride (CsCl, 48 mmol) in 10 g of water is added to the filtrate, and the mixture is stirred for 30 min to produce 19 g of light orange product. The product is washed with two 50-mL portions of water and dried under vacuum. Yield: 15 g (56%).

Anal. Calcd. for Cs$_6$[PV$_3$W$_9$O$_{40}$]: Cs, 24.32; V, 4.67; W, 50.04; P, 1.01. Found: Cs, 24.05; V, 4.36; W, 49.07; P, 0.91.
IR: (medium = mineral oil, range = 1600–650 cm^{-1}): 1085(s), 1053(m), 953(vs, sh), 863(m), and 789(vs, br). ^{51}V NMR: -566.1, linewidth 275 Hz, pH 1.8 and 30 °C.

E. HEXACESIUM PENTATUNGSTODIPHOSPHATE, Cs$_6$[P$_2$W$_5$O$_{23}$] AND HEPTACESIUM DECATUNGSTOPHOSPHATE, Cs$_7$[PW$_{10}$O$_{36}$]

Procedure[9]

In a 600-mL beaker, 60 g (0.24 mol) of tungstic acid (H$_2$WO$_4$) is slurried with 400 g of water.* Approximately 110 mL of 50% aqueous cesium hydroxide (CsOH) is added dropwise with vigorous stirring. The turbid solution is passed through a cake of 10 g Celite® Analytical Filter aid in a medium fritted filter, to produce a clear colorless solution. Phosphoric acid (85%H$_3$PO$_4$) is added dropwise while stirring to adjust the pH to 7.0 ($\sim$ 21 mL total), and the solution is stirred for an additional 1 h. The solution is filtered to produce $\sim$ 10 g of uncharacterized product. The filtrate is cooled in a 0° refrigerator for 24 h, and it is then filtered to give $\sim$ 90 g of white crystalline solid identified as Cs$_6$[P$_2$W$_5$O$_{23}$]·xH$_2$O.

A 250-mL round-bottomed flask fitted with a reflux condenser is charged with 150 g of water and 75 g of Cs$_6$[P$_2$W$_5$O$_{23}$]·H$_2$O, and the contents are heated at reflux for 24 h. The solution is filtered hot through a medium frit to yield $\sim$ 18 g of Cs$_7$[PW$_{10}$O$_{36}$]·H$_2$O.† The filtrate is cooled for 48 h at 0 °C; it is then filtered to recover 45 g of unconverted Cs$_6$[P$_2$W$_5$O$_{23}$], which is used again in later preparations. The IR spectrum is a clear indicator of the purity and identity of all products.

---

*The checkers note that anhydrous WO$_3$ can be used instead of the H$_2$WO$_4$, and is more easily available commercially.
†The checkers report a lower yield.

Properties

The yields of the products from the previous reactions are variable. In some instances, the yield of the initial unknown product is as high as 60 g. The anion is identical to $Cs_7Na_2[W_{10}PO_{37}]$ reported by Knoth and Harlow.[9] More recent analytical data and subsequent reaction chemistry, give the present formulation as $(Cs_{13-x}H_x[PW_8O_{33}])_n$, but the evidence is far from complete. In spite of its incomplete characterization, the IR is a clear indicator of its presence. IR (medium = mineral oil, range = 1600–650 cm^{-1}): 1127 (s), 1060 (s), 1000 (m), 950 (m), 934 (s + shoulders), 893 (s), 846 (m), 816 (m), and 684 (s). Solid state ^{31}P NMR MAS is a single line at -5.1 ppm.

The compound $Cs_6[P_2W_5O_{23}] \cdot H_2O$ has a characteristic IR spectrum. IR (medium = mineral oil, range = 1600–650 cm^{-1}): 1180 (m), 1153 (m), 1048 (s), 986 (w), 893 (vs, shoulders), 801 (w), and 686 (s). In more rapidly crystallized forms, the 1180, 1153 pair contains an intermediate line producing a broad overall peak. Solid state ^{31}P MAS shows two lines with almost identical chemical shift anisotropy at $+0.2$ and -5.1 ppm. The latter line in the spectrum is indistinguishable from that of $[PW_8O_{33}]^{13-}$. The solution ^{31}P NMR at pH 6 is a single line at -2.13 ppm, exhibiting ^{183}W satellites with $J_{P-W} = 0.90$ Hz, producing an overall 3:19:53:19:3 pattern.

The complex $Cs_7[PW_{10}O_{36}] \cdot H_2O$ also has a characteristic IR spectrum (medium = mineral oil, range = 1600 – 650 cm^{-1}): 1086 (ms), 1053 (m), 1023 (m), 965 and 952 (sh), 937 (s), 892 (s), 850 (m), 830 (m), and 735 (vs, br). Solid state ^{31}P NMR MAS is a single line at -12.5 ppm.

F. PENTACESIUM γ-DIVANADODECATUNGSTOPHOSPHATE, $Cs_5[\gamma\text{-}PV_2W_{10}O_{40}]$

Procedure[4]

Sodium metavanadate ($NaVO_3$, 1 g, 8.2 mmol) is dissolved in 40 g of water in a 150-mL beaker by heating to 70 °C. The solution is cooled to 25 °C, and 3 M HCl is added dropwise with vigorous stirring to reduce the pH to 0.8. The solution is pale yellow, indicating the presence of $[VO_2]^+$. If flecks of V_2O_5 are evident, the solution is passed through a fine fritted filter. Portions of $Cs_7[PW_{10}O_{36}] \cdot xH_2O$ are added slowly until a total of 12.5 g (3.6 mmol) accumulates ($\sim$ 30 min.) The solution is stirred for an additional 30 min, and it is filtered by suction to give 10.9 g (90%) of quite pure pale yellow product. Analytical quality material is produced by dissolving 0.1 g (0.8 mmol) of sodium metavanadate in 100 g of water in a 200-mL beaker and reducing the pH to 2 by dropwise addition of 3 M HCl. Crude $Cs_5[\gamma\text{-}PV_2W_{10}O_{40}]$ (5 g, 1.5 mmol) is added, and the mixture is stirred for 30 min at 25 °C. The solution is passed through a filter cake of Celite® Analytical Filter Aid (2 g) and then chilled at 0 °C for 12 h. Crystals (2.4 g, 0.7 mmol) are collected by suction filtration.

Anal. Calcd. for $Cs_5[\gamma\text{-}PV_2W_{10}O_{40}]\cdot6H_2O$: Cs, 19.6; P, 0.92; V, 3.01; W, 54.3; O, 21.8; H, 0.36; H_2O, 3.2. Found: Cs, 19.0; P, 0.72; V, 3.27; W, 54.1; O, 22.8; H, 0.44; H_2O, 3.2.

Properties

This high-energy pale yellow isomer is quite stable at pH 2 in the presence of excess vanadium. However, in the absence of $[VO_2]^+$, isomerization to the darker orange $[\beta\text{-}PV_2W_{10}O_{40}]^{5-}$ occurs. Identification of the γ-isomer is possible through ^{31}P NMR (-14.55 ppm, pH 4), ^{51}V NMR (-547.1 ppm, linewidth 112 Hz, pH 2.5) and IR (medium = mineral oil, range = 1600–650 cm^{-1}): 1096 (m), 1060 (m), 1040 (m), 1007 (sh), 985 (sh), 954 (s), and 766 (vs, br).

G. PENTACESIUM β-DIVANADODECATUNGSTOPHOSPHATE, $Cs_5[\beta\text{-}PV_2W_{10}O_{40}]$

Procedure[5]

In a 500-mL beaker, 10 g (3 mmol) of $Cs_5[\gamma\text{-}PV_2W_{10}O_{40}]$ is stirred with 300 mL of water for 60 h. The solution is cooled to 0 °C and passed through a cake (10 g) of Celite® Analytical Filter Aid. The solution is reduced in volume to ~ 75 mL by rotary evaporation and then cooled for 1 h at 0 °C to produce a precipitate. Solid $Cs_5[\beta\text{-}PV_2W_{10}O_{40}]$ (8.5 g, 70%) is collected by suction filtration. The product is dissolved in 200 mL of water at 25 °C (some turbidity remains as a result of undissolved impurity). The solution is passed through a filter cake of Celite® Analytical Filter Aid and the filtrate is stripped to dryness to yield 7 g (60% overall) of pure compound.

Anal. Calcd. for $Cs_5[\beta\text{-}PV_2W_{10}O_{40}]\cdot10H_2O$: Cs, 19.2; P, 0.90; V, 2.95; W, 53.2; O, 23.2; H, 0.58; H_2O, 5.2. Found: Cs, 18.9; P, 0.84; V, 2.99; W, 51.7; O, 24.4; H, 0.63; H_2O, 5.3.

Properties

The ^{31}P NMR (-12.85, pH 4, 30 °C), ^{51}V NMR (-544.2, -555.2 with $^2J_{V-O-V} \sim 20$ Hz, pH 3.5, 30 °C), and IR (medium = mineral oil, range =

$1600-650 \text{cm}^{-1}$): 1090 (m), 1068 (m), 1058 (m), 976 (s, sh), 960 (s), 890 (m), and 794 (vs, br) provide the necessary identification.[‡]

References

1. M. T. Pope, *Heteropoly and Isopolymetalates*, Springer-Verlag, New York, 1983.
2. P. J. Domaille, *J. Am. Chem. Soc.*, **106**, 7677 (1984).
3. P. J. Domaille and G. Watunya, *Inorg. Chem.*, **25**, 1239 (1986).
4. P. J. Domaille and R. L. Harlow, *J. Am. Chem. Soc.*, **108**, 2108 (1986).
5. P. J. Domaille and R. L. Harlow, *Inorg. Chem.* (in preparation).
6. Systematic nomenclature from Y. Jeannin and M. Fournier, *Pure Appl. Chem.*, **59**, 1529 (1987) defines: $K_5[\alpha\text{-}1,2\text{-}PV_2W_{10}O_{40}]$ as 9,12; $K_5[\alpha\text{-}1,4\text{-}PV_2W_{10}O_{40}]$ as 11,12; $Cs_6[\alpha\text{-}1,2,3\text{-}PV_3W_9O_{40}]$ as 8,9,12; and $K_6[\alpha\text{-}1,4,9\text{-}PV_3W_9O_{40}]$ as 10,11,12.
7. W. H. Knoth, P. J. Domaille, and R. D. Farlee, *Organometallics*, **4**, 62 (1985).
8. R. Massart, R. Contant, J. M. Fruchart, J. P. Ciabrini, and M. Fournier, *Inorg. Chem.*, **16**, 2916 (1977).
9. W. H. Knoth and R. L. Harlow, *J. Am. Chem. Soc.*, **103**, 1865 (1981).

18. POTASSIUM OCTADECATUNGSTODIPHOSPHATES(V) AND RELATED LACUNARY COMPOUNDS

Submitted by ROLAND CONTANT*
Checked by WALTER G. KLEMPERER[†] and OMAR YAGHI

The octadecatungstodiphosphates (α and β isomers) and related lacunary compounds form the most stable family of polyoxotungstophosphates.[1] The structure of the α-octadecatungstodiphosphate anion has been solved by Dawson.[2] It consists of two identical half-anions in which the PO_4 tetrahedron is linked to a W_3O_{13} group and three W_2O_{10} groups, the latter forming a "belt". The two half-anions are joined by the W_2O_{10} groups to give an anion of D_{3h} symmetry in which there are two types of tungsten atoms, 6 "polar" and 12 "equatorial."

The β isomer has a structure in which one W_3O_{13} group has been rotated by $\pi/3$. All the stable lacunary compounds derive from the α structure and correspond to the departure of: (a) one equatorial tungsten atom for

---

[‡]The checkers note that major differences are observed in the IR spectra of $Cs_5[PV_2W_{10}O_{40}]$ isomers in the 280 to 420-cm^{-1} range.

*Université Pierre et Marie Curie, Chimie Inorganique, 4 place Jussieu, 75252 Paris, Cedex 05, France.

[†]School of Chemical Science, University of Illinois, Urbana IL 61801.

α_1-heptadecatungstodiphosphate; (b) one polar tungsten atom for α_2-heptadecatungstodiphosphate; (c) a group of three polar tungsten atoms for pentadecatungstodiphosphate; (d) six adjacent tungsten atoms that form a longitudinal third of the polyanion, that is, two polar and four equatorial tungsten atoms, for dodecatungstodiphosphate. The octatetracontatungsto-octaphosphate is a cyclic tetramer[3] of the dodecatungstodiphosphate.

The lacunary species can act as ligands with numerous metal ions, and they are starting materials for syntheses of various mixed polyoxoanions.[4,5]

A. POTASSIUM OCTADECATUNGSTODIPHOSPHATES, α ISOMER: $K_6[P_2W_{18}O_{62}]\cdot14H_2O$; β ISOMER: $K_6[P_2W_{18}O_{62}]\cdot19H_2O$

$$18[WO_4]^{2-} + 32H_3PO_4 \longrightarrow [P_2W_{18}O_{62}]^{6-} + 30[H_2PO_4]^- + 18H_2O$$

At present the only route for synthesizing octadecatungstodiphophates is to heat at reflux a mixture of sodium tungstate and excess H_3PO_4. Three species are formed simultaneously: the two octadecatungstodiphosphate isomers and the triacontatungstopentaphosphate. The following procedure is a modified version of the Wu[6] preparation.

Procedure

In a 2-L beaker a sample of 250 g (0.76 mol) of $Na_2WO_4\cdot2H_2O$ is dissolved in 500 mL of water and 210 mL (3.09 mol) of orthophosphoric acid (85%) is added. The solution is heated at reflux for 4 h. The greenish coloration can be removed by addition of a few drops of bromine to the hot solution.

- **Caution.** *Bromine is very toxic by inhalation and causes severe burns. Work in well-ventilated fume hood and wear gloves.*

After cooling, 100 g of ammonium chloride is added, and the solution is stirred for 10 min. The pale yellow salt is removed by filtering, dissolved in 600 mL of water, and precipitated again with 100 g of ammonium chloride. After stirring for 10 min, filtration through a course frit, and suction, the precipitate is dissolved in 250 mL of warm water ($\sim 45\,°C$). If the solution is not quite clear, it is filtered on filter paper, then left to evaporate at room temperature. After 5 days the crystallization of the β-isomer ammonium salt is almost complete. (Yield: 21 g (10%). A 40 g (0.54 mol) quantity of potassium chloride is added to the filtrate, and the precipitate is collected on a filter and then dissolved in 250 mL of hot water ($\sim 80\,°C$). On slow cooling to 20 °C, white needles of $K_{14}Na\,P_5W_{30}O_{110}\cdot xH_2O$ appear and are removed by filtration after 4 to 5 h. The solution is heated to ebullition and, if necessary, filtered again after cooling. A 25-g quantity of potassium chloride is added to the filtrate to precipitate crude α-octadecatungstodiphosphate, which is collected on a filter and air dried for 2 days. Yield: 140 g (68%). For purification it is dissolved in 200 mL of warm water (~ 40 °C), acidified to pH 2 with HCl, and left to evaporate at room temperature. After several days 123 g of yellow crystals are collected. Yield: 60%.

Anal. Calcd. for $K_6P_2W_{18}O_{62} \cdot 14H_2O$: P, 1.28; W, 68.2; K, 4.84; H_2O, 5.20. Found: P, 1.29; W, 68.3; K, 4.84; H_2O, 5.31.

Purification of the *β* isomer is accomplished by recrystallization of the ammonium salt from 40 mL of water acidified to pH 2 with HCl at 45 °C. The crystals collected, after cooling to ambient temperature and allowing solvent to evaporate for several days, can be dissolved in 40 mL of water and the solution treated with 6 g of potassium chloride. The potassium salt is collected by filtering and recrystallized from 30 mL of acidified water at 45 °C as just described. Yield: 8 g; (4%).

Anal. Calcd. for $K_6P_2W_{18}O_{62} \cdot 19H_2O$: P, 1.25; W, 67.0; K, 4.75; H_2O, 6.92. Found: P, 1.27; W, 67.3; K, 4.74; H_2O, 7.04.

Alternative procedure

If the separation of the two isomers is not necessary a time saving alternative procedure is possible. The crude ammonium salt obtained from the refluxed solution is washed by stirring in a solution of 25 g of NH_4Cl in 100 mL of water for 10 min and is then redissolved in 250 mL of warm water (~ 45 °C). Potassium chloride (40 g) is added to the cold solution. The potassium salt is removed by filtering and then dissolved in 250 mL of hot water (~ 80 °C). The white needles of triacontatungstopentaphosphate that appear on cooling to 15 °C are removed by filtration. The filtrate is either treated directly with 25 g of potassium chloride to get a mixture of the two potassium octade-catungstodiphosphate isomers, or it is refluxed for 6 h to get a solution of the pure α isomer, that is then precipitated with 25 g of potassium chloride. In either case the precipitate is filtered on a sintered glass frit and air dried for 3 days. Yield: 165 g (80%). The mixture of isomers is suitable for preparation of potassium heptadecatungstodiphosphate ($α_2$ isomer) and potassium dodecatungstodiphosphate.

■ **Caution.** *Avoid using metal spatulas; these reduce solutions or wet solids of both the octadecatungstodiphosphate isomers.*

Properties

The potassium salts are efflorescent and can lose some water in dry air. Crystals and solutions of the *β* isomer are deeper yellow and turn green in the light by formation of small amounts of reduced species. Indeed, the reduction potentials of the β isomer are more positive than those of the α isomer. In molar acetic acid–sodium acetate buffer, the half-wave potentials (V versus SCE) are, respectively, as follows, for $[\alpha\text{-}P_2W_{18}O]^{6-}$, $+0.02$ (1e), -0.15 (1e), -0.52 (1e), -0.67 (1e), and -0.90 (2e); for $\beta\text{-}P_2W_{18}O_{62}{}^{6-}$, 0.05 (1e), -0.12 (1e), -0.49 (1e), -0.64 (1e), and -0.88 (2e).

The IR spectra (KBr pellet) of the two isomers are identical in the P—O region, with two bands at 1090 and 1012 cm^{-1}. Better differentiation occurs with ^{31}P NMR spectra: One resonance at -12.5 ppm (H$_3$PO$_4$ 85%) for the α isomer and two equal resonances at -11.0 and -11.7 ppm for the β isomer.

In aqueous solutions $\beta \rightarrow \alpha$ isomerization occurs above pH 2. Aqueous α solutions are stable below pH 4.5.

B. POTASSIUM α_2-HEPTADECATUNGSTODIPHOSPHATE, $K_{10}[\alpha_2\text{-}P_2W_{17}O_{61}]\cdot20H_2O$

$$[P_2W_{18}O_{62}]^{6-} + \tfrac{34}{7}HCO_3{}^-$$
$$\longrightarrow [P_2W_{17}O_{61}]^{10-} + \tfrac{1}{7}[W_7O_{24}]^{6-} + \tfrac{34}{7}CO_2 + \tfrac{34}{14}H_2O$$

Procedure

In a 1-L beaker a sample of 80 g (1.15×10^{-2} mol) of K_6 [α- or $\beta\text{-}P_2W_{18}O_{62}]\cdot xH_2O$ is dissolved in 200 mL of water, and a solution of 20 g (0.2 mol) of potassium hydrogen carbonate in 200 mL of water is added while stirring. After 1 h, the reaction is complete, and the white precipitate is filtered on a coarse sintered glass frit, dried under suction, and then redissolved in 500 mL of hot water (95 °C). The snowlike crystals that appear on cooling to ambient temperature are filtered after 3 h, dried under suction for 5 h and air-dried for 2 to 3 days. Yield: 57 g (70%).

Anal. Calcd. for $K_{10}P_2W_{17}O_{61}\cdot20H_2O$: P, 1.26; W, 63.6; K, 7.95; H$_2$O, 7.32. Found: P, 1.27; W, 63.8; K, 7.93; H$_2$O, 7.44.

Properties

The aqueous solution is stable in the 2 to 6 pH range. In molar acetic acid–sodium acetate buffer the half-wave potentials (V vs SCE) are -0.44 (2e), -0.59 (2e), and -0.85 (2e). The ^{31}P NMR spectrum exhibits two equal sharp resonances at -7.1 and -13.6 ppm (85% H$_3$PO$_4$ reference). In the IR spectrum the P—O bands are at 1084, 1050, and 1012 cm^{-1} (KBr pellet).

C. SODIUM α-PENTADECATUNGSTODIPHOSPHATE, $Na_{12}[\alpha\text{-}P_2W_{15}O_{56}]\cdot 24H_2O$

$$[P_2W_{18}O_{62}]^{6-} + 12CO_3^{2-} + 6H_2O$$

$$\longrightarrow [P_2W_{15}O_{56}]^{12-} + 3[WO_4]^{2-} + 12HCO_3^{-}$$

Procedure

In a 600-mL beaker a sample of 38.5 g (8×10^{-3} mol) of $K_6[\alpha\text{-}P_2W_{18}O_{62}]\cdot 14H_2O$ is dissolved in 125 mL of water, and 35 g (0.25 mol) of $NaClO_4\cdot H_2O$ is added. After vigorous stirring for 20 min, the mixture is cooled on an ice bath, and the potassium perchlorate is removed by filtering after ~ 3 h. A solution of 10.6 g (0.1 mol) of Na_2CO_3 in 100 mL of water is added to the filtrate. A fine white precipitate appears almost instantaneously and is decanted and then filtered on a medium porosity sintered glass frit and dried under suction for ~ 3 h. The precipitate is then washed for 1 to 2 min with a solution of 4 g of sodium chloride in 25 mL of water, dried under suction for ~ 3 h, washed for 2 to 3 min with 25 mL of ethanol, and air dried under suction for 3 h, washed again with ethanol and dried under suction, and finally air dried for 3 days. Yield: 22 g (62%).

Anal. Calcd. for $Na_{12}P_2W_{15}O_{56}\cdot 24H_2O$: P, 1.40; W, 62.3; Na, 6.24; H_2O, 9.76. Found: P, 1.39; W, 62.8; Na, 6.20; H_2O, 9.88.

Properties

The sodium salt is a white powder, little soluble in pure water, but solubilized in the presence of lithium ions. The solutions are unstable, and conversion to the $[\alpha_2\text{-}P_2W_{17}O_{61}]^{10-}$ anion is complete after several hours. In molar acetic acid–lithium acetate buffer the half-wave potentials (V vs SCE) are -0.52 (4e) and -0.78 (2e). The ^{31}P NMR spectrum of a freshly prepared solution in molar acetic acid–lithium acetate buffer exhibits two equal resonances at $+0.1$ and -13.3 ppm (85% H_3PO_4 reference). In the IR spectrum the P—O bands are at 1130, 1075, and 1008 cm^{-1} (KBr pellet). As occurs with hydrated compounds, some bands are shifted if the measurements are performed in mineral oil (1130, 1086, and 1009 cm^{-1}).

D. POTASSIUM α-DODECATUNGSTODIPHOSPHATE, $K_{12}[\alpha\text{-}H_2P_2W_{12}O_{48}]\cdot 24H_2O$

$$[P_2W_{18}O_{62}]^{6-} + 18(CH_2OH)_3-C-NH_2 + 10H_2O$$

$$\longrightarrow [H_2P_2W_{12}O_{48}]^{12-} + 6[WO_4]^{2-} + 18(CH_2OH)_3-C-NH_3^{+}$$

Procedure

In a 1-L beaker a sample of 83 g (1.7×10^{-2} mol) of $K_6[\alpha$- or β-$P_2W_{18}O_{62}]\cdot$ xH_2O is dissolved in 300 mL of water, and a solution of 48.4 g (0.4 mol) of tris(hydroxymethyl)aminomethane in 200 mL of water is added. The solution is left at room temperature for $\frac{1}{2}$ h and then 80 g of potassium chloride is added. After complete dissolution, a solution of 55.3 g (0.4 mol) of potassium carbonate in 200 mL of water is added. The solution is vigorously stirred for ∼ 15 min, and the white precipitate that appears after a few minutes is filtered on a coarse sintered glass frit, dried under suction for 12 h, washed with 50 mL of ethanol for 2 to 3 min air dried under suction for 3 h, washed again with ethanol and dried under suction, and finally air dried for 3 days. Yield: 60 g (89%).

Anal. Calcd. for $K_{12}H_2P_2W_{12}O_{48}\cdot24H_2O$: P, 1.57; W, 56.0; K, 11.91; H_2O, 11.42. Found: P, 1.53; W, 55.6; K, 11.90; H_2O, 11.45.

Properties

The potassium salt is a white crystalline powder, more soluble in lithium salt solutions than in pure water. The solution is not stable: Moderate acidification leads to a mixture of $[\alpha_1$-$P_2W_{17}O_{61}]^{10-}$ and $[P_8W_{48}O_{184}]^{40-}$ anions. The half-wave potentials (V vs SCE) in molar acetic acid–lithium acetate buffer are -0.59 (2e) and -0.69 (2e). The ^{31}P NMR spectrum of a freshly prepared solution in lithium chloride exhibits a single resonance at -8.6 ppm. In the IR spectrum, the P—O bands are at 1130, 1075, and 1012 cm^{-1} (KBr pellet).

E.   POTASSIUM α_1-LITHIUM HEPTADECATUNGSTODI-PHOSPHATE, $K_9[\alpha_1$-$LiP_2W_{17}O_{61}]\cdot20H_2O$

$$[H_2P_2W_{12}O_{48}]^{12-} + 5[WO_4]^{2-} + Li^+ + 12H_3O^+$$
$$\longrightarrow [\alpha_1\text{-}LiP_2W_{17}O_{61}]^{9-} + 19H_2O$$

*Procedure**

In a 1-L beaker, 21.2 g (0.50 mol) of lithium chloride is dissolved in 500 mL of water acidified with 10 mL of 1 *M* HCl. After the solution has been allowed

---

*It is essential that the time intervals specified be adhered to rigorously and that solution temperatures not be allowed to rise above ambient temperature if analytically pure product is to be obtained.

to stand for at least 10 min to cool to ambient temperature, 40 g (10^{-2} mol) of $K_{12}[H_2P_2W_{12}O_{48}]\cdot24H_2O$ is added with vigorous stirring. A clear solution is obtained after 2 to 3 min, and immediately thereafter, 50 mL of a 1 M lithium tungstate solution* is added over a 20-s time period. As soon as the lithium tungstate addition is complete, 110 mL of 1 M aqueous HCl is added dropwise over a time period of 2 to 3 min. The pH of the solution must stay between 4 and 5. This addition is immediately followed by the addition of 200 mL of saturated aqueous KCl solution. Solid product appears instantly as a white precipitate, which is collected within 3 to 5 min on a medium porosity sintered glass frit with suction. The product is allowed to air dry under aspiration for 4 h and is then washed by stirring in 250 mL of ethanol for 15 min. Following removal of the ethanol by suction filtration, the product is air dried in an open container for 2 to 3 days. Yield: 38 g (77%).

Anal. Calcd. for $K_9LiP_2W_{17}O_{61}\cdot20H_2O$: P, 1.27; W, 64.0; K, 7.21; Li, 0.14; H_2O, 7.37. Found: P, 1.27; W, 63.5; K, 7.18; Li, 0.14; H_2O, 7.28.

Properties

The lithium potassium salt is a white crystalline powder, soluble in water. The $[\alpha_1\text{-}P_2W_{17}O_{61}]^{10-}$ anion is unstable in aqueous solution and isomerizes to give the α_2 anion. The isomerization is slowed by lithium ions. In acetic acid–lithium acetate buffer, the half-wave potentials (V vs SCE) are -0.48 (2e), -0.60 (2e), and -0.99 (2e). The ^{31}P NMR spectrum exhibits two equal resonances at -9.0 and -13.1 ppm. In the IR spectrum the P—O bands are at 1121, 1084, and 1012 cm^{-1} (KBr pellet).

F. POTASSIUM LITHIUM OCTATETRACONTATUNGSTO-OCTAPHOSPHATE, $K_{28}Li_5H_7[P_8W_{48}O_{184}]\cdot92H_2O$

$$4[H_2P_2W_{12}O_{48}]^{12-} + 15H_3O^+ \longrightarrow [H_7P_8W_{48}O_{184}]^{33-} + 23H_2O$$

Procedure

In 950 mL of water are dissolved, successively, 60 g (1 mol) of glacial acetic acid, 21 g (0.5 mol) of lithium hydroxide, 21 g (0.5 mol) of lithium chloride, and 28 g (7×10^{-2} mol) of $K_{12}[H_2P_2W_{12}O_{48}]\cdot24H_2O$. The solution is left in a closed flask. After 1 day, white needles appear and crystallization

---

*The molar solution of lithium tungstate is obtained by dissolution of 8.4 g (0.2 mol) of lithium hydroxide and 23.2 g (0.1 mol) of tungsten(VI) oxide in 100 mL of hot water. The small amount of insoluble oxide is removed by filtration.

continues for several days. When a week has passed the crystals are collected by suction filtration on a coarse frit and air dried for 3 days. Yield: 9 g (34%).

Anal. Calcd. for $K_{28}Li_5H_7P_8W_{48}O_{184} \cdot 92H_2O$: P, 1.67; W, 59.6; K, 7.39; Li, 0.23; H_2O, 11.61. Found: P, 1.66; W, 59.4; K, 7.26; Li, 0.25; H_2O, 11.56.

Properties

The lithium potassium salt appears as white efflorescent needles. The solutions are stable over a large pH range (1–8). The ^{31}P NMR spectrum in a lithium chloride aqueous solution exhibits a single sharp resonance at − 6.6 ppm. In the IR spectrum the P—O bands are at 1140, 1090, and $1020 \, cm^{-1}$ (KBr pellet).

References

1. M. T. Pope, *Inorganic Chemistry Concepts*, Vol. **8**, Springer-Verlag, Berlin, 1983.
2. B. Dawson, *Acta Crystallogr.*, **6**, 113 (1953).
3. R. Contant and A. Tézé, *Inorg. Chem.*, **24**, 4610 (1985).
4. R. Contant and J. P. Ciabrini, *J. Inorg. Nucl. Chem.*, **43**, 1525 (1981).
4. R. G. Finke and M. W. Droege, *Inorg. Chem.*, **22**, 1006 (1983).
6. H. Wu, *J. Biol. Chem.*, **43**, 189 (1920).

19. THE AQUATRIHEXACONTAOXOBIS[TRIOXO-ARSENATO(III)]HENICOSATUNGSTATE(6 −) ANION ISOLATED AS THE ACID OR AS THE RUBIDIUM SALT

$$As_2O_3 + 21[WO_4]^{2-} + 36H^+ \longrightarrow [As_2W_{21}O_{69}(H_2O)]^{6-} + 17H_2O$$

Submitted by Y. JEANNIN* and J. MARTIN-FRERE*
Checked by JINGFU LIU[†] and M. T. POPE[†]

The title anion derives from the Keggin structure[1] by replacing the central tetracoordinated atom by an arsenic(III) atom. Because of the lone pair on this heteroatom, the T_d Keggin structure can no longer be completed, and the species can be only $[As^{III}W_9O_{33}]^{9-}$.[2] Two $As^VW_9O_{34}$ groups can be

*Laboratoire de Chimie des Métaux de Transition, UA.CNRS n°419, Université Pierre et Marie Curie 4 place Jussieu, 75252 Paris Cedex 05, France.
[†]Department of Chemistry,Georgetown University, Washington, DC 20057.

linked together to build the Dawson structure, $[As^V_2W_{18}O_{62}]^{6-}$.[3] This is not possible with $As^{III}W_9O_{33}$ because of the arsenic atom lone pair. However, if the two $As^{III}W_9O_{33}$ units are joined by three tungsten atoms, the two arsenic(III) atoms are kept apart, and the compound can be prepared.

Procedure

A. AQUAHEXAHYDROXOHEPTAPENTACONTAOXOBIS-[TRIOXOARSENATO(III)]HENICOSATUNGSTEN $H_6[As_2W_{21}O_{69}(H_2O)]\cdot nH_2O$

■ **Caution.** *Diarsenic trioxide is carcinogenic and toxic. Wear gloves and work in a well-ventilated hood.*

The reaction container is a 400-mL Pyrex beaker. Starting materials, of analytical grade, are obtained commercially.

A 66-g sample of sodium tungstate ($Na_2WO_4\cdot2H_2O$, 0.2 mol) and 1.9 g of diarsenic trioxide (As_2O_3, 0.2/21 mol) are dissolved in 70 mL of boiling water. A 32-mL volume of concentrated HCl ($d = 1.18\,g\,cm^{-3}$) is poured carefully into the boiling solution, which becomes yellow; it sometimes becomes cloudy at the end of the addition.

■ **Caution.** *The hydrogen chloride must be added with great care because the reaction is very violent. Indeed, since the initial solution has a pH of 10 and the hydrochloric acid solution is concentrated, some spattering may occur; safety goggles should be worn.*

The solution is cooled and allowed to evaporate. It first gives a light, white precipitate (the nature of which is presently unknown), and then yellow crystals form after ~ 1 day. A slight stirring brings the light precipitate into suspension. The crystals are then separated from the solution, which carries away the white precipitate, by decantation. The decantate is now filtered, and the white precipitate is discarded. The resulting clear liquid is used to wash the yellow crystals several times. The yield of yellow crystals usually ranges from 25 to 28 g.

The yellow crystals (25–28 g) are dissolved in 20 mL of water. The heteropolyacid is extracted from this solution by a concentrated HCl (5 mL, $d = 1.18\,g\,cm^{-3}$) and diethyl-ether (10 mL) mixture.[4] The heavy layer is collected, dissolved in 20 mL of water, and reextracted by the same procedure. Finally, the heavy layer is dissolved again in water, and diethyl-ether is removed by heating. By evaporating this aqueous solution, 15 g of yellow crystals of the heteropolyacid are obtained.

Elementary analysis gives As 2.61% and W 66.7%, which corresponds to W/As = 10.4 (theory, 10.5). These crystals contain a substantial amount of water, which is quite difficult to determine since the crystals are efflorescent.

B. TETRARUBIDIUM AQUADIHYDROXOHENHEXA-
CONTAOXOBIS[TRIOXOARSENATO(III)] HENICOSA-
TUNGSTATE, $H_2Rb_4[As_2W_{21}O_{69}(H_2O)]\cdot34H_2O$

Five grams of heteropolyacid crystals, (Section. A) are dissolved in 2 mL of water (the solution may be warmed gently until dissolution), and 1 g of rubidium chloride is dissolved in 2 mL of water. The two aqueous solutions are mixed at room temperature, and a yellow precipitate of the rubidium salt is obtained. It is recrystallized, first, from 40 mL of a lukewarm HCl solution ($c = 0.25$ mol L^{-1}, temp = 60 °C). After cooling, a microcrystalline product is formed. It is separated and redissolved in 100 mL of HCl solution ($c = 1$ mol L^{-1}, temp = 60 °C). The yellow needles of the rubidium salt start growing slowly after 2 days at room temperature. After 6 days the yield is 2.2 g (42% with respect to the starting heteropolyacid).

v (cm^{-1})

Fig. 1. Infrared spectrum of $H_2Rb_4[As_2W_{21}O_{69}(H_2O)]$ in KBr pellet.

TABLE I. Powder Diffraction Data for $H_2Rb_4[As_2W_{21}O_{69}(H_2O)]\cdot34H_2O$, Cu K_α Radiation

$I_{obs}\%$	$d_{obs}(\text{Å})$	$d_{calc}(\text{Å})$	h	k	l
100	14.71	14.66	0	1	0
41	11.55	11.55	0	1	1
31	9.40	9.38	0	0	2
ε	8.50	8.46	1	1	0
28	7.93	7.90	0	1	2
24	6.83	6.83	0	2	1
13	5.31	5.31	1	2	1
12	4.88	4.89	0	3	0
12	4.73	4.73	0	3	1
13	4.22	4.23	2	2	0
12	4.15	4.15	1	2	3
8	4.07	4.06	1	3	0
8	3.97	3.97	1	3	1
9	3.85	3.86	2	2	2
		3.85	0	3	3
16	3.73	3.73	1	3	2
9	3.65	3.66	0	4	0
		3.64	0	1	5
18	3.59	3.60	0	4	1
		3.58	1	2	4
30	3.41	3.41	0	4	2
		3.41	1	3	3
48	3.16	3.16	0	4	3
		3.15	1	4	1
11	3.06	3.06	0	1	6
9	3.03	3.03	1	4	2
19	2.89	2.89	0	4	4
9	2.85	2.85	1	4	3
8	2.70	2.70	3	3	2
25	2.65	2.65	0	5	3
		2.64	1	4	4
11	2.60	2.61	1	5	1
19	2.42	2.43	1	5	3
		2.42	0	6	1
6	2.27	2.28	2	5	2
		2.28	0	6	3
11	2.25	2.25	3	4	3
8	2.23	2.23	0	2	8
18	2.11	2.11	0	3	8

Anal. Calcd. for $H_2Rb_4[As_2W_{21}O_{69}(H_2O)] \cdot 34H_2O$: As, 2.46%; W, 63.4%; Rb, 5.62%. Found: As, 2.48%; W, 64.1%; Rb, 5.26%.

This corresponds to W–Rb–As:10.5–1.86–1 (theory, 10.5–2–1) Water molecules measured by thermogravimetry: 17.5 per arsenic atom.

Properties

Both compounds are yellow. The acid is extremely soluble in water, and the crystals are efflorescent. The polarogram in a 0.5-mol L^{-1} HCl solution (water–methanol 1:1 in volume) shows two waves of $2e$/molecule: $E_{1/2} = -0.42$ and -0.57 V versus SCE. The characteristic IR spectrum of the anion is given in Fig. 1. Crystals of the rubidium salt are hexagonal, with lattice constants $a = b = 16.926(7)$, $c = 18.767(8)$ Å. They contain two $H_2Rb_4[As_2W_{21}O_{69}(H_2O)]$ molecules per unit cell ($d = 4.4$ g cm^{-3} and belong to the space group $P6_3/mmc$.[5] X-ray powder diffraction data are given in Table I.

References

1. J. F. Keggin, *Proc. R. Soc. London,* **A 144**, 75 (1934).
2. A. Rosenheim and A. Wolff, *Z. Anorg. Allgem. Chem.,* **64**, 193 (1930). C. Tourné, A. Revel, G. Tourné, and M. Vendrell, *C. R. Acad. Sci. Ser. C,* **277**, 643 (1973).
3. B. Dawon, *Acta Crystallogr.,* **6**, 113 (1953).
4. E. Drechsel, Berichte, **20**, 1452 (1887).
5. Y. Jeannin and J. Martin-Frère, *J. Am. Chem. Soc.,* **103**, 1164 (1981).

20. THE SODIUM PENTAPHOSPHATO(V)-TRIACONTATUNGSTATE ANION ISOLATED AS THE AMMONIUM SALT

$$5[PO_4]^{3-} + 30[WO_4]^{2-} + Na^+ + 60H^+ \longrightarrow [NaP_5W_{30}O_{110}]^{14-} + 30H_2O$$

Submitted by Y. JEANNIN* and J. MARTIN-FRERE*
Checked by DANIEL J. CHOI† and MICHAEL T. POPE†

The title anion, first prepared by C. Preyssler,[1] is formed as a by-product in the synthesis of the $[P_2W_{18}O_{62}]^{6-}$ anion. It is one of the largest known

*Laboratoire de Chimie des Métaux de Transition, UA CNRS n°41, Université Pierre et Marie Curie, 4 place Jussieu, 75252 Paris Cedex 5, France.
†Department of Chemistry, Georgetown University, Washington, DC 20057.

polyanions and is a rare example of a chemical assembly with a true fivefold symmetry axis. The structure consists of a cyclic arrangement of five PW_6O_{22} units, each formally derived from the Keggin anion $[PW_{12}O_{40}]^{3-}$ (ref. 2) by removal of two sets of three corner shared WO_6 groups. It contains an encapsulated sodium atom located on the fivefold symmetry axis but not at the center of the anion. The sodium atom cannot be removed by ion exchange. and the heteropolyanion cannot be synthezised without it.[3]

Procedure

The following procedure is based upon the preparation given by C. Preyssler.[1]

The reaction container is a 250-mL round-bottomed flask fitted with a water cooled reflux condenser. Starting materials, of analytical grade, are obtained commercially.

Sodium tungstate ($Na_2WO_4 \cdot 2H_2O$, 50 g, 0.15 mol) is dissolved in 60 mL of boiling water, and 80 g of concentrated phosphoric acid (H_3PO_4, $d = 1.7\,g\,cm^{-3}$) is poured carefully into the boiling solution, which then becomes yellow. The mixture is heated at reflux for 5 h. During this time it may become greenish in color because of a slight reduction; it can be reoxidized by addition of nitric acid (1 mL, $d = 1.33\,g\,cm^{-3}$).

▪ **Caution.** *The orthophosphoric acid must be added with great care because spattering may occur; safety goggles should be worn.*

Ammonium chloride (NH_4Cl, 50 g) is added to the lukewarm solution (60 °C) to give a white precipitate, which is removed by filtering after cooling. The precipitate is redissolved in 150 mL of boiling water. If a slight white precipitate persists, it is removed by filtering and is discarded. Ammonium chloride (40 g) is then added to the solution. The white precipitate that forms is separated by filtration after cooling and is redissolved in 140 mL of boiling water. The solution is filtered, if necessary, and the resulting clear liquid is evaporated at room temperature.

Colorless needles of $(NH_4)_{14}[NaP_5W_{30}O_{110}] \cdot 31H_2O$ appear first, after one to several days, before the yellow crystals of $(NH_4)_6[P_2W_{18}O_{62}]$ form. They are separated by fractional crystallization, and air dried at room temperature. The yield usually ranges between 4 and 6 g (9–14% with respect to the starting tungstate). The product can be recrystallized easily from water ($1.5\,g\,mL^{-1}$). The solution is warmed gently until dissolution. Crystals are formed by cooling to room temperature. One recrystallization is usually sufficient to yield a product that is pure as shown by cyclic voltammetry and ^{31}P NMR.

Anal. Calcd. for $(NH_4)_{14}[NaP_5W_{30}O_{110}] \cdot 31H_2O$: P, 1.87%; W, 65.7%; N, 2.37%. Found: P, 1.77%; W, 65.7%; N, 2.22%.

The solution content measured by ^{23}Na NMR is 1Na atom per molecule of polyanion. Thermogravimetric analysis shows 31 water molecules per polyanion.

Properties

The compound is efflorescent and soluble in water. Reducing agents, such as TiCl$_3$, lead to a blue color. In a 1 mol L^{-1} HCl solution, cyclic voltammograms exhibit cathodic peaks at -0.22, -0.35, and -0.56 V versus SCE. The first two reduction steps correspond to 4e each.[3] The ^{31}P NMR spectrum exhibits a single peak at $\delta = -10.4$ ppm. The characteristic IR spectrum is shown in Fig. 1. The crystals are triclinic, and they belong to the space group $P\bar{1}$. Lattice constants are $a = 23.570(7)$ Å; $b = 17.82(5)$ Å; $c = 17.593(9)$ Å; $\alpha = 112.17(3)°$; $\beta = 98.10(3)°$; $\gamma = 96.84(3)°$; $Z = 2$. Density $= 4.0$ g cm^{-3}.

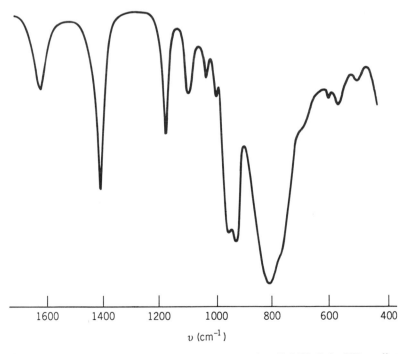

Fig. 1. Infrared spectrum of $(NH_4)_{14}[NaP_5W_{30}O_{110}]\cdot31H_2O$ in KBr pellet.

References

1. C. Preyssler, *Bull. Soc. Chim. Fr.*, **30** (1970).
2. J. F. Keggin, *Proc. R. Soc. London. Ser.*, A **144**, 75 (1934).
3. M. H. Alizadeh, S. P. Harmalker, Y. Jeannin, J. Martin-Frère, and M. T. Pope, *J. Am. Chem. Soc.*, **107**, 2662 (1985).

21. TETRACONTATUNGSTOTETRAARSENATE(III) AND ITS COBALT(II) COMPLEX

Submitted by GILBERT HERVÉ* and ANDRÉ TÉZÉ*
Checked by JINGFU LIU† and MICHAEL T. POPE†

The tetracontatungstotetraarsenate(III) ion is an especially interesting species:(a) it is a crown molecule resulting from the association of four identical AsW_9O_{33} units linked by four tungsten atoms, and (b) it presents two structurally nonequivalent types of coordination sites. One site is specific for alkali and alkaline earth metal cations, which are trapped at the center of the structure. Exchange reactions between cations are rapid in aqueous solution, and the stabilities of these inorganic cryptates depend on the radii of the cation.[1] The other site is specific for transition metal cations, and two such ions can be complexed. With Co^{2+}, for example, the polyanionic complex $[(NH_4)As_4W_{40}O_{140}(Co(H_2O))_2]^{23-}$ can be obtained and isolated as the ammonium salt. The flexibility of the crown polyanionic structure allows conformational changes induced by the Co^{2+} cation.[2] Moreover, the properties of the Co^{2+} cation in the complex, such as the intensity of the visible electronic spectrum or the substitution of a water molecule in its coordination sphere, are dependent on the nature of the alkali or alkaline earth cation in the center of the structure.[3]

A. SODIUM TETRACONTATUNGSTOTETRAARSENATE(III), $Na_{27}[Na\,As_4W_{40}O_{140}]\cdot60H_2O$

$$40[WO_4]^{2-} + 4[H_2AsO_3]^- + 28Na^+ + 56H^+ + 28H_2O$$

$$\longrightarrow Na_{27}[Na\,As_4W_{40}O_{140}]\cdot60H_2O$$

*Laboratoire de Physicochimie Inorganique, U.A. CNRS 419, Université Pierre et Marie Curie, 75252 Paris Cedex 05, France.
†Department of Chemistry, Georgetown University, Washington, DC 20057

Procedure

■ **Caution.** *As(III) compounds are toxic and should be handled with care.*

Sodium tungstate ($Na_2WO_4 \cdot 2H_2O$, 132g, 0.4 mol) and sodium meta-arsenite ($NaAsO_2$, 5.2 g, 40 mmol) are dissolved in 200 mL of distilled water at 80 °C. A 6 M HCl solution (82 mL) is added slowly with vigorous stirring. The final pH is ~ 4. At 0 °C, the solid product crystallizes slowly. One day is required for complete deposition. The white solid is collected on a filter and air dried. Yield: ~ 80 g (69%).

Anal. Calcd. for $Na_{27}[NaAs_4W_{40}O_{140}] \cdot 60H_2O$: Na, 5.54; W, 63.3; As, 2.58; H_2O, 9.3. Found: Na, 5.65; W, 62.5; As, 2.49; H_2O, 9.2.

The potassium salt of $[KAs_4W_{40}O_{140}]^{27-}$ can be obtained by the previous procedure, using a K_2WO_4 solution prepared by the action of a stoichiometric amount of potassium hydroxide on an aqueous suspension of tungsten(VI) oxide.

Properties

The compound is an air-stable, white, crystalline solid, soluble in water. Its aqueous solution is stable between pH 4 and 7.5. Its polarogram in 0.5 M tris(hydroxymethyl)aminomethane + 0.5 M NaCl buffer, pH 7.5, obtained with a dropping mercury electrode, shows two waves with half-wave potentials -1.06 and -1.16 V *versus* SCE. There is a shoulder on the electronic spectrum in the UV region at 240 nm ($\varepsilon = 1.4 \cdot 10^5 \ M^{-1} cm^{-1}$). The 250-MHz ^{183}W NMR spectrum in H_2O-D_2O (90:10) solution shows six resonance lines with relative intensities 2:1:2:2:2:1 at $\delta = -93.8$, -103.7, -104.8, -118.6, -189.8, and -196.8 ppm, respectively (reference: external 2 M Na_2WO_4 in alkaline D_2O).

The polyanion is characterized in the solid state by its IR spectrum (KBr pellet, cm^{-1}): 945, 875, 795, 700, 620, 470, 360, and 330.

B. AMMONIUM TETRACONTATUNGSTODICOBALTOTETRA-ARSENATE(III), $(NH_4)_{23}[NH_4As_4W_{40}O_{140}(Co(H_2O))_2] \cdot 19H_2O$

$$[NaAs_4W_{40}O_{140}]^{27-} + 2Co^{2+} + 24NH_4^+ + 21H_2O$$

$$\longrightarrow (NH_4)_{23}[NH_4As_4W_{40}O_{140}(Co(H_2O))_2] \cdot 19H_2O + Na^+$$

Procedure

Aqueous 1 M $Co(NO)_3$ (5 mL, 5 mmol) is added with stirring to a solution of 30 g (2.5 mmol) of the sodium salt of tetracontatungstotetraarsenate dissolved in 200 mL of water. After 10 min, 10 g of solid ammonium chloride is added. The precipitate is collected on a filter. The product is redissolved in 200 mL of water and precipitated again with 10 g of solid ammonium chloride. This operation is performed twice in order to obtain a sodium free salt.

The material is recrystallized by dissolution in 60 mL of lukewarm water. Yield: 19 g (70%).

Anal. Calcd. for $(NH_4)_{23}[NH_4As_4W_{40}O_{140}(Co(H_2O))_2]\cdot 19H_2O$: NH_4, 3.98; W, 67.8; Co, 1.09; As, 2.76; H_2O, 3.15. Found: NH_4, 3.92; W, 66.7; Co, 1.15; As, 2.84; H_2O, 3.10.

Properties

The complex is a dark green crystalline solid soluble in water. It is stable between pH 4 and 7. It is well characterized by the electronic spectrum of its solution: $\varepsilon^{max}600 = 110$, $\varepsilon^{min}544 = 46$, $\varepsilon^{max}530 = 56$, $\varepsilon^{sh}518 = 53$, $\varepsilon^{min}504 = 42$, $\varepsilon^{max}480 = 68$, $\varepsilon^{min}458 = 60$.[‡]

References

1. M. Leyrie and G. Hervé, *Nouv. J. Chim.*, **2**, 233 (1978).
2. F. Robert, M. Leyrie, G. Hervé, A. Tézé, and Y. Jeannin, *Inorg. Chem.*, **19**, 1746 (1980).
3. M. Leyrie, Ph.D thesis, Université P. et M. Curie, Paris, 1984.

22. AMMONIUM SODIUM HENICOSATUNGSTONONA-ANTIMONATE(III), $(NH_4)_{18}[NaSb_9W_{21}O_{86}]\cdot 24H_2O$

$$21[WO_4]^{2-} + \tfrac{9}{2}Sb_2O_3 + Na^+ + 41H^+ + 18NH_3 + 12.5H_2O$$
$$\longrightarrow (NH_4)_{18}[NaSb_9W_{21}O_{86}]\cdot 24H_2O$$

Submitted by GILBERT HERVÉ* and ANDRE TÉZÉ*
Checked by JINGFU LIU[†] and MICHAEL T. POPE[†]

Many heteropolyanions can be formed when tungstate is acidified in the presence of Sb(III).[1,2] The $[NaSb_9W_{21}O_{86}]^{18-}$ polyanion has been

---

[‡]The checkers report the following peak maxima: 608, 535, and 490 nm.

*Laboratoire de Physicochimie Inorganique, U.A. CNRS 419, Université Pierre et Marie Curie, 75252 Paris Cedex 05, France.
[†]Department of Chemistry, Georgetown University, Washington, DC 20057.

particularly studied because of its unusual characteristics and medical applications. Its structure is odd:It is constituted by the association of polytungstic and polyantimonic fragments.[3] Its stability is related to the presence of a sodium cation at the center of the polyanion. This position can be occupied by other cations, such as K^+, Ca^{2+}, or Sr^{2+}, but exchange reactions cannot be carried out in aqueous solution.[4] The henicosatungsto-nonaantimonate is an efficient antiviral compound, *in vitro* and *in vivo* at noncytotoxic doses.[5,6] It strongly inhibits the cellular and bacterial DNA and RNA polymerases and the reverse transcriptases of retrovirus.[7] It is used (as HPA 23) in an AIDS treatment.[8]

■ **Caution.** *The entire preparation should be carried out in a well-ventilated fume hood in order to avoid inhaling gaseous hydrogen chloride, or smoke of ammonium chloride.*

Procedure

Antimony trioxide (18.7 g, 64 mmol) is dissolved in 50 mL of concentrated HCl ($d = 1.18$) (Solution A). Sodium tungstate ($Na_2WO_4 \cdot 2H_2O$, 99 g, 0.3 mol) is dissolved in 120 mL of distilled water at 80 °C in a 0.5-L beaker containing a magnetic stirring bar (Solution B).

Solution A is added slowly to Solution B with vigorous stirring. Before the end of this addition, a yellow color appears. When a yellow color is observed, add concentrated aqueous ammonia ($d = 0.91$) in 3-mL portions, alternately with 3-mL portions of the remainder of solution A, until all of Solution A has been added. Approximately 25 mL of ammonia solution is required. The final solution is colorless and has a pH of ~ 8. During this operation a precipitate of a sodium–ammonium mixed salt of the henicosa-tungstononaantimonate appears. The solid is collected on a filter and washed with 2 M ammonium chloride solution. It is dissolved in 200 mL of water. If there is a small residue of insoluble material, it is removed by filtration on a filter paper.

A saturated solution of ammonium chloride, brought to pH ~ 8 by addition of a small amount of concentrated aqueous ammonia, is prepared, and 60 mL of this solution is added to the former one with gentle stirring. The precipitate is collected on a filter. This operation is performed twice in order to obtain a sodium free salt. The product is air dried. Yield: ~ 90 g (89%). The product may be recrystallized from a solution in its own weight of water at 30 °C.

Anal. Calcd. for $(NH_4)_{18} [NaSb_9W_{21}O_{86}] \cdot 24H_2O$: Sb, 15.4; W, 54.3; NH_4, 4.6; H_2O, 6.1. Found: Sb, 15.7; W, 55.0; NH_4, 4.8; H_2O, 6.0.

Properties

The compound is an air-stable, white, crystalline solid, soluble in water. Its aqueous solution is stable between pH 7 and 8.5. In 0.5 M tris(hydroxymethyl)aminomethane $+ 0.5 M$ NaCl buffer of pH 8.1, the polarogram, obtained with a dropping mercury electrode, of a $6 \times 10^{-5} M$ solution of the polyanion shows two waves with half-wave potentials $- 1.15$ and $- 1.26$ V *versus* SCE. The 250 MHz ^{183}W NMR spectrum in H_2O-D_2O (90:10) solution at 40 °C, shows four resonance lines with relative intensities 1:2:2:2 at $- 16.8$, $- 68.5$, $- 129.4$, and $- 224.6$ ppm (reference: external 2 M Na_2WO_4 in alkaline D_2O). The IR spectrum (KBr pellet) shows numerous bands characteristic of: (a) crystallization water molecules: 3400–3420 and 1630 cm^{-1}, (b) ammonium counterions: 3160, 3000, 2820, and 1405 cm^{-1}, and (c) vibration modes of the polyanion: 925, 865 (sh), 825, 770, 730, 680, 560, 480, 410, 350, and 320 cm^{-1}.

References

1. P. Souchay, M. Leray, and G. Hervé C. R. Acad. Sci. Ser. C, **271**, 1337 (1970).

2. C. Tourné, A. Revel, G. Tourné, and M. Vendrell, C. R. Acad. Sci. Ser. C, **277**, 643 (1973).

3. J. Fisher, L. Ricard, and, R. Weiss, J. Am. Chem. Soc., **98**, 3050 (1976).

4. M. Michelon, G. Hervé, and M. Leyrie, J. Inorg. Nucl. Chem., **42**, 1583 (1980).

5. C. Jasmin, J. C. Chermann, G. Hervé, A. Tézé, P. Souchay, C. Boy-Loustau, N. Raybaud, F. Sinoussi, and M. Raynaud. J. Natl. Cancer Inst., **53**, 469 (1974).

6. J. Blancou, H. Tsiang, J. C. Chermann, and L. Andral, Curr. Chemoth., **1981**, 1070.

7. M. Hervé, F. Sinoussi-Barre, J. C. Chermann, G. Hervé, and C. Jasmin, Biochem. Biophys. Res. Commun., **116**, 222 (1983).

8. W. Rozenbaum, D. Dormont, B. Spire, E. Vilmer, M. Gentilini, C. Griscelli, L. Montagnier, F. Barre-Sinoussi, and J. C. Chermann, Lancet, **1985**, 450.

23.   BIS(μ_5-ORGANOPHOSPHONATO-O, O′, O″)-PENTA-μ-OXO-PENTAKIS(DIOXOMOLYBDATES, -TUNGSTATES)-(4 −)(PENTAMETALLOBISPHOSPHONATES)

$$5[MO_4]^{2-} + 2RPO_3^{2-} + 10H^+$$

$$\longrightarrow [(RPO_3)_2M_5O_{15}]^{4-} + 5H_2O \qquad (M = Mo \text{ and } W)$$

**Submitted by WON SUK KWAK*, MICHAEL T. POPE[†]
and PURISAI R. SETHURAMAN[†]
Checked by XIAO SUN[‡] and JON ZUBIETA[‡]**

The pentamolybdobis(phosphonate) anions[1] were the first examples of organic derivatives of heteropolyanions to be structurally characterized,[2] although it is probable that analogous glycerophosphate[3] and phosphite[4] molybdate complexes had been prepared earlier. The heteropolyanions are readily formed in aqueous solutions at pH 2 to 6, but the isolation of pure salts is not always straightforward because of the possibility of partial hydrolytic dissociation in solution and the consequent contamination of the product with isopolyanion salts. The latter complication can be minimized with salts that are recrystallized from nonaqueous solvents or by the appropriate choice of the countercation.[5] Although the above equation represents the formation of the heteropolyanion from the tetraoxometallate anion, in some instances it is convenient to begin with an isopolyanion or molybdic oxide. Examples of the various approaches are given.

Procedures

A.   PENTAMOLYBDOBIS(PHOSPHONATE)

Ammonium salt, $(NH_4)_4[(HPO_3)_2Mo_5O_{15}]\cdot4H_2O$: Phosphorous acid (4.1 g, 50 mmol) is dissolved in 70 mL of water. After 7 mL of 15 M ammonia (105 mmol) is added to the phosphorous acid solution, it is heated to boiling. To the boiling solution is added molybdic oxide (anhy MoO_3, 14.4 g, 100 mmol) in small portions. When the powder is dissolved (40–45 min), the solution is filtered. The clear filtrate is reduced to 20–25 mL by boiling and then allowed to cool to room temperature. The crystalline powder formed

*PPG Industries, Inc., Chemical Group, Barberton, OH 44203.
[†]Georgetown University, Washington, DC 20057.
[‡]State University of New York at Albany, Albany, NY 12222.

is collected by filtration, washed with 7 mL of ice-cold water, and then air dried. Yield: 16.4 g (78%).

Anal. Calcd. for $H_{26}N_4P_2Mo_5O_{25}$:H, 2.51; N, 5.37; P, 5.93; H_2O, 6.90. Found: H, 2.56; N, 5.60; P, 6.02; H_2O, 7.10.

^{1}H NMR   7.15δ, $J_{(^1H-^{31}P)} = 683$ Hz.   UV   (0.7 mM,   pH   4.5),   nm (M^{-1} cm^{-1}): 245(3.1 × 10^4), 213(3.2 × 10^4).

B.   PENTAMOLYBDOBIS(METHYLPHOSPHONATE)

Ammonium salt, $(NH_4)_4[(CH_3PO_3)_2Mo_5O_{15}]\cdot2H_2O$: Molybdic oxide (anhy MoO_3, 7.2 g, 50 mmol) is dissolved with heating in 30 mL of water containing 2.7 mL of 15 M ammonia (40 mmol). After methylphosphonic acid (1.92 g, 20 mmol) (Aldrich) is dissolved in the reaction solution, it is boiled for 15–20 min. Then, the hot solution is filtered, and the filtrate is allowed to evaporate in the air. In a few days the crystals formed are collected by filtration, washed twice with 4-mL volumes of cold water and dried in the air. Yield: 7.7 g (75%). This product is recrystallized from the minimum (∼ 10 mL) of boiling water.

Anal. Calcd. for $C_2H_{26}N_4P_2Mo_5O_{23}$: C, 2.36; H, 2.58; N, 5.52; P, 6.10; H_2O, 3.54. Found: C, 2.70; H, 3.27; N, 5.20; P, 6.56; H_2O, 4.18.

^{1}H NMR   1.68δ, $J_{(^1H-^{31}P)} = 16.9$ Hz.   UV   (0.7 mM,   pH   4.5),   nm (M^{-1} cm^{-1}): 251 (3.4 × 10^4), 211(3.2 × 10^4).

Tetramethylammonium sodium salt, $[(CH_3)_4N]_2Na_2[(CH_3PO_3)_2Mo_5O_{15}]\cdot$ $3H_2O$: Molybdic oxide (anhy MoO_3, 7.2 g, 50 mmol) is dissolved with heating in 20 mL of water containing 1.6 g of NaOH (40 mmol). Then, methylphosphonic acid (1.92 g, 20 mmol) is dissolved in the molybdate solution, which is boiled for 15–20 min. After 4.4 g of tetramethylammonium chloride (40 mmol) dissolved in 6 mL of warm water is added, the reaction solution is heated for another 10–15 min and filtered. When the filtrate is allowed to stand overnight in the air, the salt is formed as large crystals. It is collected by filtration, washed twice with cold water, and dried in the air. Yield: 6.5 g (56%). The product is recrystallized from the minimum (∼ 10 mL) of boiling water.

Anal. Calcd. for $Na_2C_{10}H_{36}N_2P_2Mo_5O_{24}$: Na, 3.98; C, 10.39; H, 3.14; N, 2.42; P, 5.36; H_2O, 4.68. Found: Na, 3.79; C, 10.21; H, 3.62; N, 2.38; P, 5.04; H_2O, 4.95.

The NMR and UV as for the ammonium salt.

C. PENTAMOLYBDOBIS(ETHYLPHOSPHONATE)

Ammonium salt, $(NH_4)_4[(C_2H_5PO_3)_2Mo_5O_{15}]$: Ethylphosphonic acid (0.88 g, 8 mmol) (Aldrich) is dissolved in 25 mL of water containing 4.83 g of sodium molybdate dihydrate (20 mmol). The solution is acidified with 4 mL of 6 M HCl (24 mmol) and then boiled for 15–20 min. To the reaction solution is added ammonium chloride (0.86 g, 16 mmol) dissolved in 3 mL of warm water. The volume is reduced to 15 to 20 mL by heating, and the resulting solution is filtered while hot. When the filtrate is allowed to crystallize in the air, chunky crystals are formed in a few days. They are collected, washed with 2 mL of cold water, and air dried. Yield: 3.7 g (92%). This product is recrystallized from the minimum of hot water.

Anal. Calcd. for $C_4H_{26}N_4P_2Mo_5O_{21}$: C, 4.77; H, 2.60; N, 5.56; P, 6.15; H_2O, 0.0. Found: C, 4.93; H, 2.81; N, 5.47; P, 6.14; H_2O, 0.9.

^1H NMR 0.76–2.21δ (m, 5H). UV (0.7 mM, pH 4.5), nm $(M^{-1}cm^{-1})$: 252(3.2 × 10⁴), 210(3.2 × 10⁴).

D. PENTAMOLYBDOBIS(PHENYLPHOSPHATE)

Ammonium salt, $(NH_4)_4[(C_6H_5PO_3)_2Mo_5O_{15}] \cdot 5H_2O$: Ammonium heptamolybdate $((NH_4)_6Mo_7O_{24} \cdot 4H_2O$, 10 g, 8.1 mmol) is dissolved in 60 mL of water with the addition of 15 M ammonia (3 mL, 45 mmol). To this solution is added phenylphosphonic acid (4.2 g, 26.6 mmol) (Aldrich), and the powder is dissolved. The solution is acidified to pH 4.6–5.0 (test paper) with a few drops of 6 M HCl and then heated until the volume is reduced to 40–45 mL. After standing to cool to room temperature, the slightly turbid solution is filtered. When the clear solution is allowed to evaporate in the air, large crystals are formed. They are collected, washed quickly twice with 5-mL portions of water, and then dried in the air. Yield: 12.1 g (89%). The product is recrystallized from the minimum (∼ 10 mL) of boiling water.

Anal. Calcd. for $C_{12}H_{36}N_4P_2Mo_5O_{26}$: C, 12.07; H, 3.04; N, 4.69; P, 5.19; H_2O, 7.54. Found: C, 12.14; H, 3.27; N, 4.59; P, 5.61; H_2O, 7.81.

^1H NMR 7.50 (m, 3H), 2.75 (m, 2H)δ. UV (0.7 mM, pH 4.5), nm$(M^{-1}cm^{-1})$: 248(2.9 × 10⁴), 208(5.1 × 10⁴).

Tetramethylammonium sodium salt $[(CH_3)_4N]_3Na[(C_6H_5PO_3)_2Mo_5O_{15}] \cdot 6H_2O$: Sodium molybdate dihydrate (6.05 g, 20 mmol) and phenylphosphonic acid (1.58 g, 10 mmol) are dissolved in 30 mL of water. The mixture is acidified with 5 mL of 6 M HCl (30 mmol) and then boiled for 15–20 min. Tetramethylammonium chloride (2.2 g, 20 mmol) dissolved in 3 mL of water is added to the reaction solution. After filtration, the solution is concentrated to

15–20 mL by boiling. As it is transferred to a Petri dish in a shallow layer and allowed to evaporate slowly with a loose cover, the compound is precipitated as thin, platelike crystals. The crystals are collected by filtration, washed twice very quickly with cold water and then dried in the air. Yield: 6.4 g (92%).

Anal. Calcd. for $NaC_{24}H_{58}N_3P_2Mo_5O_{27}$: Na, 1.66; C, 20.81; H, 4.22; N, 3.03; P, 4.47; H_2O, 7.80. Found: Na, 2.05; C, 21.02; H, 4.19; N, 3.03; P, 4.45; H_2O, 7.19.

The NMR and UV as for ammonium salt.

E. PENTAMOLYBDOBIS[(2-AMINOETHYL)PHOSPHONATE)]

Tetramethylammonium   sodium   salt, $[(CH_3)_4N]Na[(NH_3C_2H_4PO_3)_2$-$Mo_5O_{15}]\cdot5H_2O$: Sodium molybdate dihydrate (2.42 g, 10 mmol) and (2-aminoethyl)phosphonic acid (0.50 g, 4 mmol) (Sigma) are dissolved in 20 mL of water. To this solution is added 6 M HCl (2.7 mL, 16 mmol) with good mixing. The solution is boiled for 15–20 min and filtered. A clear filtered solution of tetramethylammonium chloride (0.88 g, 8 mmol) dissolved in 4 mL of water is added to the reaction solution. The pentahydrate salt is formed as the reaction solution is boiled to a small volume (~ 20–25 mL) and then allowed to stand overnight at ambient temperature. It is filtered, washed with 2–3 mL of cold water and dried in the air. Yield: 2.0 to 2.1 g (86–91%).

Anal. Calcd. for $NaC_8H_{36}N_3P_2Mo_5O_{26}$: Na, 1.99; C, 8.32; H, 3.14; N, 3.64; P, 5.36; H_2O, 7.79. Found: Na, 1.91; C, 8.39; H, 3.11; N, 3.50; P, 5.35; H_2O, 7.97.

UV (0.7 mM, pH 4.5), nm (M^{-1} cm^{-1}): 252(3.5 × 10^4), 213(3.2 × 10^4).

F. PENTAMOLYBDOBIS[(p-AMINOBENZYL)PHOSPHONATE)]

Ammonium   salt, $(NH_4)[(NH_3C_6H_4CH_2PO_3)_2Mo_5O_{15}]\cdot5H_2O$: Sodium molybdate dihydrate (1.21 g, 5 mmol) and (p-aminobenzyl)phosphonic acid (0.37 g, 2 mmol) (Sigma) are dissolved in 70 mL of water. To this solution is added 2.0 mL of 3 M HCl (6 mmol). The solution is boiled for 30 min and then filtered. When ammonium chloride (0.22 g, 4.2 mmol) dissolved in 2 mL of water is added to the reaction solution, and the mixture is allowed to evaporate slowly at room temperature, a pale yellow crystalline powder is formed. It is collected, washed twice, quickly, with 1 mL of cold water, and then air dried. Yield: 1.15 g (94%).

Anal. Calcd. for $C_{14}H_{36}N_4P_2Mo_5O_{26}$: C, 13.80; H, 2.98; N, 4.60; P, 5.09; H_2O, 7.39. Found: C, 13.45; H, 3.11; N, 4.99; P, 4.99; H_2O, 9.16.

Tetramethylammonium salt, $[(CH_3)_4N]_2[(NH_3C_6H_4CH_2PO_3)_2Mo_5O_{15}]$·
$4H_2O$: This salt is obtained from a reaction solution prepared as described
for the ammonium salt. When tetramethylammonium chloride (0.44 g,
4 mmol) dissolved in 3 mL of water is added to the filtrate, a white powder
is formed. After the suspension is stirred for 5–10 min at room temperature,
the powder is removed by filtering, washed quickly with 2 mL of ice-cold
water, and then air dried. Yield: 1.21 g (92%).

Anal. Calcd. for $C_{22}H_{50}N_4P_2Mo_5O_{25}$: C, 20.14; H, 3.81; N, 4.27; P, 4.72;
H_2O, 5.49. Found: C, 19.19; H, 4.04; N, 4.46; P, 4.47; H_2O, 5.59.

UV (0.7 mM, pH 4.5), nm (M^{-1} cm^{-1}): 245(3.0 × 10^4), 210(5.5 × 10^4).

G. PENTATUNGSTOBIS(PHENYLPHOSPHONATE)

Tributylammonium salt, $[(C_4H_9)_3NH]_4[(C_6H_5PO_3)_2W_5O_{15}]$: To a solution
of 3.3 g (10 mmol) of $Na_2WO_4 \cdot 2H_2O$ in 30 mL of water is added 1.6 g
(10 mmol) of phenylphosphonic acid (Aldrich), and the mixture is stirred for
15 min. The pH of the resulting solution is adjusted to 5.0 with acetic acid,
and a solution of 5 g of tributylamine in 3 mL of glacial acetic acid is added.
The heteropoly salt precipitates immediately, is removed by filtering, washed
several times with water, and recrystallized from ∼ 30 mL of acetone–benzene
(1:1).

■ **Caution.** *Benzene is a carcinogen. It should be handled with gloves in
a well-ventilated hood.*
Yield: 3.8 g (85%).

Anal. Calcd. for $C_{60}H_{122}N_4P_2W_5O_{21}$: C, 32.50; H, 5.51; N, 2.53; P, 2.80; W,
41.50. Found: C, 31.83; H, 5.66; N, 2.44; P, 2.65; W, 41.78.

NMR(CD$_3$CN) ^{1}H 8.20–8.54 (m, 2H), 7.35–7.45 (m, 3H)δ, cation resonances at 9.34, 3.06–2.94, 1.79–1.11, and 0.98–0.83δ; ^{31}P, + 16.38 ppm *vs*
85% H_3PO_4. The corresponding salt of the molybdo complex can be prepared
in an analogous manner.[6] The NMR ^{31}P, + 16.14 ppm.

Properties

These heteropoly salts are all white solids with similar IR spectra and
well-resolved medium to strong peaks at 1130–970 cm^{-1} (P—O stretch) and
950–890 and 730–660 cm^{-1} (M—O terminal and bridging stretches). The
ammonium and tetramethylammonium (sodium double) salts are soluble in
water; tributylammonium pentatungstobis(phenylphosphonate) is insoluble
in water and methanol but soluble in acetonitrile, *N,N*-dimethylformamide,
dimethyl sulfoxide, and dichloromethane. Proton NMR spectra of the

phosphonato moieties of the complexes are similar to, but shifted slightly downfield (0.1–0.2 ppm) from those of the uncomplexed phosphonates. In aqueous solution, the molybdo complexes are stable between pH 2 and 6. The tungsto complexes appear to be more prone to dissociation in aqueous solution: at pH 3.6 a 0.04 M solution of $K_4[(C_6H_5P)_2W_5O_{21}]$ is $\sim 15\%$ dissociated into the free phosphonate (^{31}P, $+14.1$ ppm), whereas the corresponding molybdate is undissociated at 0.01 M.

The heteropolyanions have a structure of C_2 symmetry,[2] but attempts to resolve enantiomers have been unsuccessful, and the anion $[(C_6H_5P)_2W_5O_{21}]^{4-}$ has been shown to be fluxional in acetonitrile solution.[7]

References

1. W. Kwak, M. T. Pope, and T. F. Scully, *J. Am. Chem. Soc.*, **97**, 5735 (1975).
2. J. K. Stalick and C. O. Quicksall, *Inorg. Chem.*, **15**, 1577 (1976).
3. P. Fleury, *Bull. Soc. Chim. Fr.*, **51**, 657 (1932).
4. A. Rosenheim and M. Schapiro, *Z. Anorg. Chem.*, **129**, 196 (1923).
5. J. Fuchs and I. Bruedgam, *Z. Naturforsch. Teil B*, **32**, 403 (1977).
6. M. A. Leparulo, Ph.D. Thesis, Georgetown University, Washington, DC, 1984.
7. P. R. Sethuraman, M. A. Leparulo, M. T. Pope, F. Zonnevijlle, C. Brévard, and J. Lemerle, *J. Am. Chem. Soc.*, **103**, 7665 (1981).

24. TETRAKIS(TETRABUTYLAMMONIUM) μ_3-[η^5-CYCLOPENTADIENYL)TRIOXOTITANATE(IV)]-A-β-1, 2, 3-TRIVANADONONATUNGSTOSILICATE(4 −), $[(C_4H_9)_4N]_4[A-\beta-(\eta^5-C_5H_5)TiSiW_9V_3O_{40}]$

Submitted by R. G. FINKE,* C. A. GREEN,* and B. RAPKO*
Checked by R. CONTANT,† R. THOUVENOT,† and N. AMMARI†

Polyoxoanions have received attention as models for oxide–supported transition metals,[1,2] as novel catalytic materials,[3,4] as imaging or labeling reagents for electron microscopy,[5] and as inorganic drugs.[6] The most difficult and slow step in such applications is often the synthesis and the full and unambiguous characterization of the desired polyoxoanion.

The title $[SiW_9V_3O_{40}]^{7-}$ supported Ti^{4+} complex is of general interest as a previously unknown composition of matter, and is of specific interest,

*Department of Chemistry, University of Oregon, Eugene, OR 97403.
†Laboratoire de Physicochimie Inorganique, Université Pierre et Marie Curie, 75252 Paris Cedex 05, France.

for example, as a discrete analog of Ti^{4+} supported on solid oxides.[7] The Keggin-type $[SiW_9V_3O_{40}]^{7-}$ is synthesized by the reaction of the lacunary sodium A-β-nonatungstosilicate, $[SiW_9O_{34}]^{10-}$, with dioxovanadium(V) ions, $[VO_2]^+$.[8] The product is initially isolated as the readily recrystallized potassium salt, and then transformed into the tetrabutyl-ammonium salt, followed by deprotonation with $(C_4H_9)_4NOH$ to yield $[C_4H_9)_4N]_7[SiW_9V_3O_{40}]$. The final product is then formed in a reaction with $[(\eta^5-C_5H_5)Ti]^{3+}$ (in CH_3CN under a dry, inert atmosphere).[9] Although a detailed account of the synthesis and characterization of $[SiW_9V_3O_{40}]^{7-}$ has appeared,[8] a refined procedure is reported herein.

These complexes have been previously characterized by elemental analyses, fast atom bombardment mass spectroscopy (FABMS), NMR spectroscopy, and other techniques.[8-10] The FABMS unequivocally establishes that the composition of the trisubstituted Keggin anion is $[SiW_9V_3O_{40}]^{7-}$. The overall C_{3v} symmetry of this anion is supported by the two resonances (relative integration of 2:1) observed in the ^{183}W NMR spectrum and the single resonance observed in the ^{51}V NMR spectrum. Spectroscopic studies of $[H_xSiW_9V_3O_{40}]^{x-7}$ by ^{51}V NMR and 1- and 2-D ^{183}W NMR are also available.[8] Elemental analysis and FABMS unequivocally establish the molecular formula of the κ^3-O supported (η^5-cyclopentadienyl)Ti^{3+} complex. The ^{51}V and ^{183}W NMR spectra are consistent with the C_s symmetry of this complex, and a 2-D INADEQUATE ^{183}W NMR spectrum establishes the W to W connectivity and, hence, the overall structure of the final product.

A. HEXAPOTASSIUM HYDROGEN A-β-1, 2, 3-TRIVANADONONATUNGSTOSILICATE(7 −) TRIHYDRATE, $K_6H[A-\beta-SiW_9V_3O_{40}]\cdot3H_2O$

$$SiO_3^{2-} + 9[WO_4]^{2-} + 11H^+ \longrightarrow A-\beta-\{H[SiW_9O_{34}]\}^{9-} + 5H_2O$$

$$3[VO_3]^- + 6H^+ \longrightarrow 3[VO_2]^+ + 3H_2O$$

$$A-\beta-\{H[SiW_9O_{34}]\}^{9-} + 3[VO_2]^+ \longrightarrow \{H[A-\beta-SiW_9V_3O_{40}]\}^{6-}$$

$$SiO_3^{2-} + 9[WO_4]^{2-} + 3[VO_3]^- + 17H^+ \longrightarrow \{H[A-\beta-SiW_9V_3O_{40}]\}^{6-} + 8H_2O$$

The preparation of the lacunary sodium nonatungstosilicate is dependent on adding an acid solution to the mixture of tungstate and silicate ions in a manner such that the aggregation process favors formation of the lacunary structure. Although the preparation described here (based on the previously reported synthesis)[11] produces impure material (by elemental analysis), this material reacts with the dioxovanadium(V) ions to form the trisubstituted complex in good yield. The structure of a single crystal of the lacunary product has been determined by X-ray crystallography.[12]

Procedure

Sodium metasilicate ($Na_2SiO_3 \cdot 9H_2O$, 60 g, 0.21 mol, an excess is used due to silica gel-forming side reaction during acidification) is dissolved in 500 mL of distilled water in a 1-L beaker with rapid stirring. Sodium tungstate ($Na_2WO_4 \cdot 2H_2O$, 362 g, 1.1 mol) is dissolved in the solution, followed by addition of 200 mL of 6 M HCl (1.2 mol) *in 1 to 2 min* with vigorous stirring (during which time a white gelatinous precipitate forms). Following completion of the HCl addition, the mixture is rapidly stirred for an additional 10 min and then filtered through an ~ 0.5-in. thick pad of Celite® on Whatman No. 1 filter paper (12.5 cm) using a Büchner funnel. (Alternatively, filteration through a 600-mL course sintered glass frit covered with an 0.5-in. thick pad of either Celite® or glass wool may be utilized.) After the filtrate has been refrigerated at 4 °C for 2–3 days (with brief stirring to enhance precipitation once solids appear; alternatively, 3–5 days with no stirring may be used), the white crystalline precipitate is collected on a coarse, sintered glass filter frit and air dried for 1–2 days at room temperature. Yield: 71 g (21%) of crude $Na_9H[A\text{-}\beta\text{-}SiW_9O_{34}] \cdot 23H_2O$.

A solution containing dioxovanadium(V) ions is prepared in a 2-L beaker by dissolving 6.4 g of sodium metavanadate, $NaVO_3$, (52 mmol) in 900 mL of hot (85 °C) water and cooling the solution to 20 °C, during which time 22.6 mL of 6 M HCl (136 mmol, ~ 8.0 equivalents) is added. After stirring the pale yellow solution (at pH < 1.5)* for 15 min,[†] the solution is vigorously stirred while 48 g of crude $Na_9H[A\text{-}\beta\text{-}SiW_9O_{34}] \cdot 23H_2O$ (17 mmol) is added. A cherry red solution forms as the lacunary sodium tungstosilicate dissolves. After stirring for 15 min, 60 g of KCl is added; the solution should remain homogeneous and cherry red. Next, 900 mL of methanol is added, precipitating the product as an orange-red solid. The product is filtered from the solution using a coarse, sintered glass frit and air dried at room temperature (12–24 h). Yield: 34.5 g (74%). The product can be recrystallized by dissolving 30 g in a mixture containing 100 mL of 0.03 M HCl and 50 mL of methanol at 55 °C, and then slowly cooling the solution to 4 °C. After 6 h, the crystals are collected and air dried as before. Yield: 22.0 g (73%).

---

*The pH is typically measured through the use of a pH meter, with the system calibrated immediately prior to use using commercially available solutions of pH 4.0, 7.0, and 10.0, respectively.

[†]At room temperature the conversion to $[VO_2]^+$ is slow at pH 1.5 but is accelerated at pH 0.7 to 0.8; see footnote 11 in ref. 8.

Anal. (following drying under vacuum overnight at 25 °C) Calcd. for $K_6H[SiW_9V_3O_{40}] \cdot 3H_2O$: K, 8.48; Si, 1.02; W, 59.8; V, 5.53; O, 24.9; H_2O, 2.0. Found: K, 8.39; Si, 1.01; W, 59.4; V, 5.70; O, 24.3; H_2O, 1.7 (H_2O by thermogravimetric analysis (TGA) to 250 °C).

Properties

Previous studies have shown that, in aqueous solution, the lacunary $\{H[A$-β-$SiW_9O_{34}]\}^{9-}$ complex rapidly dissociates and recombines to form mixtures containing predominantly $[SiO_3]^{2-}$, $[WO_4]^{2-}$, and $[SiW_{11}O_{39}]^{8-}$ (the relative ratios being pH dependent).[11,13]

The potassium salt of the $[A$-β-$HSiW_9V_3O_{40}]^{6-}$ is very soluble in water and slightly soluble in dimethyl sulfoxide (DMSO) and *N, N*-dimethylformamide (DMF). It is only very slightly soluble in methanol. It appears to be quite stable in aqueous solution (room temperature at pH 1.5), since a sample kept > 2 months showed no decomposition by ^{51}V NMR spectroscopy.

B. TETRAKIS(TETRABUTYLAMMONIUM) TRIHYDROGEN A-β-1, 2, 3-TRIVANADONONATUNGSTOSILICATE(7-), $[[(C_4H_9)_4N]_4H_3[A$-β-$SiW_9V_3O_{40}]$

$$[H[A\text{-}\beta\text{-}SiW_9V_3O_{40}]\}^{6-} + 4(C_4H_9)_4N^+ + 2H^+$$

$$\longrightarrow [(C_4H_9)_4N]_4H_3[A\text{-}\beta\text{-}SiW_9V_3O_{40}]$$

This complex is readily precipitated from acidic aqueous solution because of the very low solubility of the tetrabutylammonium salt in water.

Procedure

In a 250-mL beaker, a solution containing crude (unrecrystallized) $K_6H[A$-β-$SiW_9V_3O_{40}] \cdot 3H_2O$ (30 g, 10 mmol) in 150 mL of 0.03 *M* HCl is prepared and then added over 5 to 10 min to a 600-mL beaker containing a well-stirred solution of tetrabutylammonium bromide (15 g, 46 mmol) in 75 mL of 0.03 *M* HCl. An orange precipitate forms immediately, and the pH is reset to 1.5 by the addition of 6 *M* HCl. The precipitate is isolated from the colorless filtrate by filtration through a coarse sintered glass frit and then washed four times with 20 mL portions of 0.03 *M* HCl (pH 1.5). The product is air dried on the filter (4–6 h) and then dried overnight at 50–60 °C. Yield: 34.5 g (67%). The product is recrystallized by dissolving 20 g in 30 mL of warm (55 °C) acetonitrile in a 150-mL beaker, adding 75 mL of warm (55 °C) chloroform, and placing the beaker containing the solution inside a tightly capped jar containing ~ 100 mL of warm chloroform.

- **Caution.** *Chloroform is a suspected carcinogen and should be used only in a well-ventilated fume hood. Gloves should be worn at all times.*

After standing for 1 week at 40–45 °C, the mixture is cooled to ice temperature, and the red crystals are collected on a coarse, sintered glass frit. Air drying causes the crystals to lose much of their crystallinity; the material isolated is red-orange. Yield: 15.0 g (75%).

Anal. (following drying in vacuum overnight at 25 °C) Calcd. for $[C_4H_9)N]_4H_3[SiW_9V_3O_{40}]$: C, 22.3; H, 4.27; N, 1.62; Si, 0.81; W, 48.0; V, 4.43; O, 18.6. Found: C, 22.1; H, 4.38; N, 1.67; Si, 0.79; W, 47.7; V, 4.28; O, 17.9 (no H_2O by TGA).

Properties

The complex is soluble in dipolar solvents such as acetonitrile, DMF, and DMSO. It is slightly soluble to insoluble in chloroform, dichloromethane, alcohols, aromatic hydrocarbons, aliphatic hydrocarbons, and water.

C. TETRAKIS(TETRABUTYLAMMONIUM) μ_3-[(η^5-CYCLOPENTADIENYL)TRIOXOTITANATE(IV)]-A-β-1, 2, 3-TRIVANADONONATUNGSTOSILICATE(4-), $[(C_4H_9)_4N]_4[A\text{-}\beta\text{-}(\eta^5\text{-}C_5H_5)TiSiW_9V_3O_{40}]$

$[(C_4H_9)_4N]_4H_3[A\text{-}\beta\text{-}SiW_9V_3O_{40}] + 3(C_4H_9)_4N^+OH^-$

$\longrightarrow [(C_4H_9)_4N]_7[A\text{-}\beta\text{-}SiW_9V_3O_{40}] + 3H_2O$

$(\eta^5\text{-}C_5H_5)TiCl_3 + 3Ag^+ \longrightarrow [(\eta^5\text{-}C_5H_5)Ti]^{3+} + 3AgCl$

$[(\eta^5\text{-}C_5H_5)Ti]^{3+} + [(C_4H_9)_4N]_7[A\text{-}\beta\text{-}SiW_9V_3O_{40}]$

$\longrightarrow [(C_4H_9)_4N]_4[A\text{-}\beta\text{-}(\eta^5\text{-}C_5H_5)TiSiW_9V_3O_{40}] + 3(C_4H_9)_4N^+$

$[(C_4H_9)_4N]_4H_3[A\text{-}\beta\text{-}SiW_9V_3O_{40}] + 3(C_4H_9)_4N^+OH^- + (\eta^5\text{-}C_5H_5)TiCl_3$

$+ 3Ag^+ \longrightarrow [(C_4H_9)_4N]_4[A\text{-}\beta\text{-}(\eta^5\text{-}C_5H_5)TiSiW_9V_3O_{40}]$

$+ 3(C_4H_9)_4N^+ + 3AgCl + 3H_2O$

Procedure

The compound $[(C_4H_9)_4N]_4H_3[A\text{-}\beta\text{-}SiW_9V_3O_{40}]$ (15.0 g, 4.4 mmol) is placed in a 500-mL round-bottomed flask and dissolved in 150 mL of acetonitrile. While this solution is stirred rapidly, 13.1 mL of $1.0\,M\,(C_4H_9)_4NOH$ in methanol (13.1 mmol, Aldrich) is added, and the resulting solution is stirred for 15 min. The solvent is then removed under vacuum at room temperature (12–16 h) to yield the crude $[(C_4H_9)_4N]_7[A-\beta-SiW_9V_3O_{40}]$ as a light orange-brown powder, which contains decomposition products plus $[(C_4H_9)_4N]_6H[A-\beta-SiW_9V_3O_{40}]$, by ^{29}Si and ^{51}V NMR.[8]

The following steps are carried out in a Vacuum Atmospheres inert atmosphere box under N_2 [the nitrogen was obtained from the boil-off from liquid N_2 and further purified by passing it first through 4-Å molecular sieves and then through a reduced Cu catalyst (BASF)]. All glassware and Celite® used in this preparation should be dried in an oven at 150 °C overnight and cooled under vacuum in the dry box ante-chamber before use. Acetonitrile is dried by distillation over CaH_2 under N_2 followed by standing for at least 48 h over $\sim 30\%$ by volume 3-Å molecular sieves previously activated under vacuum at 170 °C.[8]

A solution of $[(\eta^5-cyclopentadienyl)Ti]^{3+}$ is prepared by dissolving 0.79 g of sublimed $(\eta^5-cyclopentadienyl)TiCl_3$[14] (3.6 mmol)* in 30 mL of dry[8] CH_3CN (150-mL beaker), adding dropwise a solution of $AgNO_3$ (1.83 g, 10.8 mmol) in 20 mL of dry CH_3CN (50-mL beaker), and stirring the resulting mixture for 15 min. The mixture is then filtered through a medium glass frit to remove the precipitated AgCl, with the yellow filtrate passing directly into a stirred solution of the $[A-\beta-SiW_9V_3O_{40}]^{7-}$ heteropolyanion. The latter is prepared by dissolving crude $[(C_4H_9)_4N]_7[SiW_9V_3O_{40}]$ (15.0 g, 3.6 mmol) in 50 mL of dry CH_3CN (150-mL beaker) and filtering the resulting suspension through a coarse, sintered glass frit (covered with a 0.5-in. thick pad of Celite®) into a 250-mL round-bottomed flask in order to remove an unidentified white, insoluble, decomposition product. The resulting dark orange solution is gently heated at reflux for 1 h. The solvent is completely removed under vacuum by rotary evaporation. The reaction product is then removed from the dry box. The residue is slurried in 150 mL of chloroform for 15 min, and the insoluble product is isolated by filtration through a medium sintered glass frit. The solid product is purified by dissolving it in ~ 20 mL of acetonitrile (100-mL beaker), filtering the solution through a medium sintered glass frit, and then adding the filtrate to a well-stirred solution of 200 to 250 mL of $CHCl_3$ (400-mL beaker) to precipitate the product. The light brown to dark red-orange precipitate is isolated by filtration on a medium sintered glass frit. It is then suspended in 100 mL of absolute ethanol, refiltered, and then reprecipitated by dissolving the solid in ~ 15 mL of CH_3CN followed by slow addition (over 1 h) to a well-stirred

---

*The checkers report that commercial $(\eta^5-C_5H_5)TiCl_3$ (Alfa) is unsuitable for this synthesis.

solution containing 150 mL of $CHCl_3$. This ethanol wash and precipitation* are repeated (usually 1–2 times) until the sample is pure by ^{51}V NMR spectroscopy.[9] The product is then dried overnight at 25 °C under vacuum. Yield: 7.6 g (59%).[†]

Anal. Calcd. for $[(C_4H_9)_4N]_4[(C_5H_5)TiSiW_9V_3O_{40}]$: C, 23.3; H, 4.22; N, 1.57; Ti, 1.35; Si, 0.79; W, 46.5; V, 4.26; O, 18.0. Found: C, 23.1; H, 4.25; N, 1.65; Ti, 1.09; Si, 0.78; W, 46.4; V, 4.60; O, 18.2.

Properties

The complex is soluble in acetonitrile, DMSO, and DMF. It is slightly soluble in acetone and methanol. Although the $[(\eta^5\text{-cyclopentadienyl})Ti]^{3+}$ ion is quite sensitive to even atmospheric moisture, once supported, $[(\eta^5\text{-cyclopentadienyl})TiSiW_9V_3O_{40}]^{4-}$ is stable to the atmosphere for > 1 month, either in the solid state or in CH_3CN solution (by solution ^{51}V and ^{183}W NMR). An orange CH_3CN solution of $[(C_4H_9)_4N]_4[(\eta^5\text{-}C_5H_5)\text{-}TiSiW_9V_3O_{40}]$ passes unaltered (^1H NMR, IR) through an Amberlyst® A-27 cation-exchange column (Ⓟ-SO_3^- Bu_4N^+), whereas a control shows that organometallic cations like $[(\eta^5\text{-cyclopentadienyl})Ti]^{3+}$ are retained at the top of the column. Conversely, the anionic $[(\eta^5\text{-}C_5H_5)TiSiW_9V_3O_{40}]^{4-}$ is completely retained at the top of an Amberlyst® A-27 anion-exchange column in the Cl^- form (Ⓟ-$NR_3^+Cl^-$), with no visible elution of $(\eta^5\text{-cyclo-pentadienyl})Ti^{3+}$ from this band.[9] The ^{51}V and 2-D ^{183}W INADEQUATE spectra have been described elsewhere.[9]

References

1. V. W. Day and W. G. Klemperer, *Science* **228**, 533 (1985), and references therein.
2. (a) R. G. Finke, M. W. Droege, J. R. Hutchinson, and O. Gansow, *J. Am. Chem. Soc.*, **103**, 1587 (1981). (b) R. G. Finke and M. W. Droege, *J. Am. Chem. Soc.*, **106**, 7274 (1984).
3. For a listing of lead references, see ref. 4 and 5 in R. G. Finke, M. W. Droege, J. C. Cook, and K. S. Suslick, *J. Am. Chem. Soc.*, **106**, 5750 (1984).
4. R. G. Finke, D. K. Lyon, K. Nomiya, and S. Sur, Inorg. Chem., in press.
5. (a) See ref. 7 in ref. 3 cited above. (b) J. F. W. Keana, M. C. Ogan, Y. Lü, M. Beer, and J. Varkey, *J. Am. Chem. Soc.*, **107**, 6714 (1985); **108**, 7956 (1986).
6. W. Rozenbaum, D. Dormont, B. Spire, E. Vilmer, M. Gentilini, C. Griscelli, L. Montagnier, F. Barre-Sinoussi, and J. C. Chermann, *Lancet*, **1985**, 450.

---

*Crystallization can also be done from acetonitrile–chloroform or acetonitrile–benzene systems over several days using vapor diffusion techniques at 40 to 45 °C (ref. 9)

[†]The checkers yield was 4.6 g.

7. For example, Ti^{4+} on SiO_2 is an excellent olefin oxidation catalyst: R. A. Sheldon and J. K. Kochi, *Metal-Catalyzed Oxidations of Organic Compounds*, Academic Press, New York, 1981, Chapter 9.

8. R. G. Finke, B. Rapko, R. J. Saxton, and P. J. Domaille, *J. Am. Chem. Soc.*, **108**, 2947 (1986).

9. R. G. Finke, B. Rapko, and P. J. Domaille, *Organometallics*, **5**, 175 (1986).

10. R. G. Finke, M W. Droege, J. C. Cook, and K. S. Suslick, *J. Am. Chem. Soc.*, **106**, 5750 (1984)

11. G. Hervé and A. Tézé, *Inorg. Chem.*, **16**, 2115 (1977). A. Tézé and G. Hervé, *J. Inorg. Nucl. Chem.*, **39**, 999 (1977).

12. F. Robert and A. Tézé, *Acta Crystallogr. Sect. B*, **B37**, 318 (1981).

13. D. L. Kepert and J. H. Kyle, *J. Chem. Soc. Dalton Tans.*, 137 (1978).

14. C. R. Lucas and M. L. H. Green, *Inorg. Synth.*, **16**, 238 (1976).

Chapter Four

LANTHANIDE AND ACTINIDE COMPLEXES

25. LANTHANIDE TRICHLORIDES BY REACTION OF LANTHANIDE METALS WITH MERCURY(II) CHLORIDE IN TETRAHYDROFURAN

Submitted by GLEN B. DEACON,* TRAN D. TUONG,*
and DALLAS L. WILKINSON*†
Checked by TOBIN MARKS‡

Anhydrous lanthanide trihalides, particularly the trichlorides, are important reactants for the formation of a variety of lanthanide complexes, including organometallics. Routes for the syntheses of anhydrous lanthanide trihalides generally involve high temperature procedures or dehydration of the hydrated halides.[1-5] The former are inconvenient and complex for small scale laboratory syntheses, while dehydration methods may also be complex[4] and have limitations, for example, use of thionyl chloride.[1,5] Moreover, the products from these routes may require purification by vacuum sublimation at elevated temperatures.[3,4] Redox transmetalation between lanthanide metals and mercury(II) halides was initially carried out at high temperatures.[2,3] However, this reaction can be carried out in tetrahydrofuran (THF, solvent) to give complexes of lanthanide trihalides with the solvent.[6] These products are equally as suitable as reactants for synthetic purposes as the uncomplexed

*Chemistry Department, Monash University, Clayton, Victoria, Australia, 3168.
†The submitters are grateful to the Australian Research Council for support and to Rare Earth Products for a gift of REACTION ytterbium.
‡Department of Chemistry, Northwestern University, Evanston, IL 60201.

trihalides. Other workers have used this transmetalation reaction for activation of the lanthanide metals.[7]

Detailed syntheses of four representative lanthanide trihalide–tetrahydrofuran (thf, ligand) complexes, $YbCl_3(thf)_3$, $ErCl_3(thf)_{3.5}$, $SmCl_3(thf)_2$, and $NdCl_3(thf)_{1.5}$ by redox transmetalation are described. The first three compounds have previously been prepared by direct reaction of anhydrous lanthanide trihalides with THF,[8,9] but the composition of the last differs slightly from that reported,[9] namely, $NdCl_3(thf)_2$, for the product from reaction of $NdCl_3$ with THF. Although the method below describes isolation of the complex trichlorides, the THF solutions can be used *in situ* for further reactions.[7] The method can also be used for other trichlorides[7,10] and other trihalides.[6]

General Procedure

The lanthanide trichlorides described here are moisture sensitive, but they can be stored indefinitely under purified nitrogen or argon at room temperature. All operations are carried out under nitrogen or argon, which is purified by passage through BASF R3/11 oxygen removal catalyst and molecular sieves (see ref. 11 for a discussion of inert atmosphere techniques). Tetrahydrofuran is dried by distillation from sodium benzophenone under nitrogen.

■ **Caution.** *Only fresh, peroxide-free THF should be distilled.*

The solvent is stored under nitrogen in Schlenk vessels equipped with high-vacuum Teflon taps (e.g., Young or Rotaflo). Syringes (preflushed with nitrogen) are used for transfers of solvent. For reliable results, an oven-dried greaseless Schlenk apparatus incorporating polytetrafluoroethylene O-rings and taps (Young or Rotaflo) should be used, and the lanthanide metal powder should be stored, weighed, and handled under nitrogen. The metal powder can be obtained from Research Chemicals or REACTON distilled lanthanide metals from Rare Earth Products, can be crushed under nitrogen to a powder. Dried reagent grade mercury(II) chloride can be used without purification.

The apparatus used in all reactions is shown in Fig. 1. Mercury(II) chloride (1.084 g, 4.00 mmol), an excess of the lanthanide metal powder (6.00 mmol), and a magnetic stirring bar are placed in the lower 100-mL Schlenk flask. Onto this flask are attached (in order) a condenser,* a Schlenk filter [covered with a Whatman microfibre glass filter paper, a layer at least 6 mm thick of dried diatomaceous earth (Sigma grade 1, D-3877) and a second glass filter paper], and a preweighed 100-mL Schlenk flask. The Schlenk apparatus is purged of air, and an inert atmosphere is established by evacuation of the

---

*If heating is carefully controlled, the condenser may be omitted.

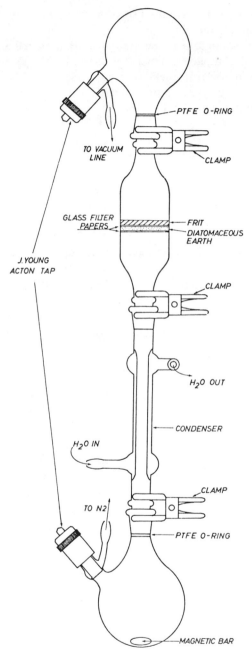

Fig. 1. Reaction apparatus.

apparatus to 10^{-3} torr and backfilling with nitrogen or argon at least three times.* Tetrahydrofuran (25 mL) is then added, stirring is commenced, and the reaction mixture is heated to 65 °C. After stirring and heating for 1.5 h the apparatus is inverted, and the reaction mixture is filtered into the second Schlenk flask under reduced pressure. The residue on the Schlenk filter is washed with THF (25 mL). Evaporation of the filtrate and washings (10^{-3} torr, room temperature) gives the lanthanide trichloride, which is dried under vacuum (10^{-3} torr) for 1–2 h at room temperature and weighed.

- **Caution.** *Finely divided metal residues from the Schlenk filter can be pyrophoric.*

The composition of the THF complexes of the lanthanide trichlorides is established by dissolution of a weighed amount of each compound in water and determination of chloride potentiometrically with aqueous silver nitrate.[12] Analysis for the lanthanide is effected by titration with disodium ethylenediaminetetraacetate using xylenol orange indicator after appropriate buffering of the solution.[13,14] The IR spectra of all the trichlorides show intense bands at ~ 1020 and 880 cm^{-1} (see below), indicative of coordinated THF.[15]

A. YTTERBIUM TRICHLORIDE–TETRAHYDROFURAN (1/3)

$$2Yb + 3HgCl_2 + 6THF \xrightarrow{\text{THF}} 2YbCl_3(thf)_3 + 3Hg\downarrow$$

- **Caution.** *Mercury(II) chloride and mercury metal are toxic. Protective rubber gloves should be used. Operations should be carried out in a well-ventilated hood.*

Procedure

Mercury(II) chloride (1.084 g, 4.00 mmol) and ytterbium metal (1.038 g, 6.00 mmol) are allowed to react as described previously, giving the white title compound. Yield: 1.16 g (88%).

Anal. Calcd. for $C_{12}H_{24}Cl_3O_3Yb$: Cl, 21.5; Yb, 34.9. Found: Cl, 21.4; Yb, 34.9%.

IR (Nujol mull, prepared under N_2), 1055 (sh), 1018 (s), 925 (m), 880 (s), 860 (sh), and 735 (w) cm^{-1}.

---

*Vacuum–nitrogen (argon) lines are connected to both Schlenk flasks. Initially the reaction flask is attached to the nitrogen line and the product flask is attached to the vacuum line.

B.   ERBIUM TRICHLORIDE–TETRAHYDROFURAN (2/7)

$$2Er + 3HgCl_2 + 7THF \xrightarrow{THF} 2ErCl_3(thf)_{3.5} + 3Hg\downarrow$$

- ■ **Caution.**   *See Section A.*

Procedure

Mercury(II) chloride (1.084 g, 4.00 mmol) and erbium metal (1.004 g, 6.00 mmol) are allowed to react as indicated above to give the pink title compound. Yield: 1.22 g (87%).

Anal. Calcd. for $C_{28}H_{56}Cl_6Er_2O_7$: Cl, 20.2; Er, 31.8. Found: Cl, 20.1; Er, 31.9%.

IR (Nujol mull, prepared under N_2): 1191 (m), 1021 (s), 931 (m), 871 (s), and 737 (w) cm^{-1}.

C.   SAMARIUM TRICHLORIDE–TETRAHYDROFURAN (1/2)

$$2Sm + 3HgCl_2 + 4THF \xrightarrow{THF} 2SmCl_3(thf)_2 + 3Hg\downarrow$$

- ■ **Caution.**   *See Section A.*

Procedure

Mercury(II) chloride (1.084 g, 4.00 mmol) and samarium metal (0.902 g, 6.00 mmol) are allowed to react as described above, giving the white title compound. Yield: 0.520 g (49%).

Anal. Calcd. for $C_8H_{16}Cl_3O_2Sm$: Cl, 26.5; Sm, 37.5. Found: Cl, 25.6; Sm, 37.5%.

IR (Nujol mull, prepared under N_2): 1020 (s), 926 (m), 870 (s), and 746 (m) cm^{-1}.

D.   NEODYMIUM TRICHLORIDE–TETRAHYDROFURAN (2/3)

$$2Nd + 3HgCl_2 + 3THF \xrightarrow{THF} 2NdCl_3(thf)_{1.5} + 3Hg\downarrow$$

- ■ **Caution.**   *See Section A.*

Procedure

Mercury(II) chloride (1.084 g, 4.00 mmol) and neodymium metal (0.865 g, 6.00 mmol) are allowed to react as described above, giving the pale blue title compound. Yield: 0.560 g (58%).

Anal. Calcd. for $C_{12}H_{24}Cl_6Nd_2O_3$: Cl, 29.6; Nd, 40.2. Found: Cl, 29.0; Nd, 39.8%.

IR (Nujol mull, prepared under N_2): 1022 (s), 930 (m), 871 (s), and 736 (m) cm^{-1}.

References

1. M. D. Taylor, *Chem. Rev.*, **62**, 503 (1962).
2. L. B. Asprey, T. K. Keenan, and F. H. Kruse, *Inorg. Chem.*, **3**, 1137 (1964).
3. F. L. Carter and J. F. Murray, *Mater. Res. Bull.*, **7**, 519 (1972).
4. K. E. Johnson and K. R. MacKenzie, *J. Inorg. Nucl. Chem.*, **32**, 43 (1970).
5. J. H. Freeman and M. L. Smith, *J. Inorg. Nucl. Chem.*, **7**, 224 (1958).
6. G. B. Deacon and A. J. Koplick, *Inorg. Nucl. Chem. Lett.* **15**, 263 (1979).
7. G. Z. Suleimanov, T. Kh. Kurbanov, Yu. A. Nuriev, L. F. Rybakova, and I. P. Beletskaya, *Dokl. Chem.*, **265**, 254 (1982).
8. K. Rossmanith and C. Auer-Welsbach, *Monatsh. Chem.*, **96**, 602 (1965).
9. K. Rossmanith, *Monatsh. Chem.*, **100**, 1484 (1969).
10. A. A. Pasynskii, I. L. Eremenko, G. Z. Suleimanov, Yu. A. Nuriev, I. P. Beletskaya, V. E. Shklover, and Yu. T. Struchkov, *J. Organomet. Chem.* **266**, 45 (1984).
11. J. J. Eisch, *Organometallic Syntheses*, Vol. 2 *Nontransition-Metals*, J. J. Eisch and R. B. King (eds.) Academic Press, New York, 1981.
12. A. I. Vogel, *A Text Book of Quantitative Inorganic Analysis*, 3rd ed., Longmans, Green, London, **1961**, p. 950.
13. J. Korbl and R. Pribil, *Chem. Anal*, **45**, 102 (1956).
14. J. L. Atwood, W. E. Hunter, A. L. Wayda, and W. J. Evans, *Inorg. Chem.*, **20**, 4115 (1981).
15. J. Lewis, J. R. Miller, R. L. Richards, and A. Thompson, *J. Chem, Soc.*, **1965**, 5850. R. J. H. Clark, J. Lewis, D. J. Machin, and R. S. Nyholm, *J. Chem. Soc.*, **1963**, 379.

26. BIS(PHENYLETHYNYL)YTTERBIUM(II) FROM YTTERBIUM METAL AND BIS(PHENYLETHYNYL)MERCURY*

Submitted by GLEN B. DEACON,[†] CRAIG M. FORSYTH,[†]
and DALLAS L. WILKINSON[†]
Checked by ANDREA L. WAYDA[‡]

Organolanthanide chemistry is dominated by the trivalent compounds.[1-5] Compounds in oxidation state (II) are restricted to derivatives of europium, samarium, and ytterbium, but they have considerable importance in both organic and organometallic syntheses because of their reducing properties.[1-6] Redox transmetalation reactions of organomercurials with lanthanide metals provide convenient syntheses of a number of diorganolanthanides, for example, R_2M, $R = C_6F_5$ or $PhCC$, $M = Yb$ or Eu.[7-10]

$$R_2Hg + M \rightarrow R_2M + Hg\downarrow$$

Details of the redox transmetalation synthesis of bis(phenylethynyl)-ytterbium(II) are presented here.[8,10] This diorganoytterbium(II) compound has a rich reaction chemistry, which is currently being developed.[11,12] The compound is highly air and moisture sensitive and must be handled under purified nitrogen or argon. Bis(phenylethynyl)ytterbium(II) can be stored indefinitely in an inert atmosphere at room temperature.[8,10]

General Procedure

All operations are carried out under nitrogen or argon, which is purified by passage through BASF R3/11 oxygen removal catalyst and molecular sieves. Tetrahydrofuran (THF) is purified by distillation from sodium benzophenone under nitrogen.

- ■ **Caution.** *Only fresh, peroxide-free THF should be distilled.*

The solvent is stored under nitrogen in Schlenk vessels equipped with high-vacuum Teflon taps (*e.g.*, Young or Rotaflo). All transfers of solvent and solutions are carried out with syringes that have been preflushed with nitrogen. For reliable results, an over-dried greaseless Schlenk apparatus incorporating polytetrafluoroethylene O-rings and taps (Young or Rotaflo) should be used. Bis(phenylethynyl)mercury is prepared by mercuration of

*The submitters are grateful to the Australian Research Grants Scheme for support, and Rare Earth Products for a gift of REACTON ytterbium.
[†]Chemistry Department, Monash University, Clayton, Victoria, Australia, 3168.
[‡]AT & T Bell Laboratories, Murray Hill, NJ 07974.

phenylacetylene with K_2HgI_4 under basic conditions.[13,14] Ytterbium powder can be obtained from Research Chemicals, or REACTON distilled ytterbium from Rare Earth Products, can be crushed under nitrogen to a powder. Although metal from both sources reacts readily, we find that a higher $Yb-R_2Hg$ ratio is needed with the latter to ensure complete reaction (see below). The ytterbium metal should be stored, weighed, and handled in an inert atmosphere.

A.   BIS(PHENYLETHYNYL)YTTERBIUM(II)

$$(PhCC)_2Hg + Yb \xrightarrow{THF} (PhCC)_2Yb + Hg\downarrow$$

■ **Caution.** *Mercury compounds and the metal are toxic. Protective rubber gloves should be used. Operations should be carried out in a well-ventilated area.*

Bis(phenylethynl)mercury (1.21 g, 3.0 mmol) and ytterbium metal (1.04 g, 6.0 mmol Research Chemicals powder or 2.08 g, 12.0 mmol, crushed RE-ACTON distilled) are placed in a 100-mL Schlenk flask containing a magnetic stirring bar. To this flask are attached a Schlenk filter [covered with a Whatman microfibre glass filter paper, a layer at least 6 mm thick of dried diatomaceous earth (Sigma grade 1, D-3877), and a second glass filter paper], and another preweighed, Schlenk flask (Fig. 1). After assembly, the Schlenk apparatus is purged of air by evacuation to 10^{-3} torr and backfilling with purified nitrogen or argon three times.* Purified THF (20 mL) is then added and stirring commenced. After an initial induction period of 2–5 min, the solution changes color to a deep purple, mercury metal is deposited, and the solution becomes warm.† If initiation is slow a small amount of mercury metal is added. After completion of reaction (4 h), the solution is filtered and the filter washed with THF (2×5 mL).‡ Evaporation of the solvent to dryness yields the title compound as a purple-black solid, which is dried at 50–60 °C under vacuum for 4 h to remove THF. Yield: 1.07 g (95%) in preweighed flask; 0.95 g (85%) scraped out and weighed in a dry box.

Anal. Calcd. for $C_{16}H_{10}Yb$: Yb, 46.1. Found: Yb, 45.8%.

---

*Vacuum–nitrogen (argon) lines are connected to both Schlenk flasks (Fig. 1). Initially the reaction flask is connected to the nitrogen line and the product flask to the vacuum line.

†It is undesirable to carry out the reaction at a significantly higher concentration or a purple solid coprecipitates with mercury and the excess of ytterbium.

‡The checker performed the reaction in an Erlenmeyer flask inside a recirculating atmosphere glove box. Slow gravity filtration of the reaction solution through a D-porosity frit obviated the need for using filter aid.

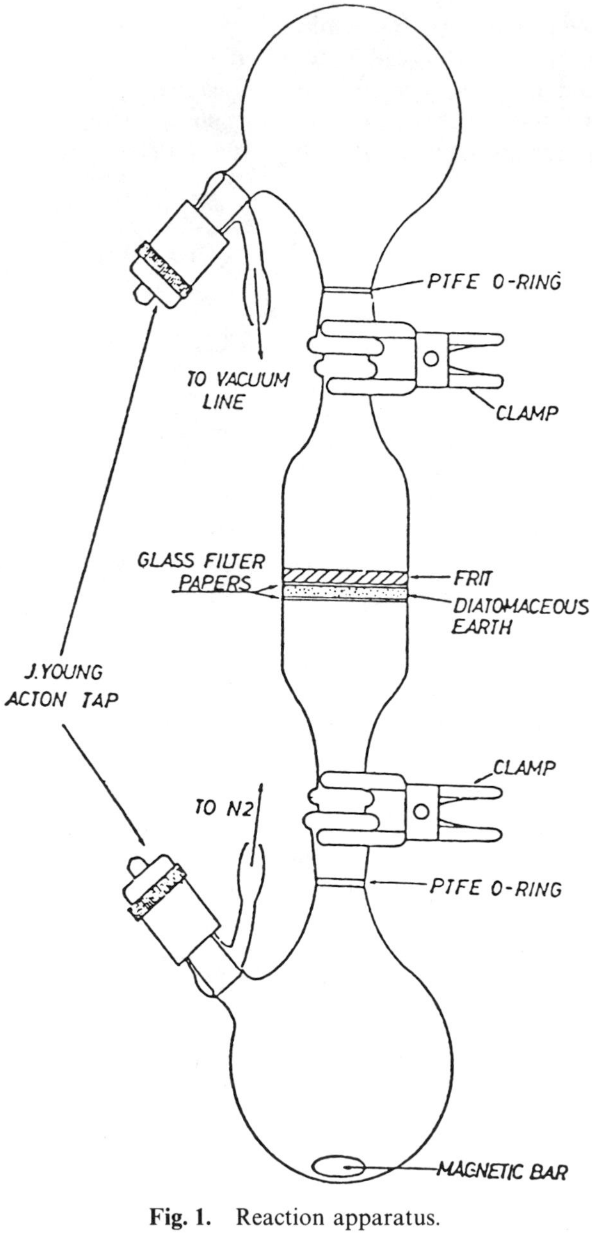
PTFE O-RING
CLAMP
TO VACUUM LINE
GLASS FILTER PAPERS
FRIT
DIATOMACEOUS EARTH
J.YOUNG ACTON TAP
CLAMP
TO N2
PTFE O-RING
MAGNETIC BAR

Fig. 1. Reaction apparatus.

The complex can be recrystallized from THF–light petroleum ether if required.

■ **Caution.** *Finely divided metal residues from the Schlenk filter can be pyrophoric. The residues can be used as an activated ytterbium metal source for reactions such as transmetalation, where the presence of mercury can be tolerated.*

Properties

Bis(phenylethynyl)ytterbium(II) is very sensitive to oxygen and water both in solution and the solid state. It is indefinitely stable under purified nitrogen or argon at room temperature. Thermal decomposition occurs at 200 °C. Dissolution occurs in THF, in which it is trimeric–tetrameric at the boiling point,[10] and in 1, 2-dimethoxyethane, but it is insoluble in less polar solvents. It is considered to have a polymeric solid state structure with at least four coordination for ytterbium.[10] ^{1}H NMR spectrum (C_4D_8O): 7.12 (br), ~ 4H, H3, 5; 7.38 (br), ~ 2H, H4; 7.71 (br), ~ 4H, H2, 6 downfield from Me$_4$Si. Electronic spectrum (300–1000 nm, THF): 325 ($\varepsilon = 900$), 543 (500) nm. On brief exposure to air, an ytterbium(III) absorption at 978 nm appears. The IR (Nujol mull): 2045 and 2020 (vw) $[v_{(C \equiv C)}]$, 1481 (s), 1439 (s), 1023 (s), 750 (vs), and 687 (vs) cm^{-1}.

References

1. T. J. Marks and R. D. Ernst, "Scandium, Yttrium and the Lanthanides and Actinides," in *Comprehensive Organometallic Chemistry*, Vol. 3, G. Wilkinson, F. G. A. Stone, and E. W. Abel, (eds), Pergamon, Oxford 1982, Chapter 21.

2. W. J. Evans, *Adv. Organomet. Chem.*, **24**, 131 (1985).

3. H. Schumann, *Angew. Chem. Int. Ed. Engl.*, **23**, 474 (1984).

4. J. H. Forsberg and T. Moeller, "Organometallic Compounds," in *Gmelin Handbook of Inorganic Chemistry*, RE Main Volume D6, 1983 pp. 137–302.

5. H. Schumann and W. Genthe, "Organometallic Compounds of the Rare Earths," in *Handbook of the Physics and Chemistry of the Rare Earths*, K. A. Gschneidner and I. Eyring, (eds.), Elsevier, Amsterdam, 1984, Chapter 53.

6. H. B. Kagan and J. L. Namy, "Preparation of Divalent Ytterbium and Samarium Derivatives and their use in Organic Chemistry," in *Handbook of the Physics and Chemistry of the Rare Earths*, K. A. Gschneidner and I. Eyring, (eds.), Elsevier, Amsterdam, 1984, Chapter 50.

7. G. B. Deacon, W. D. Raverty and D. G. Vince, *J. Organomet. Chem.*, **135**, 103 (1977).

8. G. B. Deacon and A. J. Koplick, *J. Organomet. Chem.*, **146**, C43 (1978).

9. G. B. Deacon, A. J. Koplick, W. D. Raverty, and D. G. Vince, *J. Organomet. Chem.*, **182**, 121 (1979).

10. G. B. Deacon, A. J. Koplick, and T. D. Tuong, *Aust. J. Chem.*, **35**, 941 (1982).

11. G. B. Deacon and T. D. Tuong, *J. Organomet. Chem.*, **205**, C4 (1981); G. B. Deacon and R. H. Newnham, *Aust. J. Chem.*, **38**, 1757 (1985).

12. Yu. F. Rad'kov, E. A. Fedorova, S. Ya. Khorshev, G. S. Kalimina, M. N. Bochkarev, and G. A. Razuvaev, *J. Gen. Chem. USSR*, **55**, 1911 (1985).

13. J. R. Johnson and W. C. McEwan, *J. Am. Chem. Soc.*, **48**, 469 (1926).

14. R. E. Dessy and J.-Y. Kim, *J. Am. Chem. Soc.*, **83**, 1167 (1961); G. Eglinton and W. McCrae, *J. Chem. Soc.*, 2295 (1963).

15. J. L. Atwood, W. E. Hunter, A. L. Wayda, and W. J. Evans, *Inorg. Chem.*, **20**, 4115 (1981).

27. BIS[BIS(TRIMETHYLSILYL)AMIDO]BIS(DIETHYL ETHER)YTTERBIUM AND (DIETHYL ETHER)BIS(η^5-PENTAMETHYLCYCLOPENTADIENYL)YTTERBIUM

$$Yb(s) + 2NH_4I + NH_3(l) \rightarrow YbI_2 + H_2$$

$$YbI_2 + 2NaN(SiMe_3)_2 \xrightarrow{Et_2O} Yb[N(SiMe_3)_2]_2[OEt_2]_2 + 2NaI$$

$$YbI_2 + 2NaC_5Me_5 \xrightarrow{Et_2O} Yb(\eta^5\text{-}C_5Me_5)_2(OEt_2) + 2NaI$$

Submitted by T. D. TILLEY, J. M. BONCELLA, D. J. BERG, C. J. BURNS, and R. A. ANDERSEN*
Checked by G. A. LAWLESS, M. A. EDELMAN, and M. F. LAPPERT†

Divalent ytterbium compounds with ligands such as halide, acetate, cyclopentadienyl, and so on, are insoluble in hydrocarbon solvents and not especially volatile.[1] This chemical fact is related to the large size of the metal ion and therefore the propensity for maximizing the coordination number by forming coordination polymers. One way to prevent polymerization is to prepare compounds with sterically hindered ligands, which decreases the coordination number of the metal atom and yield hydrocarbon soluble compounds. Application of this strategy is found in our preparation of two divalent ytterbium compounds, $Yb[N(SiMe_3)_2]_2[Et_2O]_2$ (ref. 2) and $Yb(\eta^5\text{-}Me_5C_5)_2$(thf) tetrahydrofuran = thf (ligand)],[3] as well as the diethyl ether complex of the latter.[4] All three compounds are best obtained from reaction of YbI_2, prepared by the method of Howell and Pytlewski,[5] and $NaN(SiMe_3)_2$ (ref. 6) or NaC_5Me_5.[7] We generally use the sodium reagents rather than the lithium or potassium ones, primarily because of the solubility properties of NaC_5Me_5, which is soluble in tetrahydrofuran (THF, solvent), though insoluble in diethyl ether, whereas the lithium and potassium analogs are insoluble in both of these ethers. Furthermore, $NaN(SiMe_3)_2$ is superior

*Chemistry Department, University of California Berkeley, CA 94720.
†School of Chemistry and Molecular Sciences, University of Sussex, Brighton BN1 9QJ, United Kingdom.

to LiN(SiMe$_3$)$_2$ as an amide-transfer reagent. The method of synthesis used to prepare Yb(η^5-Me$_5$C$_5$)$_2$(OEt$_2$) may be extended to Eu(η^5-Me$_5$C$_5$)$_2$(OEt$_2$) and Sm(η^5-Me$_5$C$_5$)$_2$(OEt$_2$).

Procedure

All reactions are performed under nitrogen. All solvents are distilled from sodium benzophenone under nitrogen. Ammonium iodide is dried by heating the commercial material to 120 °C under dynamic vacuum ($\sim 10^{-2}$ mm) for 12 to 14 h. After the white solid has cooled to room temperature, it is washed with diethyl ether (3 × 50 mL) to remove iodine that formed during heating.

■ **Caution.** *Anhydrous ammonia is toxic by inhalation and attacks the cornea. It must be handled in a well-ventilated fume hood, and eye protection and gloves should be worn.*

A. YTTERBIUM DIIODIDE

Ammonium iodide (12.7 g, 0.0875 mol) is transferred under a nitrogen flush to an oven-dried, 500-mL Schlenk flask equipped with a magnetic stirring bar. The flask is fitted with a Dry Ice–ethanol condenser, which is connected to a nitrogen manifold on a conventional Schlenk vacuum line and vented through a mineral oil bubbler that is located inside a fume hood. A mineral oil bubbler is preferred to a mercury bubbler since the back pressure is substantially less in the former. The flask is cooled to −70 °C and ~250 to 300 mL of anhydrous ammonia is condensed onto the ammonium iodide directly from a cylinder without additional drying. The mixture is stirred to assist dissolution of the ammonium iodide. If the stirring bar sticks and cannot be broken loose with gentle agitation, the reflux condenser can be removed under a brisk flow of nitrogen and the ammonium iodide broken into small pieces with a glass stirring rod. Ytterbium chips (we have used various suppliers with good results) (7.95 g, 0.0459 mol) are weighed into a small nitrogen-filled Schlenk flask (any large chips may be cut into small pieces with wire cutters). It is essential that a slight excess (~ 5%) of the metal is used in order to ensure that all of the ammonium iodide is consumed. With both flasks under a brisk nitrogen flush, the ytterbium metal is transferred, piece by piece over ~ 30 min, with the aid of a spatula or tweezers to the homogeneous solution of ammonium iodide in liquid ammonia. Dissolution of the ytterbium is accompanied by evolution of heat, hydrogen, and ammonia, and the ytterbium diiodide separates from solutions as an orange-yellow powder. When all of the metal has been added, the joint is stoppered, and the contents of the flask are stirred at −70 °C for 1 h. The cold bath is removed, and the ammonia is allowed to boil away through the nitrogen manifold with continual stirring.* The orange-yellow residue is scraped off the sides of the flask and crushed to a powder with the aid of a spatula. Ammonia fumes are still evolved at this stage. The flask is stoppered, cautiously exposed to a dynamic vacuum, and then heated in a sand bath at 200 °C for 16–20 h. The latter operation removes all of the coordinated ammonia. The yield of solvent-free, light yellow powder is essentially quantitative.

B. BIS[BIS(TRIMETHYLSILYL)AMIDO]BIS(DIETHYL ETHER)-YTTERBIUM

Ytterbium diiodide (3.18 g, 0.00745 mol) is weighed into a 500-mL, oven-dried Schlenk flask. Diethyl ether (50 mL) is added and the slurry is cooled to 0 °C with an ice bath. Sodium bis(trimethylsilyl)amide (2.56 g, 0.0140 mol)[6] is weighed into a 250-mL, oven-dried Schlenk flask, and the contents of the flask are dissolved in diethyl ether (150 mL). The solution is cooled to 0 °C, and the contents of the flask are added to the YbI_2 by way of a cannula. It is important that a slight excess of YbI_2 ($\sim 5\%$) is present, otherwise the yield of the desired compound is greatly reduced. The mixture is stirred for 1 h at 0 °C, at which point the mixture consists of a blue-green solution above a yellow-green precipitate. The mixture is allowed to warm to 20 °C. The color turns to light orange during this time, and the stirring is continued for 3 h. The contents of the flask are allowed to settle, the light orange solution is filtered, and the filtrate is concentrated to ~ 30 mL by exposing the contents of the flask to dynamic vacuum. The flask is returned to 1 atm with nitrogen, and then it is stored at -70 °C overnight. The large yellow prisms are collected by filtration and dried under reduced pressure (10^{-2} mm, 2 h). The prisms are orange at 20 °C. The diethyl ether mother liquor is concentrated to ~ 5 mL; cooling to -70 °C produces an additional crop of crystals that are collected as above. Yield: 3.2 g (71%).[†]

C. (DIETHYL ETHER)BIS(η^5-PENTAMETHYLCYCLOPENTA-DIENYL)YTTERBIUM

Sodium amide[8] (prepared from 8.6 g of sodium metal)[‡] in a 1-L Schlenk flask is suspended in THF (300 mL). With rapid stirring, Me_5C_5H (ref. 9) (42.6 g,

---

*The checkers report that after addition of ytterbium metal (Johnson Matthey) was complete, the color of the solution changed from orange to dark green. This color was discharged upon removal of the NH_3.

[†]The checkers obtained a yield of 58%.

[‡]The checkers used commercial sodium amide (Aldrich). They obtained a 97% yield of NaC_5Me_5 as a fine white powder.

48.9 mL, 0.312 mol) dissolved in THF (300 mL) is added by cannula to the suspension of $NaNH_2$. Ammonia is evolved during the addition, and the suspension becomes warm. The mixture is stirred for 8 h. The stirring is stopped, and the mixture is allowed to settle for 1–2 days, then filtered. If the filtrate is cloudy it is refiltered. The THF is removed by exposing and contents of the flask to a dynamic vacuum. When all of the solvent is evaporated, the off-white sticky precipitate is exposed to vacuum for 2–3 h, then returned to 1-atm pressure with nitrogen. The residue is washed with diethyl ether (3 × 200 mL). The white powder is exposed to vacuum for ∼ 1 h. Yield: NaC_5Me_5 39 g (80%). It is used without further purification.

Ytterbium diiodide (3.56 g, 0.00834 mol) and NaC_5Me_5 (2.52 g, 0.0159 mol) are weighed into an oven-dried 500 mL Schlenk flask equipped with a magnetic stirrer. Diethyl ether (200 mL) is added, and the green slurry is stirred at 20 °C for 16 h. The stirring is stopped, the bright green solution is filtered, and the volume of the filtrate is reduced to ∼ 50 mL by exposing the contents of the flask to dynamic vacuum. The flask is returned to 1 atm with nitrogen, then cooled to −25 °C. The bright green needles are isolated by filtration and dried under reduced pressure. A second crop of crystals can be obtained by concentrating the mother liquor to ∼ 10 mL and cooling to −25 °C. The combined yield is 2.9 g (70%).*

Properties

The divalent ytterbium compounds are air and moisture sensitive. They may be stored for months without noticeable signs of decomposition in the absence of air and moisture. Both complexes are readily soluble in aliphatic and atomatic hydrocarbons, and the diethyl ether is readily displaced by a variety of Lewis bases.[2–4,10] The compound $Yb[N(SiMe_3)_2][OEt_2]_2$ melts at 69–71 °C, and it is thermochroic, being yellow at − 70 °C and orange at 20 °C. It dissolves in benzene to give a red-orange solution. The ^1H NMR spectrum (C_6D_6, 26 °C) shows resonances at δ 3.51 (q), $J = 7$ Hz, 8H; 1.06 (t), $J = 7$ Hz, 12H and 0.41 (s), 36H. The IR spectrum as a Nujol mull between CsI plates has absorptions at 1249 (s), 1187 (m), 1151 (m), 1122 (m), 1091 (w), 1040 (m, br), 968 (s), 862 (m, sh), 830 (s, br), 774 (m), 753 (w, sh), 732 (w, sh), 663 (w, sh), 662 (m), 609 (m), 592 (w), 503 (w), 391 (m), and 372 (m) cm^{-1}. The compound $(\eta^5\text{-}Me_5C_5)_2Yb(OEt_2)$ melts at 204–207 °C. The ^1H NMR spectrum (C_6D_6, 35 °C) shows resonances at δ 3.15 (q), $J = 7$ Hz, 4H; 2.16 (s), 30H; and 1.04 (t), $J = 7$ Hz, 6H. The IR spectrum as a Nujol mull between CsI plates has absorptions at 2720 (w), 1633 (w), 1284 (w), 1180 (w, sh), 1163 (w, sh), 1149 (s), 1123 (w), 1097 (w, sh), 1077 (s, br), 1041 (m), 1019 (w), 980 (w, sh), 948 (w),

---

*The checkers obtained a yield of 2.2 g (53%).

929 (m), 839 (s), 829 (w, sh), 797 (w), 592 (m), 552 (w), 482 (w), 443 (w), 355 (m, br), 303 (m), and 268 (s, br) cm^{-1}.

References

1. D. A. Johnson, *Adv. Inorg. Chem. Radiochem.*, **20**, 1 (1977); T. J. Marks, *Prog. Inorg. Chem.*, **24**, 51 (1978).
2. T. D. Tilley, R. A. Andersen, and A. Zalkin, *Inorg. Chem.*, **23**, 2271 (1984).
3. T. D. Tilley, R. A. Andersen, B. Spencer, H. Ruben, A. Zalkin, and D. H. Templeton, *Inorg. Chem.*, **19**, 2999 (1980).
4. T. D. Tilley, R. A. Andersen, and A. Zalkin, *Inorg. Chem.*, **22**, 856 (1983). The single crystal X-ray structure of this material is known, P. L. Watson, personal communication, 1981.
5. J. K. Howell and L. L. Pytlewski, *J. Less Common Met.*, **18**, 437 (1969).
6. C. R. Krüger and H. Niederprum, *Inorg. Synth.*, **8**, 15 (1966).
7. J. E. Bercaw, R. H. Marvich, L. G. Bell, and H. H. Brintzinger, *J. Am. Chem. Soc.*, **94**, 1219 (1972).
8. K. W. Greenlee and A. L. Henne, *Inorg. Synth.*, **2**, 128 (1946).
9. R. S. Threlkel and J. E. Bercaw, *J. Organomet. Chem.*, **136**, 1 (1977); J. M. Manriquez, P. J. Fagan, L. D. Schertz, and T. J. Marks, *Inorg. Synth.*, **21**, 181 (1982).
10. T. D. Tilley, R. A. Andersen, A. Zalkin, and B. Spencer, *Inorg. Chem.*, **21**, 2647 (1982); T. D. Tilley, R. A. Andersen, and A. Zalkin, *J. Am. Chem. Soc.*, **104**, 3725 (1982).

28. CYCLOOCTATETRAENE LANTHANIDE COMPLEXES. Lu(C_8H_8)Cl(thf) AND Lu(C_8H_8)[o-$C_6H_4CH_2N(CH_3)_2$](thf)

$$LuCl_3 + K_2(C_8H_8) \xrightarrow{THF} Lu(C_8H_8)Cl(thf) + 2KCl$$

$$Lu(C_8H_8)Cl(thf) + Li[o\text{-}C_6H_4CH_2N(CH_3)_2]$$
$$\longrightarrow Lu(C_8H_8)[o\text{-}C_6H_4CH_2N(CH_3)_2](thf) + LiCl$$

tetrahydrofuran = THF (solvent) and thf (ligand)

Submitted by ANDREA L WAYDA*
Checked by JOHN T. RIGSBEE† **and ANDREW STREITWIESER, Jr.**†

In marked contrast to the extensive chemistry known for the cyclopentadienyl derivatives of the lanthanide elements,[1,2] little or no interest has been directed toward the development of the monocyclooctatetraene chemistry of these

*AT & T Bell Laboratories, Murray Hill, NJ 07974.
†Department of Chemistry, University of California, Materials and Chemical Sciences Division, Lawrence Berkeley Laboratory, Berkeley, CA 94720.

same elements.[3] This is surprising since these compounds should readily lend themselves to metathetical derivatization (as exploited to great advantage in the cyclopentadienyl systems),[1] and should allow the facile chemical study of a complete homologous series of lanthanide complexes.

We have found that the late metal congeners of this important compound class are easily synthesized and can be derivatized to yield aryl monocyclooctatetraene lanthanide complexes if the metal is small (Er—Lu) and the ligand bulky and chelating.[4] The synthesis of these compounds clears the way for comparative reactivity studies (with cyclopentadienyl lanthanide complexes) and allows investigations of fundamental chemical reactivity to be conducted for this ligand system.

Procedure

■ **Caution.** *Tetrahydrofuran is extremely flammable and forms explosive peroxides; only fresh, peroxide-free material should be used. N, N-dimethylbenzylamine causes skin and eye burns and is harmful if inhaled or absorbed through the skin. It should be manipulated in a well-ventilated fume hood; protective gloves and goggles should be worn.*

Weighing of reagents, some reactions and final product isolation and characterization are conducted in the recirculated argon atmosphere of a Vacuum Atmospheres HE-43 Dri-Lab. Manipulations performed on an argon-containing Schlenk line or vacuum line are so indicated. All solvents are Aldrich Sure-Seal grade. They are distilled on the Schlenk line under argon for sodium benzophenone (THF and toluene) or sodium benzophenone–tetraglyme (pentane). Hydrated lutetium trichloride is obtained from Research Chemicals, and is dehydrated on the vacuum line by the method of Taylor and Carter.[5,*] Cyclooctatetraene (Aldrich) is dried over activated 4-Å molecular sieves and degassed by vacuum transfer. N, N-dimethylbenzylamine (Aldrich) is heated at reflux over, and distilled from, sodium under argon on the Schlenk line. It is converted to the lithium salt, $Li[o-C_6H_4CH_2N(CH_3)_2]$ by metalation with BuLi (Aldrich)[6,†] on the Schlenk line.

---

*The checkers obtained anhydrous $LuCl_3$ from Cerac and used it without further purification.
†The checkers prepared $Li[o-C_6H_4CH_2N(CH_3)_2]$ as follows: N, N-Dimethylbenzylamine (2.53 g, 0.019 mol) is added to a flask containing 25 mL of hexane with stirring under argon. Butyllithium (18 mL containing 0.023 mol) is added via syringe, and the flask is fitted with a condenser. The mixture is refluxed under argon for 14 h. The initially pale yellow solution turns orange on heating, and after 2 h of reflux, a sand-colored precipitate begins to form. The solution is concentrated under vacuum but is not taken to dryness. The flask is removed to an inert atmosphere glove box, where the precipitate is filtered off and washed repeatedly with small portions of hexane. The sandy brown solid is dried on a Schlenk line, resulting in 2.25 g of o-lithio-N, N-dimethylbenzylamine (0.016 mol, 84% yield).

A. Lu(C$_8$H$_8$)Cl(thf)

The compound Lu(C$_8$H$_8$)Cl(thf) is prepared by using the method of Streitwieser and coworkers,[7] with minor modification. The entire procedure is conducted in the dry box. The compound K$_2$C$_8$H$_8$ is prepared by slowly adding 0.972 g (9.33 mmol) dry and degassed C$_8$H$_8$ to a stirred suspension of 0.730 g (18.66 mmol) of freshly scraped potassium metal (Alfa) in 30 mL of THF contained in a 125-mL Erlenmeyer flask equipped with a magnetic stirring bar.

■ **Caution.** *Potassium metal ignites on contact with moisture and may also form dangerous superoxides. Scraps should be disposed of by reaction with 1-butanol or a higher alcohol.*

The reaction is conducted at ambient temperature. Slow addition of the tetraene (over a 15-min period) immediately produces the intense deep brown color of the cyclooctatetraene dianion. The reaction mixture is stirred overnight or until all of the potassium metal has been consumed. The solution is then slowly added to a magnetically stirred slurry of 2.625 g (9.33 mmol) anhydrous LuCl$_3$ in 30 mL of THF contained in a 125-mL Erlenmeyer flask. Upon addition, the deep brown color of the potassium salt is immediately quenched by reaction with the lutetium trichloride. After addition is complete (15 min), the resultant light ginger ale-colored slurry is stirred for 24 h to ensure complete reaction.

The crude product is isolated and initially purified by removal of solvent by rotary evaporation followed by extraction with three small washes of THF (3 × 10 mL) to remove the extremely soluble by-product, K[Lu(C$_8$H$_8$)$_2$].* Final purification and removal of KCl are achieved by transfer of the crude product (∼4.3 g) to a Kontes Schlenk Soxhlet extraction apparatus. The apparatus is removed from the dry box and the residue is extracted with THF (60 mL) over a 24-h period on a Schlenk line under argon. During extraction, the pure white product precipitates in the Soxhlet receiver flask. It is isolated in 50% yield (1.73 g) by vacuum filtration in the dry box.† Cooling of the light yellow filtrate to −40 °C does not yield additional microcrystalline product. In practice, it is normally discarded (by removal from the dry box and careful quenching with 2-propanol) since it contains K[Lu(C$_8$H$_8$)$_2$], which has not been removed by the previous purification step.

Anal. Calcd. for LuC$_{12}$H$_{16}$ClO: Lu, 45.25. Found: Lu, 45.61.

---

*The checkers transferred the reaction mixture to a 100-mL Schlenk flask, which was then taken from the glove box and attached to a Schlenk line. The solvent was removed by vacuum transfer, then the dry material was returned to the glove box for the washes with THF.

†The checkers isolated the product by gravity filtration in the glove box, followed by drying on a Schlenk line. It was necessary to rinse the material with several small portions of THF to remove completely the pale yellow of the filtrate. Yield: 40%.

B.   Lu(C$_8$H$_8$)[o-C$_6$H$_4$CH$_2$N(CH$_3$)$_2$](thf)

In the dry box, 0.603 g (1.56 mmol) of Lu(C$_8$H$_8$)Cl(thf) is suspended in 50 mL of THF in a 100-mL Schlenk flask equipped with a stirring bar. A 100-mL Schlenk addition funnel containing a light gold-colored solution of Li[o-C$_6$H$_4$CH$_2$N(CH$_3$)$_2$] (0.220 g, 1.56 mmol) in 20 mL of THF is attached to this flask, and the apparatus is removed to an argon-containing Schlenk line. The receiver flask is cooled to −78 °C (Dry Ice–2-propanol), and the lithium reagent is added slowly (15 min). No change in the appearance of the reaction slurry is noted. The reaction solution is then allowed to reach ambient temperature slowly, while being stirred.* During warm-up, reaction occurs to produce a clear yellow solution. After a 24-h reaction time, the apparatus is removed to the dry box where solvent is removed from the reaction solution by rotary evaporation.† A clear yellow oil is obtained, which is triturated with 20 mL of pentane to produce an off-white powder. Exhaustive extraction of this crude powder with toluene leaves behind insoluble LiCl. Removal of solvent from the filtrate gives microcrystalline Lu(C$_8$H$_8$)[o-C$_6$H$_4$CH$_2$N(CH$_3$)$_2$](thf). Yield: 0.37 g (50%).

Anal. Calcd. for LuC$_{21}$H$_{28}$NO. C, 51.96; H, 5.81; N, 2.88; Lu, 36.04; O, 3.29. Found: C, 51.62; H, 5.68; N, 2.78; Lu, 36.45; O, 3.47 (by difference).

Properties

■ **Caution.** *Particular care must be taken when working with dry powders of these materials. They have been observed to ignite spontaneously and violently in air. Partially oxidized samples are particularly dangerous and have exploded without warning (partially decomposed material adhering to the walls of glassware and frits can be particularly dangerous). In practice, samples of these*

---

*Alternatively, we have conducted the reaction at ambient temperature in the dry box by adding the lithium reagent to a magnetically stirred slurry of Lu(C$_8$H$_8$)Cl(thf) in THF contained in a 125-mL Erlenmeyer flask. In this variation, the suspension of the lanthanide reagent slowly disappears as the lithium reagent is added. Upon completion of addition, a dark cloudy yellow-gold reaction solution is obtained. After stirring for 24 h at ambient temperature, solvent is removed from the slightly cloudy medium yellow solution (with green highlights) to produce a sticky froth. This material is triturated with pentane to produce a gray-green powder. The powder is then washed with a small quantity of toluene (10–20 mL) to remove very soluble impurities. The remaining powder is exhaustively extracted with toluene to produce a light gold solution. Removal of solvent by rotary evaporation yields the product in ∼50% yield. Using this modification, recrystallization is necessary to produce spectroscopically pure material. We prefer the synthesis described in the main body of the text for small scale preparations since it produces analytically pure product without further treatment. However, the alternate procedure is more convenient and may be preferable for large scale syntheses.

†The checkers removed the solvent by vacuum transfer prior to removal to the glove box.

complexes are destroyed in small quantities in THF solution by slowly exposing them to ambient air. All manipulations are performed in a well-protected fume hood.

Pure white $Lu(C_8H_8)Cl(thf)$ is insoluble in aromatic or alphatic solvents and is only slightly soluble in THF. It is characterized by IR [Nujol mull, cm^{-1}: 1020(s), 890(s), and 700(s)] and ^1H NMR [THF-d_8: δ 6.33(s)] spectroscopy. The absence of by-product $K[Lu(C_8H_8)_2]$ is determined by X-ray fluorescence analysis for potassium. White $Lu(C_8H_8)[o\text{-}C_6H_4CH_2N(CH_3)_2](thf)$ is soluble in aromatic and coordinating solvents. It is routinely characterized by IR [Nujol mull, cm^{-1}: 1420(m), 1380(m), 1360(m), 1305(m), 1235(m), 1170(s), 1095(m), 1040(m), 1010(m), 990(m), 940(m), 890(s), 855(s), 750(s), and 700(s)] and ^1H NMR (benzene-d_6: δ 8.15, 7.01 [m, $C_6H_4CH_2N(CH_3)_2$], 6.71(s, C_8H_8), 3.01, 1.00(m, α- and β-THF, respectively), 2.79 [s, $C_6H_4CH_2N(CH_3)_2$], 1.82 [s, $C_6H_4CH_2N(CH_3)_2$)] spectroscopy. It is definitively characterized by single crystal X-ray diffraction analysis.[8]

References

1. For recent reviews and descriptions of general chemical reactivity, see (a) H. Schumann, *Angew. Chem., Int. Ed. Engl.*, **23**, 474 (1984). (b) W. J. Evans, *The Chemistry of the Metal–Carbon Bond*, F. R. Hartley and S. Patai (eds.), Wiley-Interscience, New York, 1982, Chapter 12. (c) T. J. Marks, *Prog. Inorg. Chem.*, **24**, 51 (1978).

2. For representative examples of parent and substituted cyclopentadienyl complexes, see ref. 1 and (a) A. L. Wayda and W. J. Evans, *Inorg. Chem.*, **19**, 2190 (1980); T. D. Tilley and R. A. Andersen, *Inorg. Chem.*, **20**, 3267 (1981); P. L. Watson, J. Whitney, and R. L. Harlow, *Inorg. Chem.*, **20**, 3271 (1981). (b) J. N. John and M. Tsutsui, *Inorg. Chem.*, **20**, 1602 (1981), M. Tsutsui, Li-Ban Chen, D. E. Bergbreiter, and T. K. Miyamoto, *J. Am. Chem. Soc.*, **104**, 855 (1982). (c) M. F. Lappert, A. Singh, J. L. Atwood, and W. E. Hunter, *J. Chem. Soc. Chem. Commun.*, 1190 (1981).

3. Only one derivative of this type, $Lu(C_8H_8)(C_5H_5)(thf)$, has been reported. See J. D. Jamerson, A. P. Masino, and J. Takats, *J. Organomet. Chem.*, **65**, C33 (1974).

4. A. L. Wayda, *Organometallics*, **2**, 565 (1983).

5. M. D. Taylor and C. P. Carter, *J. Inorg. Nucl. Chem.*, **24**, 387 (1962).

6. A. C. Cope and R. N. Gourley, *J. Organomet. Chem.*, **8**, 527 (1967).

7. F. Mares, K. O. Hodgson, and A. Streitwieser, Jr., *J. Organomet. Chem.*, **28**, C24 (1971), K. O. Hodgson, F. Mares, D. F. Starks and A Streitwieser, Jr. *J. Am. Chem. Soc.*, **95**, 8650 (1973).

8. A. L. Wayda and R. D. Rogers, *Organometallics*, **4** 1440 (1985).

29. BIS(η⁵-PENTAMETHYLCYCLOPENTADIENYL)-BIS(TETRAHYDROFURAN)SAMARIUM(II)

$$SmI_2(thf)_2 + 2KC_5Me_5 \longrightarrow Sm(C_5Me_5)_2(thf)_2 + 2KI$$

thf (ligand) and THF (solvent) = tetrahydrofuran

Submitted by WILLIAM J. EVANS* and TAMARA A. ULIBARRI
Checked by HERBERT SCHUMANN† and SIEGBERT NICKEL

In recent years, the organometallic chemistry of the lanthanide metals in low oxidation states has been actively investigated. These low-valent studies have involved the zerovalent metals in the elemental state, using metal–vapor techniques, as well as the complexes of the three lanthanide metals that have divalent states readily accessible under "normal" solution reaction conditions, that is, Eu, Yb, and Sm. Although Sm(II) is the most reactive of these divalent lanthanides $[Sm(III) + e \rightarrow Sm(II): -1.5 \text{ V} \text{ vs} \text{ NHE}]$,[1] its chemistry in organometallic systems had not been previously investigated because the only known divalent organosamarium complexes, $[Sm(C_5H_5)_2(thf)_x]_y$,[2,3] and $[Sm(CH_3C_5H_4)_2(thf)_x]_y$,[4] are insoluble. Recently, however, the first soluble organosamarium(II) complex, $Sm(C_5Me_5)_2(thf)_2$, was synthesized by metal–vapor techniques.[5,6] Although the original synthesis was achieved on a preparative scale, a rotary metal vaporization reactor was required. The following solution synthesis[7] of the title compound is a more generally available route to this soluble divalent organosamarium complex.

General Procedure

The complexes described below are extremely air and moisture sensitive. Therefore, the syntheses are conducted under nitrogen with rigorous exclusion of air and water by using Schlenk, vacuum line, and glove box techniques.[8]

■ **Caution.** *Tetrahydrofuran and toluene are harmful if inhaled or absorbed through the skin. They should be handled in a well-ventilated fume hood, and gloves should be worn. THF forms explosive peroxides; only fresh, peroxide-free material should be distilled. Potassium hydride removed from oil suspension and KC_5Me_5 are pyrophoric; they should be manipulated in an inert atmosphere only.[8] The compound $1,2\text{-}C_2H_4I_2$ is heat and light sensitive, as well as sublimable.*

Toluene, hexane, and THF are distilled under nitrogen from sodium/

---
*Department of Chemistry, University of California, Irvine, Irvine, CA 92717.
†Institut für Anorganische und Analytische Chemie der Technischen, Universitat, Berlin, D-1000 Berlin 12, Federal Republic of Germany.

benzophenone. The compound C_5Me_5H [Strem, 95%] is dried with 4 Å molecular sieves and degassed by repeated freeze–pump–thaw cycles. The compound KC_5Me_5 is prepared by slowly adding ∼5 to 7 g of C_5Me_5H (3% molar excess) to a vigorously stirring suspension of the appropriate amount of KH in ∼40 mL of THF. Immediately upon addition of the C_5Me_5H, H_2 evolution is evident. The reaction mixture is stirred for 10 h or until H_2 evolution stops. The solution is filtered through a medium porosity fritted funnel (10–20 μm) to isolate the insoluble white KC_5Me_5. The KC_5Me_5 is washed with three 5-mL aliquots of hexane and dried by rotary evaporation.*
The complex $SmI_2(thf)_2$ is prepared from excess Sm metal (Research Chemicals) and $1,2$-$C_2H_4I_2$ in a THF solution.[3,7] In the air in a fume hood, $1,2$-$C_2H_4I_2$ (Aldrich, 97%) is dissolved in diethyl ether, washed with a sodium thiosulfate solution, and then washed with water. The ether solution is dried with $MgSO_4$, the ether is removed by rotary evaporation, and $1,2$-$C_2H_4I_2$ is recovered as a white powder. To remove all traces of H_2O, the $1,2$-$C_2H_4I_2$ is placed in a round-bottomed flask and dried on a vacuum line. Excess samarium pieces are placed in a 500-mL round-bottomed flask equipped with a Teflon stirring bar. Approximately 2 to 3 g of $1,2$-$C_2H_4I_2$ and 50 to 75 mL of THF are added to the samarium. The solution is stirred until it is a homogeneous dark blue color without any trace of the insoluble, yellow, samarium triiodide. The solution is then filtered through a medium porosity fritted funnel to remove the excess metal pieces, which can be saved and recycled. The dark blue solution is dried by rotary evaporation to yield the dark blue solid, $SmI_2(thf)_2$. This reaction is quantitative in $1,2$-$C_2H_4I_2$. The compound $SmI_2(thf)_2$ is very sensitive to oxidation. Solid $SmI_2(thf)_2$ is stable if stored under N_2. Solutions should be stored with a small amount of metal present.[†]

Procedure

In the glove box,[‡] KC_5Me_5 (2.01 g, 11.5 mmol) and $SmI_2(thf)_2$ (3.06 g, 5.58 mmol) are placed in a 125-mL Erlenmeyer flask equipped with a Teflon stirring bar. The powders are mixed thoroughly, and while stirring, 25 mL of THF is added. The mixture is allowed to stir for ∼12 h. The purple solution is filtered through a medium porosity fritted funnel (10–20 μm). The potassium

---

*The checkers prepared KC_5Me_5 from C_5Me_5H and KNH_2 in THF or with potassium in $1,2$-dimethoxyethane (DME).
[†]The checkers recommended that the $SmI_2(thf)_2$ be freshly prepared.
[‡]All of the following procedures were done in the nitrogen atmosphere of a Vacuum/Atmospheres HE-553 Dri-Lab glove box equipped with an oxygen- and water-getter and an atmosphere recirculation system. The checkers used Schlenk technique with 100-mL flasks for the entire procedure.

salts are washed with 5 mL of THF and discarded. The 5 mL of wash solution is combined with the filtrate and the THF is removed by rotary evaporation to yield a purple solid. The purple solid is dissolved in 30 mL of toluene, and the resulting solution is allowed to stir for 6 h. The purple solution is filtered through a fine porosity fritted funnel (4–8 μm). The purple insoluble materials are washed with two 5-mL portions of toluene. The toluene washes are combined with the filtrate, and the toluene is removed by rotary evaporation to yield a greenish-purple solid. The toluene extraction procedure is repeated (this is important to insure that all iodide containing species are excluded). The resulting solid is dissolved in ~ 5 mL of THF. The THF is removed from the resulting purple solution by rotary evaporation to yield the purple solid, $Sm(C_5Me_5)_2(thf)_2$. Yield: 2.29 g (73%).

Anal. Calcd. for $SmO_2C_{28}H_{46}$: Sm, 26.61. Found: 26.1.

Properties

The compound $Sm(C_5Me_5)_2(thf)_2$ is a purple air-sensitive solid that is soluble in both aromatic hydrocarbons and ethers. The monosolvate, $Sm(C_5Me_5)_2(thf)$, is a green solid; it is obtained by repeated evaporation of toluene solutions of the disolvate. The degree of solvation is easily monitored by integration of the absorptions in the ^1H NMR spectrum in benzene-d_6. The ^1H NMR shifts of these samarium(II) complexes are concentration and solvent dependent. The C_5Me_5 signal is typically located between 2.0 and 3.0 ppm in solutions of $Sm(C_5Me_5)_2(thf)_2$ in benzene-d_6. The two THF signals are found between 12.0 to 16.0 and 1.0 to 2.0 ppm. The most common impurity formed as a result of partial decomposition is $[Sm(C_5Me_5)_2]_2(\mu\text{-}O)$,[9] which has a distinctive ^1H NMR signal in benzene-d_6 at 0.05 to 0.06 ppm.

References

1. L. R. Morss, *Chem. Rev.*, **76**, 827 (1976).
2. G. W. Watt and E. W. Gillow, *J. Am. Chem. Soc.*, **91**, 775 (1969).
3. J. L. Namy, P. Girard, H. B. Kagan and P. E. Caro, *Nouv. J. Chim.*, **5**, 479 (1981).
4. W. J. Evans and H. A. Zinnen, unpublished results.
5. W. J. Evans, I. Bloom, W. E. Hunter, and J. L. Atwood, *J. Am. Chem. Soc.*, **103**, 6507 (1981).
6. W. J. Evans, I. Bloom, W. E. Hunter, and J. L. Atwood, *Organometallics*, **4**, 112 (1985).
7. W. J. Evans, J. W. Grate, H. W. Choi, I. Bloom, W. E. Hunter, and J. L. Atwood, *J. Am. Chem. Soc.*, **107**, 941 (1985).
8. D. F. Shriver and M. A. Drezdzon, *The Manipulation of Air-Sensitive Compounds*, 2nd. ed. Wiley, New York, 1986.
9. W. J. Evans, J. W. Grate, I. Bloom, W. E. Hunter, and J. L. Atwood, *J. Am. Chem. Soc.*, **107**, 405 (1985).

30.   *tert*-BUTYLBIS(η^5-CYCLOPENTADIENYL)-(TETRAHYDROFURAN)NEODYMIUM

$$Nd(C_5H_5)_3(thf) + LiC(CH_3)_3 \longrightarrow Nd(C_5H_5)_2[C(CH_3)_3](thf) + LiC_5H_5$$

tetrahydrofuran = thf (ligand) and THF (solvent)

Submitted by H. SCHUMANN* and G. JESKE*
Checked by G. B. DEACON† and P. I MacKINNON†

Bis(η^5-cyclopentadienyl)lanthanide alkyl derivatives, $Ln(C_5H_5)_2R(thf)$, are usually prepared from a bis(η^5-cyclopentadienyl)lanthanide chloride and an alkyllithium compound or a Grignard reagent in THF.[1,2] However, when R is a bulky alkyl ligand like *sec*-butyl or *tert*-butyl, the high reactivity of the respective alkyllithium compound allows a simpler synthesis starting from the tris(cyclopentadienyl)lanthanide complex $Ln(C_5H_5)_3(thf)$. In contrast to the bis(η^5-cyclopentadienyl)lanthanide chloride,[3] these complexes are available for the whole series of the lanthanide metals. The example given here describes the synthesis of $Nd(\eta^5\text{-}C_5H_5)_2(t\text{-}C_4H_9)(thf)$. The corresponding La, Sm, Yb, and Lu derivatives can be prepared in a similar way. The air and moisture sensitivity of the starting materials as well as of the products require working in an atmosphere of argon dried and freed from oxygen by passing over a BASF oxygen removal catalyst and P_4O_{10} (Sicapent, Merck) and Schlenk tube[4] and/or dry box techniques using glassware flame dried under vacuum and allowed to cool to room temperature under argon.‡

■   **Caution.**   *Solutions of $LiC(CH_3)_3$ in pentane burn in contact with air. $Ln(C_5H_5)_2[C(CH_3)_3](thf)$ compounds smolder in air and react violently with water. A nearby fire extinguisher should be ready to use.*

Procedure

A 2.2 g (5.3 mmol) quantity of $Nd(C_5H_5)_3(thf)$[5] is suspended in 50 mL of THF (predried over KOH flakes, further dried over Na wire, and finally distilled under argon from potassium) in a 250-mL three-necked flask equipped with a stirring bar, a 10-mL pressure equalizing funnel, a reflux condenser, a connection to a second 250-mL three-necked flask via a glass filter frit, and a stopcock (Fig. 1). The suspension is then cooled, using an acetone–Dry Ice

*Institut für Anorganische und Analytische Cemie, Technische Universität Berlin, D-1000 Berlin 12, Federal Republic of Germany.
†Department of Chemistry, Monash University, Clayton, Victoria, 3168, Australia.
‡The checkers used nitrogen purified by passage through BASF R 3/11 oxygen removal catalyst and molecular sieves.

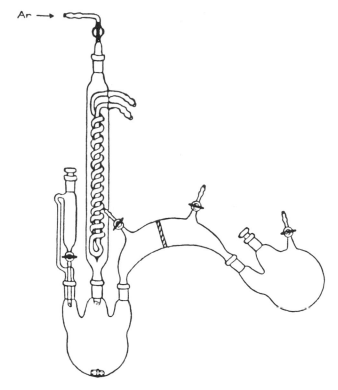

Fig. 1.   Reaction set up for synthesis of Nd(η^5-C$_5$H$_5$)$_2$[C(CH$_3$)$_3$](thf).

bath, to $-78\,°$C, and 3.6 ml of a 1.5 M solution of *tert*-butyllithium (Janssen Chimica) in pentane is added via the dropping funnel.* The reaction mixture is allowed to warm to room temperature within 6 h and stirred for a further 6 h. The solution is then concentrated to 10 mL at room temperature in vacuum, treated with 70 mL of diethyl ether (dried over sodium wire and finally distilled under argon from sodium benzophenone), and filtered via the frit into the second flask.† The insoluble LiC$_5$H$_5$ remains on the frit. The clear solution is cooled to $-30\,°$C. A 1.6 g (74%) yield of the product crystallizes as green needles, which remain after decanting the solvent carefully

---

*Checkers added the *tert*-butyllithium solution over 0.5 h. They report that faster addition led to the formation of a metallic precipitate, presumably Nd metal.
†Checkers evaporated the reaction mixture to dryness, redissolved the residue in THF and added ether (70 mL). The product must be filtered cold or it redissolves. A yield of 48% was obtained. Repeating the extraction process recovered a further 26% in one case.

at $-30\,°C$. The residual solvent is then removed under reduced pressure. The residue is pure at this stage.

Anal. Calcd. for $C_{18}H_{27}NdO$: C, 53.56; H, 6.74; Nd, 35.74. Found: C, 53.72; H, 6.83; Nd, 35.26.

Properties

The compound $Nd(\eta^5\text{-}C_5H_5)_2[C(CH_3)_3](C_4H_8O)$, a green crystalline material, reacts violently with water. The $^1H\,NMR$ spectrum of $Nd(C_5H_5)_2[C(CH_3)_3](C_4D_8O)$, recorded in C_4D_8O at $25\,°C$, exhibits only two singlet signals in an area ratio of 9:10, with the smaller *tert*-butyl signal at $\delta\,8.84$ and the larger cyclopentadienyl signal at $\delta\,-2.76$. The IR spectrum of the complex (KBr pellet) shows bands at 3092 (m), 2980 (s), 2930 (s), 2900 (s), 2870 (s), 2740 (vs), 2720 (vs), 2660 (s), 1450 (m), 1385 (m), 1370 (w), 1345 (w), 1295 (w), 1180 (w), 1152 (w), 1120 (w), 1090 (m), 1040 (s), 1010 (s), 915 (m), 890 (s), 760 (vs), 667 (w), 470 (w), and 400 (w). The compound melts at $42\,°C$ with decomposition.[6,7].

References

1. H. Schumann, W. Genthe, N. Bruncks, and J. Pickardt, *Organometallics*, **1**, 1194 (1982).
2. W. J. Evans, J. H. Meadows, A. L. Wayda, W. E. Hunter, and J. L. Atwood, *J. Am. Chem. Soc.*, **104**, 2008 (1982).
3. H. Schumann, *Angew. Chem.*, **96**, 475 (1984); *Angew. Chem., Int. Ed. Engl.*, **23**, 474 (1984).
4. H. Lux, *Anorganisch-chemische Experimentierkunst*, Ambrosius Barth-Verlag, Leipzig, 1970, p. 257; R. B. King and J. J. Eisch, *Organometallic Syntheses*, Vol. 1, Academic Press, New York, 1981.
5. (a) G. B. Deacon, A. J. Koplick, and T. D. Tuong, *Aust. J. Chem.*, **37**, 517 (1984). (b) G. B. Deacon, G. N. Pain, and T. D. Tuong, *Inorg. Synth.*, **26**, 17 (1989).
6. H. Schumann and G. Jeske, *Angew. Chem.*, **97**, 208 (1985); *Angew. Chem., Int. Ed. Engl.*, **24**, 225 (1985)
7. H. Schumann and G. Jeske, *Z. Naturforsch. Teil B*, **40**, 1490 (1985).

31. BIS(η^5-CYCLOPENTADIENYL)(TETRAHYDROFURAN)-[(TRIMETHYLSILYL)METHYL]LUTETIUM AND BIS(η^5-CYCLOPENTADIENYL)(TETRAHYDROFURAN)-*p*-TOLYLLUTETIUM

$$Lu(\eta^5\text{-}C_5H_5)_2Cl(thf) + LiCH_2Si(CH_3)_3$$
$$\longrightarrow Lu(\eta^5\text{-}C_5H_5)_2[CH_2Si(CH_3)_3](thf) + LiCl$$

$$Lu(\eta^5\text{-}C_5H_5)_2Cl(thf) + Li\text{-}p\text{-}C_6H_4CH_3$$

$$\longrightarrow Lu(\eta^5\text{-}C_5H_5)_2[p\text{-}C_6H_4CH_3](thf) + LiCl$$

tetrahydrofuran = thf (ligand) and THF (solvent)

Submitted by HERBERT SCHUMANN*, WOLFGANG GENTHE,*
and EFTIMIOS PALAMIDIS*
Checked by G. B. DEACON† and S. NICKEL†

The title compounds are obtained by stoichiometric exchange reactions between chlorobis(η⁵-cyclopentadienyl)lutetium and [(trimethylsilyl)methyl]-lithium or p-tolyl lithium, respectively. The procedure is based on the original preparation.[1,2] It can be easily adapted to the synthesis of the corresponding derivatives of Sc, Y, La, and the other lanthanide metals as well as to compounds of the type Ln(η⁵-C₅H₅)₂R(thf) with more or less bulky alkyl or aryl groups like $R = t\text{-}C_4H_9$,[3] $CH_2C(CH_3)_3$,[2] or $CH_2C_6H_5$.[2] The air and moisture sensitivity of the starting materials, as well as of the products, require working in an argon atmosphere using Schlenk tube and/or dry box techniques.

▪ **Caution.** *The lithium alkyl and aryl compounds and the resulting organolanthanides smolder in contact with air and react violently with water. A fire extinguisher should be ready to use nearby.*

A. Lu(η⁵-C₅H₅)₂[CH₂Si(CH₃)₃](thf)

Procedure

A 3.33-g (11.8 mmol) quantity of $LuCl_3$ (ref. 4) is suspended in 50 mL of dry THF (predried over KOH flakes, further dried over Na wire, and finally distilled under argon from potassium) in a 250-mL three-necked flask equipped with a stirring bar, a 20-mL pressure equalizing funnel, a reflux condenser, a connection to a second 250-mL three-necked flask via a glass filter frit (D4, 20-mm diameter), and a stopcock (see diagram in the synthesis in Section 30).[5] An atmosphere of argon, purified by passage through BASF R3-11 and molecular sieves/P_4O_{10}, is used. The glassware is flame dried under vacuum and allowed to cool to room temperature under predried argon. This procedure is repeated one time. A 9.2-mL volume of a 2.578 M solution of NaC_5H_5 (23.7 mmol)[6] in THF is added to the suspension at room

*Institut für Anorganische und Analytische Chemie, Technische Universität Berlin, D-1000 Berlin 12, Federal Republic of Germany.
†Department of Chemistry, Monash University, Clayton, Victoria, 3168, Australia.

temperature. The mixture is stirred for 2 h and then cooled to $-78\,°C$ using an acetone–Dry Ice bath. After addition of 16 mL of a $0.756\,M$ solution of $LiCH_2Si(CH_3)_3$ (12 mmol)[7] in pentane (dried by refluxing over NaH for 2 days and finally distilled from NaH under argon) and stirring for 6 h at $-78\,°C$, the solution is allowed to warm to room temperature. Volatile materials are removed by vacuum, and the residue is extracted three times with 200-mL portions of toluene,* washing the compound over the frit in the second flask. The toluene is evaporated at reduced pressure after each extraction, and the residue is washed four times with 200 mL of dry pentane–diethyl ether mixture (20:1). A yellowish crystalline powder results: 4.05 g (74% yield), which gives colorless crystals after recrystallization from 12 mL of diethyl ether at $-30\,°C$.

Anal. Calcd. for $C_{18}H_{29}LuOSi$: C, 46.54; H, 6.29; Lu, 37.67. Found: C, 46.27; H, 6.09; Lu, 37.96.

Properties

The compound $Lu(\eta^5\text{-}C_5H_5)_2[CH_2Si(CH_3)_3](C_4H_8O)$ reacts violently with water and it smolders in air. The ^1H NMR spectrum of the compound, recorded in C_6D_6 at $25\,°C$, exhibits three singlets with an area ratio of 10:2:9 at $\delta\,6.15$ (C_5H_5), $\delta\,-0.68$ (CH_2), and $\delta\,0.51$ $[Si(CH_3)_3]$ along with the THF multiplets at $\delta\,3.16$ and 1.11. The ^{13}C NMR data (in C_6D_6 at $25\,°C$) are $\delta\,110.8$ (C_5H_5), 28.3 (CH_2), 5.9 (CH_3), 72.6 (THF), and 26.0 (THF). The IR spectrum of the complex [in Nujol–poly(chlorotrifluoroethylene) oil] shows bands at 3750 (m), 3090 (w), 2945 (s), 2900 (sh), 2865 (m), 2815 (m), 2740 (m), 1775 (br), 1662 (br), 1562 (br), 1447 (s), 1370 (m), 1350 (m), 1308 (w), 1253 (m), 1240 (m), 1180 (w), 1040 (sh), 1020 (s), 920 (w), 860 (s), 785 (vs), 728 (m), 672 (m), 578 (w), 510 (m), 450 (w), 408 (m), 390 (m), and 250 (w). The compound decomposes between 100 and $110\,°C$. The crystals are orthorhombic, space group $P2_12_12_1$, with $a = 17.381(8)\,Å$, $b = 12.268(3)\,Å$, $c = 9.170(3)\,Å$.

B. $Lu(\eta^5\text{-}C_5H_5)_2(C_6H_4CH_3\text{-}p)(thf)$

Procedure

The procedure is exactly the same as for $Lu(\eta^5\text{-}C_5H_5)_2[CH_2Si(CH_3)_3](thf)$ using 2.94 g (10.44 mmol) $LuCl_3$ in 50 mL of THF, 8.2 mL of a $2.54\,M$ solution of NaC_5H_5 (20.8 mmol) in THF, and 22.7 mL of a $0.46\,M$ solution of p-tolyl lithium (10.44 mmol) (Ventron) in diethyl ether. Yield: 3.62 g (74%) of colorless crystals.

*Checkers used 50 mL of toluene per extraction.

Anal. Calcd. for $C_{21}H_{25}LuO$: C, 53.85; H, 5.38; Lu, 37.36. Found: C, 53.80; H, 5.18; Lu, 36.65.*

Properties

The compound $Lu(\eta^5\text{-}C_5H_5)_2(C_6H_4CH_3\text{-}p)(C_4H_8O)$ reacts violently with water and it smolders in air. The ^1H NMR spectrum of the compound, recorded in C_6D_6 at 25 °C, exhibits two singlets at δ 6.25 (C_5H_5) and δ 2.78 (CH_3) and two doublets at δ 7.93 (H_β) and δ 7.39 (H_γ) [$^3J_{(HH)} = 7\,Hz$] in an area ratio of 10:3:2:2, along with the THF multiplets at δ 3.21 and 1.10. The ^{13}C NMR data (in C_6D_6 at 25 °C are δ 111.8 (C_5H_5), 184.2 (C_α), 142.2 (C_β), 129.7 (C_γ), 134.3 (C_δ), 22.6 (CH_3), 73.1 (THF), and 25.9 (THF). The IR spectrum of the complex [in Nujol–poly(chlorotrifluoroethylene) oil] shows bands at 3575 (m), 3098 (w), 3080 (w), 3030 (m), 2995 (s), 2950 (s), 2900 (s), 2840 (sh), 2720 (m), 2422 (w), 2278 (w), 2080 (w), 1905 (w), 1770 (br), 1660 (br), 1590 (w), 1554 (br), 1445 (sh), 1432 (sh), 1420 (m), 1350 (m), 1300 (w), 1250 (w), 1240 (w), 1228 (m), 1195 (s), 1175 (m), 1140 (w), 1070 (w), 1040 (sh), 1015 (vs), 938 (w), 920 (w), 870 (w), 860 (s), 850 (s), 795 (sh), 785 (vs), 775 (sh), 680 (w), 630 (w), 580 (w), 552 (m), 510 (m), 482 (s), 450 (br), 390 (w), and 250 (w) cm^{-1}. The compound decomposes between 110 and 120° C. The crystals are orthorhombic, space group *Pnma*, with $a = 15.814(36)$ Å, $b = 12.511(11)$ Å, $c = 9.467(19)$ Å.

References

1. H. Schumann, W. Genthe, and N. Bruncks, *Angew. Chem.*, **93**, 126 (1981); *Angew. Chem., Int. Ed. Engl.*, **20**, 120 (1981).

2. H. Schumann, W. Genthe, N. Bruncks, and J. Pickardt, *Organometallics*, **1**, 1194 (1982).

3. W. J. Evans, A. L. Wayda, W. E. Hunter, and J. L. Atwood, *J. Chem. Soc. Chem. Commun.*, 292, **1981**.

4. J. D. Corbett, *Inorg. Synth.*, **22**, 39 (1983).

5. H. Schumann and G. Jeske, *Inorg. Synth.*, **27**, 158 (1990).

6. R. B. King and F. G. A. Stone, *Inorg. Synth.*, **7**, 99 (1963).

7. C. T. Youngs and O. T. Beachley, *Inorg. Synth.*, **24**, 95 (1985).

*Checkers worked at 60% of the suggested scale. They dried their glassware with Me_3SiCl followed by purging with argon. The product obtained by toluene extraction and evaporation was oily. It required washing with 5 × 120 mL of pentane–diethyl ether (20:1, v/v) to achieve solidification. The crystals from ether were initially colorless, but after vacuum drying gave the title compound as a yellowish-white powder (51%).

32. HYDROCARBON-SOLUBLE HOMOLEPTIC BULKY ARYL OXIDES OF THE LANTHANIDE METALS: [Ln(OAr^R)$_3$]

$$2LnCl_3 + 3[Li(\mu\text{-}OAr^R)(OEt_2)]_2 \xrightarrow[\text{(2)sublimation}]{\text{(1)THF,reflux}} 2[Ln(OAr^R)_3] + 6LiCl$$

$$(1)$$

$$[Ln\{N(SiMe_3)_2\}_3] + 3Ar^ROH \xrightarrow[\text{(2)sublimation}]{\text{(1)toluene,reflux}} [Ln(OAr^R)_3] + 3NH(SiMe_3)_2$$

$$(1)$$

$$Ar^R = C_6H_2t\text{-}Bu_2\text{-}2, 6\text{-}R\text{-}4$$

Submitted by MICHAEL F. LAPPERT,* ANIRUDH SINGH,*
and RICHARD G. SMITH*
Checked by HILMAR A. STECHER† and AYUSMAN SEN†

The chemistry of Sc(III), Y(III), and the 4f-metal(III) [collectively abbreviated as Ln(III)] alkoxides is well documented.[1] The compounds [{Ln(OR′)$_3$}$_n$] (R′ = an alkyl group) are often hydrocarbon soluble, volatile solids or liquids and are believed to be tetramers or hexamers, involving bridging alkoxo ligands, although few X-ray data are available. However, until recently, the known corresponding Ln(III) aryl oxides were involatile solids, insoluble in hydrocarbons, and were presumed to be high polymers.[2] In 1983, it was recognized that using the greatly hindered 2, 6-di-*tert*-butylphenoxo ligands $^-$OAr^R (Ar^R = C$_6$H$_2t$-Bu$_2$-2, 6-R-4; R = H, Me, or t-Bu), monomeric, volatile, hydrocarbon soluble complexes [Ln(OAr^R)$_3$] (1) could be prepared.[3]

The compounds (1) are of interest for various reasons. First, they provide rare examples of Ln^{3+} complexes having the remarkably low metal coordination number of three; the only other examples are the alkyls [Ln{CH(SiMe$_3$)$_2$}$_3$] (ref. 4) and the bis(trimethylsilyl)amides [Ln{N(SiMe$_3$)$_2$}$_3$].[5] Second, it is clear that the bulky $^-$OAr^R ligands confer lipophilicity upon the complexes (1); their volatility is also noteworthy. These ligands were introduced into coordination chemistry within the last decade for additionally *inter alia* corresponding aryl oxides of Li, Na, Be, Mg, Al, Ti(III), Ti(IV), Zr(IV), Hf(IV), Ta(V), Cr(II), W(VI), Mn(II), Rh(I) (this afforded the η^5-cyclohexadienonyl isomer), Cu(I), Ge(II), Sn(II), Pb(II), P(II), Th(IV), and U(IV) (*cf.*, ref. 6). These metal aryl oxides are prepared directly or indirectly from the parent phenols Ar^{R}OH; the latter are readily and cheaply available because of their use as antioxidants.

*School of Chemistry and Molecular Sciences, University of Sussex, Brighton BN1 9QJ, United Kingdom.
†Department of Chemistry, The Pennsylvania State University, University Park, PA 16802.

General Procedures

All operations, including those of Section B, are carried out in an atmosphere of pure dry argon [used directly from a cylinder (Air Products)]. Solvents are of reagent grade or better, and are rigorously freed from oxygen and moisture (by heating with sodium benzophenone under reflux), and are freshly distilled under nitrogen [white spot; used directly from a cylinder (Air Products)] and freeze degassed prior to use. Glassware is successively oven-dried, flame-dried *in vacuo*, and sealed under argon. All manipulative procedures are performed using standard Schlenk techniques.[7] Each of the phenols Ar[R]OH is purified by passage of a hexane or pentane solution through a basic alumina (Aldrich, "Brockmann I"-150 mesh, 58 Å) column (50-cm length, 3-cm diameter), and removal of solvent from the eluent by heating *in vacuo*.

The sublimation apparatus is illustrated in Fig. 1. It comprises a horizontal Pyrex tube divided into three compartments A–C, terminating in a B14 socket D. There is an attached side arm and tap E. The sample is introduced into A using Schlenk techniques. Similarly, a glass wool plug F is inserted through D; its function is partly to clean the walls of Sections B and C of the tube, and also to retain the sample in A when sublimation *in vacuo* is in progress. The tube is placed in a horizontal furnace G, with the collection area exposed to ambient temperature. The sample gradually collects into this collection area during the course of the sublimation. When sublimation is complete, the tube is removed from the furnace and while still under vacuum is allowed to reach ambient temperature. Argon is introduced through the stopcock E and the tube is cut midway along the constricted section B; A is discarded. Using Schlenk techniques under argon, under ambient conditions, the solid sample is scraped through Section B, by removal of the stopper at D, into a Schlenk tube placed around B.

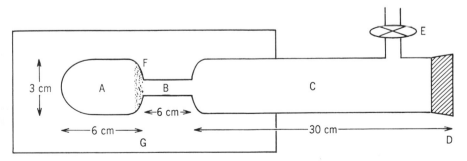

Fig. 1. Sublimation apparatus.

A. FROM A LITHIUM ARYL OXIDE

Procedure

Anhydrous samarium(III) chloride [*Rare Earth Products, Johnson Matthey*] (2.40 g, 9.40 mmol) is placed in a 200-mL Schlenk tube containing a magnetic stirring bar. Tetrahydrofuran (THF) (100 mL) is introduced via a cannula at 20 °C. The mixture is stirred at 20 °C for 2 h. Solid (2, 6-di-*tert*-butylphenoxo)-lithium-diethyl etherate[8] (8.13 g, 28.4 mmol) is added to this slurry at 20 °C. A reflux condenser is then fitted to the Schlenk tube, and the reaction mixture is heated under reflux for $\sim$ 8 h. After $\sim \frac{1}{2}$ h, the solid dissolves to yield a clear yellow-green solution. The solvent is removed at $\sim$ 25 °C and 10^{-2} torr, leaving a yellow-green solid, which is scraped from the sides of the Schlenk tube with a spatula and transferred to a sublimation tube. Sublimation at 255–260 °C and 10^{-3} torr affords yellow crystals of tris(2, 6-di-*tert*-butyl-phenoxo)samarium. Yield: 4.46 g (62%).*

Anal. Calcd. for $C_{42}H_{63}O_3Sm$: C, 65.8; H, 8.29. Found: C, 65.9; H, 8.20.
  ^{1}H NMR (external SiMe$_4$): δ 8.45 and $-$ 0.81 (broad).

B. FROM A PHENOL

Procedure

Tris[bis(trimethylsilyl)amido]lanthanum[9] (3.82 g, 6.16 mmol) in toluene (20 mL) is slowly added during $\sim$ 15 min from a dropping funnel fitted with a pressure equalization attachment to 2, 6-di-*tert*-butyl-4-methylphenol (4.18 g, 18.97 mmol) in toluene (50 mL) at 20 °C in a 200-mL Schlenk tube. The dropping funnel is replaced by a reflux condenser, and the mixture is heated under reflux for $\sim$ 48 h.† Volatiles are removed at 10^{-3} torr. Sublimation of the white residue at 250–260 °C and 10^{-3} torr affords white crystals of tris(2,6-di-*tert*-butyl-4-methylphenoxo)lanthanum. Yield: 3.30 g (63%).

Anal. Calcd. for $C_{45}H_{69}O_3La$: C, 67.8; H, 8.73. Found: C, 67.4; H, 8.68.
  ^{1}H NMR: δ 1.49, 2.23, and 6.97 (multiplet).

---

*The checkers tested this procedure with Ce instead of Sm, and obtained a 69% yield of product with purity >95%; they found that prestirring prior to addition of the lithium aryl oxide was unnecessary.

†The checkers found that refluxing was not required, and that the product was purified by heating the solid to 120 to 140 °C for 8 to 12 h under vacuum.

Preparation of Related Compounds

Procedures A and B can be employed to prepare a wide range of compounds [Ln(OAr^R)$_3$]. The lithium aryl oxide procedure has the advantage that [Li(μ-OAr^R)(OEt$_2$)]$_2$ is readily prepared[8] from Ar^{R}OH and LiBu (available as a hexane solution) in diethyl ether. The lanthanide metal(III) amide route has the advantage of greatest ease of removal of by-product, in this case of the volatile NH(SiMe$_3$)$_2$; however, because these metal(III) amides are highly hindered, substantial reflux times may be required. Data on some selected aryl oxides are summarized in Table I.

Properties

Compounds **1** are exceedingly sensitive to air or moisture. In an inert atmosphere they are thermally stable, melting without decomposition at high temperatures (Table I), and subliming at $\sim 250\,^\circ$C and 10^{-3} torr. They are soluble in hydrocarbons and are monomers in benzene (cryoscopy).

The structures of crystalline [Ln(OC$_6$H$_2$$t$-Bu$_2$-2, 6-R-4)$_3$], determined by single crystal X-ray diffraction, confirm their structure, with LnO$_3$ in a

TABLE I   Yields, Physical Properties, and Analyses for [Ln(OC$_6$H$_2$$t$-Bu$_2$-2, 6-Me-4)$_3$] (1a) and [Ln(OC$_6H_3$$t$-Bu$_2$-2, 6)$_3$] (1b)

Ln in **1**	Yield (%)^a	Melting point (°C)	Color	Calculated (%) C	H	Found (%) C	H
1a							
Sc	89	150–152	White	76.9	9.89	76.8	9.93
Y	68	178–180	White	72.4	9.31	72.1	9.42
La	40	195–197	White	67.8	8.73	67.4	8.68
Pr	52	200–204	Pale yellow	67.7	8.71	67.1	8.50
Nd	42	220–221	Blue-violet	67.4	8.67	67.0	8.25
Dy	68	214–216	White	65.0	8.48	65.4	8.24
Ho	70	198–200	Pale yellow	65.7	8.45	65.4	8.41
Er	68	180–182	Pink	65.5	8.43	65.2	8.38
Yb	70	215–218	Maroon	65.0	8.37	65.0	8.28
1b							
Sc	85	140–142	White	76.3	9.61	76.1	9.56
Y	72	188–192	White	71.6	9.01	71.8	9.17
La	69	242–245	White	66.8	8.41	67.6	8.83
Sm	62	200–205	Yellow	65.8	8.29	65.9	8.20

^{a}The quoted yields refer to sublimed material.

planar trigonal environment [for $Ln = Sc$ $(R = Me)^3$ or Y $(R = H)^{10}$] or pyramidal [for $Ln = Ce$ $(R = Me)^{11}$]; compare pyramidal LnC_3 or LnN_3 in $[Ln\{CH(SiMe_3)_2\}_3]$ $(Ln = La$ or $Sm)^4$ or $[Ln\{N(SiMe_3)_2\}_3].^5$

References

1. D. C. Bradley, R. C. Mehrotra, and D. P. Gaur, *Metal Alkoxides*, Academic Press, New York, 1978.
2. K. C. Malhotra and R. L. Martin, *J. Organomet. Chem.*, **239**, 159 (1982).
3. P. B Hitchcock, M. F. Lappert, and A. Singh, *J. Chem. Soc. Chem. Commun.*, **1983**, 1499.
4. P. B. Hitchcock, M. F. Lappert, R. G. Smith, R. A. Bartlett, and P. P. Power, *J. Chem. Soc. Chem. Commun.*, **1988**, 1007.
5. M. F. Lappert, P. P. Power, A. R. Sanger, and R. C. Srivastava, *Metal and Metalloid Amides*, Horwood-Wiley, Chichester, 1980.
6. A. W. Duff, R. A. Kamarudin, M. F. Lappert, and R. J. Norton, *J. Chem. Soc. Dalton Trans.*, **1986**, 489.
7. Compare, D. F. Shriver and M. A. Drezdzon, *The Manipulation of Air-Sensitive Compounds*, 2nd ed., Wiley, New York, 1986.
8. B. Cetinkaya, I. Gümrükcü, M. F. Lappert, J. L. Atwood, and R. Shakir, *J. Am. chem. Soc.*, **102**, 2086 (1980).
9. D. C. Bradley and R. G. Copperthwaite, *Inorg. Synth.*, **18**, 112 (1978) (for the Sc amide).
10. P. B. Hitchcock, M. F. Lappert, and R. G. Smith, *Inorg. Chim. Acta*, **139**, 183 (1987).
11. H. A. Stecher, A. Sen, and A. L. Rheingold, *Inorg. Chem.*, **27**, 1130 (1988).

33. BIS[η^5-1,3-BIS(TRIMETHYLSILYL)CYCLOPENTA-DIENYL]CHLOROLANTHANIDE(III) COMPLEXES, $[\{LnCp''_2(\mu\text{-}Cl)\}_2]$ AND $[LnCp''_2(\mu\text{-}Cl)_2Li(thf)_2]$

$$LnCl_3 + 2LiCp'' \xrightarrow[\text{(2) 25 °C, } 10^{-2} \text{ torr}]{\text{(1) THF, } -20 °C} [LnCp''_2(\mu\text{-}Cl)_2Li(thf)_2]$$
$$(1)$$

$$2[LnCp''_2(\mu\text{-}Cl)_2Li(thf)_2] \xrightarrow[3\,h]{140\text{--}150\,°C} [\{LnCp''_2(\mu\text{-}Cl)\}_2]$$
$$\quad\quad\quad (1) \quad\quad\quad\quad\quad\quad\quad\quad\quad (2)$$

tetrahydrofuran = THF (solvent) and thf (ligand)
$Cp'' = \eta^5\text{-}C_5H_3(SiMe_3)_2\text{-}1,3$

Submitted by MICHAEL F. LAPPERT* and ANIRUDH SINGH*
Checked by JOSEF TAKATS†

*School of Chemistry and Molecular Sciences, University of Sussex, Brighton, BN1 9QJ, United Kingdom.
†Department of Chemistry, University of Alberta, Edmonton, Alberta, Canada T6G 2G2.

Among the 4f-block metals and their Group 3 congeners Sc and Y (collectively abbreviated as Ln), the metallocene(III) chlorides $[\{LnCp_2(\mu\text{-}Cl)\}_2]$ $(Cp = \eta^5\text{-}C_5H_5)$ have proved to be useful precursors to a wider range of organometallic compounds of formula $LnCp_2X$.[1] These chlorides are not available for the early lanthanide metals (Ln = La, Ce, Pr, or Nd); attempts to make them from the appropriate Ln^{3+} $(f^0, f^1, f^2,$ or $f^3)$ chloride and $2NaCp$ gave the products of redistribution, $\frac{2}{3}LnCp_3 + \frac{1}{3}LnCl_3$. The instability of these lanthanocene(III) chlorides has been attributed to kinetic factors, favored by the greater ionic character of the f^0–f^3 Ln^{3+} complex than that of their f^5–f^{14} counterparts. This led to the development of the $\eta^5\text{-}C_5H_{5-n}(SiMe_3)_n{}^-$ series of ligands: $\eta^5\text{-}C_5H_4SiMe_3{}^-$ (abbreviated as Cp'), $\eta^5\text{-}C_5H_3(SiMe_3)_2\text{-}1,3$ (Cp''^-), and $\eta^5\text{-}C_5H_2(SiMe_3)_3\text{-}1,2,4$ (Cp'''^-).

The Cp''^- ligand is sterically more demanding than $\eta^5\text{-}C_5Me_5{}^-$,[2] and often confers lipophilicity on its metal complexes.

General Procedures

All experiments are performed in an atmosphere of pure dry argon [used directly from a cylinder (Air Products)], or under vacuum, using Schlenk techniques.[3] Glassware is successively oven-dried, flame-dried, and cooled under argon. Solvents are of reagent grade, or better, and are rigorously freed from oxygen or moisture, by heating under reflux with $Li[AlH_4]$ (hydrocarbons) or sodium benzophenone.

A. THE BIS[η^5-1,3-BIS(TRIMETHYLSILYL)CYCLOPENTA-DIENYL]DI-μ-CHLORO-LANTHANIDE(III)-BIS(TETRAHYDROFURAN)LITHIUM COMPLEXES, $[LnCp''_2(\mu\text{-}Cl)_2Li(thf)_2]$ (1)

Procedure

■ **Caution.** *All operations must be carried out under strictly anhydrous oxygen-free conditions. Organolithium compounds in hydrocarbon solvents are spontaneously flammable in air.*

Butyllithium (33.75 mL of a 1.6 M solution in hexane, 54 mmol [Lithium Corporation]) is added dropwise with stirring during $\sim\frac{1}{2}$ h to bis(trimethylsilyl)cyclopentadiene $(Cp''H)$[4] (11.25 g, 54 mmol) in diethyl ether (500 mL) at 0 °C. (Because of the gelatinous nature of the initially formed cyclopentadienyllithium, a large excess of solvent is used.) The mixture is stirred at ~ 20 °C for 12 h. The white precipitate of $LiCp''$ is allowed to settle during ~ 12 h at 20 °C. The precipitate is filtered off and is thoroughly washed with hexane

(3 × 50 mL) in order to remove any unreacted LiBu. It is dried *in vacuo* to yield [η^5-1, 3-bis(trimethylsilyl)cyclopentadienyl]lithium [Yield: 11.25 g (96%)] as a free-flowing white powder.

Anal. Calcd. for $C_{11}H_{21}LiSi_2$: C, 61.1; H, 9.78. Found: C, 61.0; H, 9.62%. ^{1}H NMR (C_5H_5N): $\delta = 0.368$(s) (SiMe$_3$), 6.704(d) [$^4J_{(^1H^1H)} = 1.8$ Hz], 6.803(t) [$^4J_{(^1H^1H)} = 1.8$ Hz]. Anhydrous ytterbium(III) chloride [Rare Earth Products, Johnson Matthey] (6.48 g, 24.48 mmol) is added slowly ($\sim\frac{1}{2}$ h) to a solution of LiCp″ (10.60 g, 49.0 mmol) in THF (150 mL) at $-20\,°C$, with gradual color changes from colorless, through yellow and orange to orange-red. This reaction mixture is stirred at $\sim 20\,°C$ for ~ 24 h. Volatile substances are removed at $25\,°C$ and 10^{-2} torr. The residue is recrystallized from n-C_6H_{14} (50 mL) at $-30\,°C$ to yield orange-red crystals of [YbCp″$_2$(μ-Cl)$_2$Li(thf)$_2$] (**1**, Ln = Yb). (Yield: 16.9 g (85%); mp 102–105) $°C$ (dec.)

Anal. Calcd. for $C_{30}H_{58}Cl_2LiO_2Si_4Yb$: C, 44.3; H, 7.19%. Found: C, 44.0; H, 7.30%.

Preparation of Related Compounds

The previous procedure can be employed to prepare the related compounds of formula [LnCp″$_2$(μ-Cl)$_2$Li(thf)$_2$], where Ln = Sc (72%, colorless), Y (78%, colorless), La (colorless, 80%), Ce (yellow, 75%), Pr (pale green, 70%), or Nd (blue-green, 75%).

Properties

The compounds of formula (**1**) are exceedingly sensitive to air and moisture and are somewhat thermally unstable.[6] Thus, attempted sublimation under reduced pressure results in loss of THF and formation of [{LnCp″$_2$(μ-Cl)}$_2$] (**2**) (see Section B). The thf molecules are readily displaced by treatment with Me$_2$NCH$_2$CH$_2$NMe$_2$ (tmeda), yielding [LnCp″$_2$(μ-Cl)$_2$Li (tmeda)].

The structure of crystalline [NdCp″$_2$(μ-Cl)$_2$Li(thf)$_2$], determined by X-ray diffraction, confirms the formulation as (**1**) (Ln = Nd).[6]

B. THE TETRAKIS[η^5-1,3-BIS(TRIMETHYLSILYL)CYCLOPENTA-DIENYL]-DI-μ-CHLORODILANTHANIDE(III) COMPLEXES, [{LnCp″$_2$(μ-Cl)}$_2$] (2)

Procedure

■ **Caution.** *Note Caution in Section* **A**
n-Butyllithium (35.1 mL of a 1.6 *M* solution in hexane, 56.01 mmol) is add-

ed dropwise to bis(trimethylsilyl)cyclopentadiene (11.79 g, 56.01 mmol)[4] in hexane (250 mL) at 0 °C. The mixture is stirred at ∼ 25 °C for 12 h. The white precipitate is collected and THF (100 mL) is slowly added. This solution is slowly (during ∼ 1 h) added to a cooled (0 °C) slurry of neodymium(III) chloride [Rare Earth Products, Johnson Matthey] (7.02 g, 28.01 mmol) in THF (100 mL). The resulting clear blue-violet reaction mixture is stirred at ∼ 20 °C for ∼ 48 h. After removal of volatile substances at 25 °C and 10^{-2} torr, the residue is heated at 100 °C for 1 h and then at 140–150 °C for 3 h. The green-blue residue is sublimed at 240–260 °C and 10^{-3} torr (see Fig. 1 and general procedure in Section **32**) to afford green-blue crystals of [{NdCp″₂(μ-Cl)}₂]. Yield: 12.6 g (75%).

Anal. Calcd. for $C_{22}H_{42}ClNdSi_4$: C, 50.9; H, 7.47. Found: C, 50.8; H, 7.36. The compound has a mp 295–297 °C.

Preparation of Related Compounds

The previous procedure can be employed to prepare the following related compounds [lanthanocene(III) chlorides], as summarized in Table I. Instead

TABLE I   Yields, Physical Properties, and Analysis for [{LnCp″₂(μ-Cl)}₂] (2)

Ln in [{LnCp″₂(μ-Cl)}₂]	Yield (%)[a]	Melting Point (°C)	Color	Calculated (%) C	H	Found (%) C	H
Sc	69	198–200	White	52.9	8.48	52.5	7.98
Y	78	325–327	White	48.6	7.79	48.6	7.89
La	50	238–240	White	44.5	7.14	44.4	7.32
Ce	55	265–268	Yellow	44.4	7.12	44.1	6.58
Pr	68	280–282	Yellow	44.4	7.11	44.3	7.17
Nd	75	295–297	Green-blue	44.1	7.07	44.2	6.89
Sm	69	302–304	Yellow	43.7	7.00	43.7	6.97
Eu	45	(115–120)[b]	Brown-violet	43.6	6.98	42.7	7.11
Gd	65	295–297	White	43.2	6.92	43.3	6.78
Tb	70	312–315	White	43.1	6.90	43.0	7.01
Dy	74	310–312	White	42.8	6.86	42.6	6.70
Ho	70	333–335	Pink (pale)	42.7	6.84	42.6	6.96
Er	78	322–323	Pink	42.5	6.81	42.2	6.52
Tm	72	295–297	Yellow	42.4	6.79	42.4	6.77
Yb	82	285–287	Maroon	42.1	6.75	42.5	6.64
Lu	65	313–315	White	42.0	6.73	41.7	6.71

[a]The quoted yields refer to sublimed material, except for La and Ce which relate to recrystallized material.
[b]This is a decomposition temperature.

of the final sublimation, an alternative purification procedure involves extracting the crude product into hot toluene, and cooling the toluene solution to $-30\,^\circ$C; this is recommended particularly for the La and Ce complexes, which undergo redistribution upon attempted sublimation.

Properties

The lanthanocene(III) chlorides are exceedingly sensitive to air and moisture but are thermally stable. They are stable in hot toluene.

The structures of crystalline $[\{LnCp''_2(\mu\text{-Cl})\}_2]$ $[(2)$ Ln $=$ Sc,[7a] Pr,[7a] Nd,[7b] and Yb[7a]], determined by single crystal X-ray diffraction, confirm the formulation (2).

References

1. M. F. Lappert and A. Singh, *J. Organomet. Chem.*, **239**, 133 (1982).
2. P. C. Blake, M. F. Lappert, R. G. Taylor, J. L. Atwood, W. E. Hunter, and H. Zhang, *J. Chem. Soc. Chem. Commun.*, **1986**, 1394.
3. Compare D. F. Shriver and M. A. Drezdzon, *The Manipulation of Air-Sensitive Compounds*, 2nd ed., Wiley, New York, 1986.
4. E. W. Abel and S. Moorhouse, *J. Organomet. Chem.*, **29**, 227 (1971). The starting material for $C_5H_4(SiMe_3)_2$, which is a mixture of isomers, is $C_5H_5SiMe_3$, conveniently prepared according to ref. 5.
5. C. S. Kraihanzel and M. L. Losee, *J. Am. Chem. Soc.*, **90**, 4701 (1968).
6. M. F. Lappert, A. Singh, J. L. Atwood, and W. E. Hunter, *J. Chem. Soc. Chem. Commun.*, **1981**, 1191.
7. (a) M. F. Lappert, A. Singh, J. L. Atwood, and W. E. Hunter, *J. Chem. Soc. Chem. Commun.*, **1981**, 1190. (b) *Idem*, unpublished work.

34. BIS[η^5-1,3-BIS(TRIMETHYLSILYL)CYCLOPENTA-DIENYL]HALOURANIUM(IV) AND -THORIUM(IV), [MCp''$_2$Cl$_2$] AND [UCp''$_2$X$_2$]

$$MCl_4 + 2LiCp'' \xrightarrow{\text{THF, }20\,^\circ\text{C}} [MCp''_2Cl_2] + 2LiCl$$

$$3[UCp''_2Cl_2] + 2BX_3 \xrightarrow[20\,^\circ\text{C}]{n\text{-}C_6H_{14}} 3[UCp''_2X_2] + 2BCl_3$$

$$Cp'' = \eta^5\text{-}C_5H_3(SiMe_3)_2\text{-}1,3 \qquad M = Th \text{ or } U \qquad X = Br \text{ or } I$$
tetrahydrofuran = THF (solvent)

Submitted by PAUL C. BLAKE,* MICHAEL F. LAPPERT,* and RICHARD G. TAYLOR*
Checked by PAULETTE N. HAZIN† and JOSEPH W. BRUNO†

Substituted bis(η^5-cyclopentadienyl)uranium(IV) chlorides and their thorium analogs (Th and U are collectively abbreviated as M) are of interest because of the inaccessibility of the parent compounds MCp_2Cl_2 ($Cp = \eta^5$-C_5H_5).[1] Thus, attempts to make them from $MCl_4 + 2LiCp$ gave the products of redistribution: $\frac{1}{2}MCp_3Cl + \frac{1}{2}MCpCl_3$. However, stable peralkylated complexes such as $[M(\eta^5$-$C_5Me_5)_2Cl_2]$ and $[U(\eta^5$-$C_5Me_4Et)_2Cl_2]$ have been obtained. We had previously used the η^5-1, 3-bis(trimethylsilyl)-cyclopentadienyl ligand (abbreviated as Cp''^-) to prepare the stable early lanthanocene(III) chlorides $[\{LnCp''_2(\mu$-$Cl)\}_2]$ (Ln = La, Ce, Pr, or Nd) (corresponding complexes $LnCp_2Cl$ redistributed to yield $\frac{2}{3}LnCp_3 + \frac{1}{3}LnCl_3$).[2] Using the same ligand, the complexes $[MCp''_2Cl_2]$ were obtained and used as convenient precursors to a range of complexes: $[UCp''_2(Cl)NMe_2]$ and $[UCp''_2Y_2]$ (Y = NMe_2, CH_2Ph, or CH_2SiMe_3),[3] $[\{UCp''_2(\mu$-F$)(\mu_2$-$BF_4)\}_2]$,[4] $[\{UCp''_2(\mu$-$Cl)\}_2]$,[5] and $[ThCp''_3]$.[6]

The use of boron tribromide or triiodide as a convenient reagent for converting an anhydrous metal chloride into the appropriate heavier halide was introduced in 1967.[7]

General Procedures

The experiments are all performed in an atmosphere of pure dry argon [used directly from a cylinder (Air Products)] using Schlenk techniques.[8] Glassware is successively oven-dried, flame-dried, and cooled under argon. Solvents are reagent grade, or better, and are rigorously freed from oxygen or moisture by heating under reflux with sodium benzophenone.

A. BIS[η^5-1, 3-BIS(TRIMETHYLSILYL)CYCLOPENTADIENYL]-DICHLOROTHORIUM(IV)

Procedure

■ **Caution.** *All operations must be carried out under strictly anhydrous oxygen-free conditions. Organolithium compounds in hydrocarbon solvents are spontaneously flammable in air. All uranium and thorium compounds are toxic.*

*School of Chemistry and Molecular Sciences, University of Sussex, Brighton BN1 9QJ, United Kingdom.
†Department of Chemistry, Wesleyan University, Middletown, CT 06457.

These substances should be manipulated in a well-ventilated fume hood, and gloves should be worn.

Solid [η^5-1, 3bis(trimethylsilyl)cyclopentadienyl]lithium (LiCp″, 7.26 g, 33.6 mmol, prepared as in Section 33A)[9] is added via an addition tube to a stirred, cooled (0° C) slurry of anhydrous thorium(IV) chloride (Cerac) (6.27 g, 16.8 mmol) in diethyl ether (250 mL). The mixture is stirred at 25 °C for 15 h. Volatile substances are removed at 25 °C and 10^{-2} torr. The resultant off-white foam is extracted into warm (50 °C) hexane ($\sim$ 150 mL), and the extract is filtered. The filtrate is concentrated (to $\sim$ 75 mL), then cooled to -30° C to yield colorless crystals of [ThCp″$_2$Cl$_2$] (1). Yield: 4.45 g (40%). A further crop of crystals is obtained from the mother liquor (total yield, 60%). For its characterization, see Table I.

B. BIS(η^5-1, 3-BIS(TRIMETHYLSILYL)CYCLOPENTADIENYL]-DICHLOROURANIUM(IV)

Procedure

A solution of [η^5-1, 3-bis(trimethylsilyl)cyclopentadienyl]lithium is prepared *in situ*, using the procedure described in Section 33.A, from butyllithium (18.0 mL of a 1.5 M solution in C$_6$H$_{14}$, 29 mmol) and bis(trimethylsilyl)cyclopentadiene (5.89 g, 28 mmol) in THF (100 mL). This solution is added dropwise during $\sim\frac{1}{2}$ h to a stirred, cooled (0 °C) solution of uranium(IV) chloride (5.45 g, 14 mmol) in THF (100 mL). The mixture is stirred at 25 °C for 12 h. Removal of the solvent at 25 °C and 10 torr affords a sticky brown residue; this residue is warmed to 60 °C at 10^{-2} torr. The brown involatile solid is extracted into warm toluene (100 mL), and the extract is filtered. The brown filtrate is concentrated (to $\sim$ 50 mL), then cooled to -30 °C to yield large orange-brown crystals of [UCp″$_2$Cl$_2$] (2). Yield: 6.1 g (60%). Two further crops of crystals are obtained from the mother liquor (total yield, 90%).* For its characterization, see Table I.

C. BIS[η^5-1, 3-BIS(TRIMETHYLSILYL)CYCLOPENTADIENYL]-DIBROMOURANIUM(IV)

Procedure

■ **Caution.** *Boron tribromide (like BI$_3$) reacts violently with water to yield hydrogen bromide (or HI from BI$_3$) and orthoboric acid. It must be handled with care, avoiding contact with air or moisture.*

---

*The checkers found it easier to obtain the product by crystallization from hexane. Three crops gave a total yield of 80%.

TABLE I Some Properties of [MCp″$_2$Cl$_2$] (M = Th or U) and [UCp″$_2$X$_2$] (X = Br or I)

Compound	Melting Point (°C)	Color	Calculated (%) C	Calculated (%) H	Found (%) C	Found (%) H	¹H NMR Chemical Shifts (δ)[a] SiMe$_3$	H¹ = H²	H³
[ThCp″$_2$Cl$_2$] (1)	150–152	White	36.6	5.9	36.6	6.0	+0.35 (s)	+6.93 (d)	+7.32 (t)
[UCp″$_2$Cl$_2$] (2)	140–147 (dec)	Orange-brown	36.3	5.8	36.0	5.6	−2.55 (s)	−33.4 (s)	+87.3 (s)
[UCp″$_2$Br$_2$] (3)	160–163	Red-brown	32.4	5.0	32.3	5.2	−1.78 (s)	−37.1 (s)	+99.5 (s)
[UCp″$_2$I$_2$] (4)	180–183	Dark red	29.0	4.6	29.1	4.8	+0.74 (s)	−36.2 (s)	+99.7 (s)

[a]In toluene $-d_8$ at 80 MHz at 305 K; H¹ and H² are signals from protons attached to C-4 and C-5, while H³ is attached to C-2 of the 1,3-bis(trimethylsilyl)cyclopentadienyl ligand (abbreviations: s = singlet, d = doublet, t = triplet).

175

A sample of [UCp"$_2$Cl$_2$] (2.73 g, 3.8 mmol) is placed in a 200-mL Schlenk tube containing a magnetic stirrer. Hexane (100 mL) is added from a dropping funnel, fitted with a pressure equalization attachment. Boron tribromide (0.4 mL, an approximate twofold excess) in hexane (10 mL) is introduced into the dropping funnel and is added dropwise to the solution of [UCp"$_2$Cl$_2$] in C$_6$H$_{14}$ at 20 °C. The mixture is stirred at 20 °C for 2 h and filtered. The filtrate is concentrated (to ~ 25 mL) at 20 °C and 10 torr. It is then cooled to − 30 °C to yield red-brown needles of [UCp"$_2$Br$_2$] (3). Yield: 2.67 g (87%), which are washed with cold (− 78 °C) C$_5$H$_{12}$, and dried *in vacuo*. For its characterization, see Table I.

D. BIS[η^5-1, 3-BIS(TRIMETHYLSILYL)CYCLOPENTADIENYL]-DIIODOURANIUM(IV)

Procedure

Using the procedure and apparatus described in Section C (*Note the* **Caution** in Section C), boron triiodide (1.13 g, 2.9 mmol) in hexane (10 mL) is added gradually to a stirred solution of [UCp"$_2$Cl$_2$] (2.10 g, 2.9 mmol) in C$_6$H$_{14}$ (100 mL) at 20 °C. The mixture is stirred at 20° C for 4 h, filtered, and the filtrate is concentrated to ~ 20 mL. Cooling the filtrate to − 30 °C yields red cubes of [UCp"$_2$I$_2$] (4). Yield: 2.17 g (82%), which are washed with cold (− 78 °C) C$_5$H$_{12}$ and dried *in vacuo*. For its characterization, see Table I.

Properties of the Metallocene(IV) Halides (1) to (4)

Compounds (1) to (4) are sensitive to air or moisture but are thermally stable. Some data are summarized in Table I.

The structures of crystalline [MCp"$_2$Cl$_2$] [M = Th (1) or U (2)] and [UCp"$_2$X$_2$] [X = Br (3) or I (4)], determined by single crystal X-ray diffraction, confirm them to have a distorted tetrahedral geometry around the metal atom (taking the centroid of the Cp"$^-$ ligand as occupying a single coordination site).[11]

References

1. Compare T. J. Marks and R. D. Ernst, in *Comprehensive Organometallic Chemistry*, Vol. 3, G. Wilkinson, F. G. A. Stone, and E. W. Abel (eds.), Pergamon Press, Oxford, 1982, Chapter 21.
2. M. F. Lappert, A. Singh, J. L. Atwood, and W. E. Hunter, *J. Chem. Soc. Chem. Commun.*, **1981**, 1190; M. F. Lappert and A. Singh, *Inorg. Synth.*, **27**, 168 (1990).
3. P. B. Hitchcock, M. F. Lappert, A. Singh, R. G. Taylor, and D. Brown, *J. Chem. Soc. Chem. Commun.*, **1983**, 561.

4. P. B. Hitchcock, M. F. Lappert, and R. G. Taylor, *J. Chem. Soc. Chem. Commun.*, **1984**, 1082.

5. P. C. Blake, M. F. Lappert, R. G. Taylor, J. L. Atwood, W. E. Hunter, and H. Zhang, *J. Chem. Soc. Chem. Commun.*, **1986**, 1394.

6. P. C. Blake, M. F. Lappert, J. L. Atwood, and H. Zhang, *J. Chem. Soc. Chem. Commun.*, **1986**, 1148.

7. P. M. Druce, M. F. Lappert, and P. N. K. Riley, *J. Chem. Soc. (A)*, **1967**, 486; P. M. Druce and M. F. Lappert, *J. Chem. Soc. (A)*, **1971**, 3595.

8. Compare D. F. Shriver and M. A. Drezdzon, *The Manipulation of Air-sensitive Compounds*, 2nd ed., Wiley, New York, 1986.

9. E. W. Abel and S. Moorhouse, *J. Organomet. Chem.*, **29**, 227, (1971); the starting material for $C_5H_4(SiMe_3)_2$ is $C_5H_5SiMe_3$, conveniently prepared according to ref. 10.

10. C. S. Kraihanzel and M. L. Losee, *J. Am. Chem. Soc.*, **90**, 4701, (1968).

11. P. C. Blake, M. F. Lappert, R. G. Taylor, J. L. Atwood, W. E. Hunter, and H. Zhang, *Inorg. Chim. Acta.*, **139**, 13, (1987); P. C. Blake, M. F. Lappert, R. G. Taylor, J. L. Atwood, W. E. Hunter, and H. Zhang, *J. Chem. Soc. Dalton Trans.*, in press.

35. TRIS(η⁵-CYCLOPENTADIENYL)[(DIMETHYLPHENYL-PHOSPHORANYLIDENE)METHYL]URANIUM(IV)

$$[P(CH_3)_3(C_6H_5)]I + 2CH_3Li$$

$$\longrightarrow Li(CH_2)(CH_2)P(CH_3)(C_6H_5) + 2CH_4 + LiI$$

$$U(\eta^5\text{-}C_5H_5)_3Cl + Li(CH_2)(CH_2)P(CH_3)(C_6H_5)$$

$$\longrightarrow U(\eta^5\text{-}C_5H_5)_3[CHP(CH_3)_2(C_6H_5)] + LiCl$$

Submitted by ROGER E. CRAMER,* JONG HWA JEONG,* RICHARD B. MAYNARD,* and JOHN W. GILJE*
Checked by DWIGHT D. HEINRICH† and JOHN P. FACKLER, JR.†

The only compounds yet reported that contain an *f*-element carbon multiple bond are the phosphoylide complexes $U(\eta^5\text{-}C_5H_5)_3[CHP(CH_3)_3]$, $U(\eta^5\text{-}C_5H_5)_3[CHP(CH_3)_2(C_6H_5)]$, and $U(\eta^5\text{-}C_5H_5)_3[CHP(CH_3)(C_6H_5)_2]$.[1] The most general preparative route which has been developed to date is the reaction of $U(\eta^5\text{-}C_5H_5)_3Cl$ with $Li(CH_2)(CH_2)PRR'$.[2] Obviously this method can be applied only to ylides that form well-defined lithio derivatives. Since

*Department of Chemistry, 2545 The Mall, University of Hawaii, Honolulu, HI 96822. The support of this work by the National Science Foundation through Research Grants CHE 83-10244 and CHE 85-19289 and by the donors of the Petroleum Research Fund, administered by the American Chemical Society, is gratefully acknowledged.
†Department of Chemistry, Texas A & M University, College Station, TX 77843.

many phosphoylides are prone to substituent exchange during lithiation, or undergo lithiation at several sites,[3] there are combinations of R and R' that are unsuited for the preparation, and so far only the three compounds mentioned above are well characterized.

Although the direct reaction of $H_2C=P(CH_3)_3$ with $U(\eta^5-C_5H_5)_3Cl$ does produce $U(\eta^5-C_5H_5)_3[CHP(CH_3)_3]$,

$$U(\eta^5-C_5H_5)_3Cl + 2H_2C=P(CH_3)_3$$
$$\longrightarrow U(\eta^5-C_5H_5)_3[CHP(CH_3)_3] + P(CH_3)_4Cl$$

similar reactions do not occur with less basic ylides. For example, $U(\eta^5-C_5H_5)_3[CHP(C_6H_5)_3]$ has not been isolated from the reaction of $U(\eta^5-C_5H_5)_3UCl$ with $H_2C=P(C_6H_5)_3$.[4]

Corresponding reactions have failed to produce thorium analogs,[4] even though the chemistries of $Th(\eta^5-C_5H_5)_3Cl$ and $U(\eta^5-C_5H_5)_3Cl$ are often similar.

Procedure

▪ **Caution.** *All reactants and products are very moisture and/or oxygen sensitive. Consequently, solvents and glassware must be carefully dried and deoxygenated before use, and all operations must be performed under an anhydrous nitrogen or argon atmosphere.[5] Although the physiological effects of the materials used in this synthesis are unknown, all uranium and many phosphorus compounds are toxic. The compounds $Li(CH_2)(CH_2)P(CH_3)(C_6H_5)$ and $U(\eta^5-C_5H_5)_3[CHP(CH_3)_2(C_6H_5)]$ are pyrophoric and produce fine dusts and volatile substances upon exposure to air. Care should be taken to avoid ingesting or breathing these substances, and manipulations should be carried out in a fume hood or in a well-vented dry box.[5]*

A. $Li(CH_2)(CH_2)P(CH_3)(C_6H_5)$

The compound $[(CH_3)_3P(C_6H_5)]I$ is obtained from the reaction of $(CH_3)_2P(C_6H_5)$ (ref. 6) with CH_3I using the procedure described for the preparation of $[P(CH_3)_4]Br$.[7] The lithio derivative, $Li(CH_2)(CH_2)P(CH_3)(C_6H_5)$, is conveniently prepared in a dry box by a method that appears in the literature.[8] Using an addition funnel, 28 mmol (18 mL of a 1.6 M solution) of CH_3Li in diethyl ether (Aldrich) is added dropwise to a stirred suspension of 4.0 g (14 mmol) of $[(CH_3)_3P(C_6H_5)]I$ and 50 mL of diethyl ether in a 200-mL flask.

▪ **Caution.** *In order to avoid pressure buildup resulting from the formation of CH_4 in this reaction, the apparatus should not be tightly closed.*

Since methyllithium converts the diethyl ether insoluble $[(CH_3)_3P(C_6H_5)]I$ into the diethyl ether soluble $ILi \cdot CH_2P(CH_3)_2(C_6H_5)$, a clear yellow solution results after the addition of one equivalent of CH_3Li. As the remaining CH_3Li is added, $Li(CH_2)(CH_2)P(CH_3)(C_6H_5)$ precipitates. After the CH_3Li addition is complete, the solution is stirred for 5 h. The white precipitate is separated by filtration through a fine porosity frit and washed with 300 mL of diethyl ether in order to remove any LiI. Yield: 1.8 g (80%)* of white, pyrophoric $Li(CH_2)(CH_2)P(CH_3)(C_6H_5)$. This material is extremely basic, and reacts readily with proton sources to form yellow $CH_2=P(CH_3)_2(C_6H_5)$. Samples of $Li(CH_2)(CH_2)P(CH_3)(C_6H_5)$ which have a distinct yellow coloration are partially decomposed and should not be used in the following preparation. However, tetrahydrofuran (THF) solutions of $Li(CH_2)(CH_2)P(CH_3)(C_6H_5)$ are yellow so that coloration need not indicate decomposition in such solutions.

B. $U(\eta^5-C_5H_5)_3[CHP(CH_3)_2(C_6H_5)]$

In a dry box, a pressure-equalizing addition funnel containing a solution of 0.314 g (1.99 mmol) of $Li(CH_2)(CH_2)P(CH_3)(C_6H_5)$ in 10 mL of THF is attached to a 100-mL Schlenk flask containing a stirring bar and 0.932 g (1.99 mmol) of $U(\eta^5-C_5H_5)_3Cl$ (ref. 9) dissolved in 15 mL of THF. The flask is removed from the dry box, attached to a Schlenk double-manifold vacuum line,[5] and placed in an acetone–Dry Ice bath. The $Li(CH_2)(CH_2)P(CH_3) \cdot (C_6H_5)$ solution is added dropwise over 30 min to the well cooled, stirred $U(\eta^5-C_5H_5)_3Cl$ solution. After 3 h, the mixture is allowed to warm, while volatile components are evaporated under vacuum. In order to prevent the formation of side products, the flask should not be allowed to reach room temperature until all of the solvent has been removed. The green solid residue is extracted with 25 mL of benzene. ■ **Caution.** *Benzene is a human carcinogen. It should be handled only in a well-ventilated hood, and gloves should be worn.* The benzene is then evaporated *in vacuo*, leaving 0.935 g (1.60 mmol, 80%)† of $U(\eta^5-C_5H_5)_3[CHP(CH_3)_2(C_6H_5)]$ which, by NMR, contains only minor impurities.

Further purification is achieved by stirring the crude material with 15 mL of toluene for ~ 15 min, filtering the solution, and adding 30 mL of heptane to the filtrate. Long, dark green needles of $U(\eta^5-C_5H_5)_3[CHP(CH_3)_2(C_6H_5)]$ form after this solution has been stored at ~ − 15 °C for a day. After filtration, washing with 20 mL of hexane, and drying under vacuum, 0.605 g (52%) $U(\eta^5-C_5H_5)_3[CHP(CH_3)_2(C_6H_5)]$ is obtained. After evaporation of the

---

*The checkers obtained a 61% yield.
†The checkers obtained a 70% yield.

filtrate, extraction of the residue with 5 mL of toluene, and addition of 10 mL of heptane, a second crop of product, 0.160 g (14%), is obtained by an analogous recrystallization.

Anal. Calcd. for $UPC_{24}H_{27}$: C, 49.31; H; 4.67; P; 5.30. Found: C, 49.53; H, 4.91; P; 5.22.

Properties

The complex $U(\eta^5\text{-}C_5H_5)_3[CHP(CH_3)_2(C_6H_5)]$ is a dark green solid that is very oxygen and moisture sensitive and pyrophoric. It is soluble in ethereal and aromatic solvents, but almost insoluble in aliphatic hydrocarbons. ^1H NMR (23 °C, C_6D_6, ppm from TMS): $P(C_6H_5)$, 18.0 (dd) (ortho), 11.1 (t) (meta), 9.1 (t) (para); PCH_3, -8.5 (d) $J_{PCH} = 12$ Hz; $(C_5H_5)_3U$, -12.7 (s); PCH, -117 (d), $J_{HCP} = 16$ Hz.

The preparations of $U(\eta^5\text{-}C_5H_5)_3[CHP(CH_3)_3]$ and $U(\eta^5\text{-}C_5H_5)_3\cdot[CHP(CH_3)(C_6H_5)_2]$ are exactly analogous to the one for $U(\eta^5\text{-}C_5H_5)_3[CHP(CH_3)_2(C_6H_5)]$. The ^1H NMR spectra of these compounds are, $U(\eta^5\text{-}C_5H_5)_3[CHP(CH_3)_3]$: PCH_3, -2.9 d $J_{PCH} = 12$ Hz; $(\eta^5\text{-}C_5H_5)_3U$, -13.2 (s); PCH, -108 (d) $J_{HCP} = 13$ Hz. $U(\eta^5\text{-}C_5H_5)_3[CHP(CH_3)(C_6H_5)_2]$: $P(C_6H_5)$, 10.8 (dd) (ortho), 8.4 (t) (meta), 7.5 (t) (para); $(\eta^5\text{-}C_5H_5)_3U$, -12.1 (s); PCH_3, -13.8 (d) $J_{PCH} = 12$ Hz; PCH, -131 (d), $J_{HCP} = 16$ Hz.

The purity of the $U(\eta^5\text{-}C_5H_5)_3[CHP(CH_3)_2(C_6H_5)]$ depends on careful control of the temperature and stoichiometry of the reactants. If an excess of $U(\eta^5\text{-}C_5H_5)_3Cl$ is used it will remain as an impurity that is difficult to remove from the final product. If $Li(CH_2)(CH_2)P(CH_3)(C_6H_5)$ is in excess or if the temperatures exceed those described, $U(\eta^5\text{-}C_5H_5)_2[(CH)P(CH_3)\cdot(C_6H_5)(CH_2)]_2U(\eta^5\text{-}C_5H_5)_2$ and/or $U(\eta^5\text{-}C_5H_5)[(CH_2)(CH_2)P(CH_3)\cdot(C_6H_5)]_3$ may form.[2] The presence of these materials is indicated by NMR resonances of their $(\eta^5\text{-}C_5H_5)$ groups: $U(\eta^5\text{-}C_5H_5)_3UCl$, -2.8 ppm; $U(\eta^5\text{-}C_5H_5)_2[(CH)P(CH_3)(C_6H_5)(CH_2)]_2U(\eta^5\text{-}C_5H_5)_2$, -26.8 ppm; and $U(\eta^5\text{-}C_5H_5)[(CH_2)(CH_2)P(CH_3)(C_6H_5)]_3$, -16.3 ppm. The impurities $CH_2{=}P(CH_3)_2(C_6H_5)$, $LiI{\cdot}CH_2{=}P(CH_3)_2(C_6H_5)$, C_5H_6, and/or $(C_5H_6)_2$, which will form if traces of moisture or other protic solvents are present, are indicated if the ^1H NMR spectrum contains peaks between 8 and -1 ppm. Although recrystallization of $U(\eta^5\text{-}C_5H_5)_3[CHP(CH_3)_2(C_6H_5)]$ from a 1:2 toluene–heptane mixture removes the bulk of these impurities, it is difficult to eliminate all traces of them.

The uranium–carbon multiple bond has an extensive insertion chemistry with polar unsaturated molecules including carbon monoxide,[10] nitriles,[11] isocyanides,[12] and isocyanates.[13] Metal carbonyls also insert into this bond to form metallaphosphoniumenolates,[14,15] which undergo novel reactions including C—O bond cleavage,[16] metal migrations,[15] and carbonyl coupling reactions.[17] The carbon atom is nucleophilic and can deprotonate weak acids to form Cp_3U complexes of the conjugate base of the acid.[18]

References

1. (a) R. E. Cramer, R. B. Maynard, J. C. Paw, and J. W. Gilje, *Organometallics*, **2**, 1336 (1983). (b) R. E. Cramer, R. B. Maynard, J. C. Paw, and J. W. Gilje, *J. Am. Chem. Soc.*, **103**, 3589, (1981). (c) M. A. Bruck, F. Edelmann, D. Afzal, R. E. Cramer, J. W. Gilje, and H. Schmidbaur, *Chem. Ber.* **121**, 417 (1988).

2. R. E. Cramer, R. B. Maynard, and J. W. Gilje, *Inorg. Chem.*, **20**, 2466 (1981).

3. See, for example, B. Schaub, T. Jenny, and M. Schlosser, *Tetrahedron Lett.*, **25**, 4097 (1984).

4. R. E. Cramer and J. W. Gilje, unpublished results.

5. D. F. Shriver and M. A. Drezdzon, *The Manipulation of Air Sensitive Compounds*, 2nd ed., Wiley-Interscience, New York, 1986.

6. (a) M. A. Mathur, W. H. Myers, H. H. Sisler, and G. E. Ryschkewitsch, *Inorg. Synth.*, **15**, 132 (1974). (b) C. Frajerman and R. Meunier, *Inorg. Synth.*, **22**, 133 (1983).

7. H. F. Klein, *Inorg. Synth.*, **18**, 138 (1978).

8. L. E. Manzer, *Inorg. Chem.*, **15**, 2567 (1976).

9. T. J. Marks, A. M. Seyam, and W. A. Wachter, *Inorg. Synth.*, **16**, 147 (1976).

10. R. E. Cramer, R. B. Maynard, J. C. Paw, and J. W. Gilje, *Organometallics*, **1**, 869 (1982).

11. R. E. Cramer, K. Panchanatheswaran, and J. W. Gilje, *J. Am. Chem. Soc.*, **106**, 1853 (1984).

12. R. E. Cramer, K. Panchanatheswaran, and J. W. Gilje, *Angew. Chem. Int. Ed. Engl.*, **23**, 912 (1984).

13. R. E. Cramer, J. H. Jeong, and J. W. Gilje, *Organometallics*, **6**, 2010 (1987).

14. R. E. Cramer, K. T. Higa, and J. W. Gilje, *J. Am. Chem. Soc.*, **106**, 7245 (1984).

15. R. E. Cramer, J. H. Jeong, and J. W. Gilje, *Organometallics*, **5**, 2555 (1986).

16. R. E. Cramer, K. T. Higa, and J. W. Gilje, *Organometallics*, **4**, 1140 (1985).

17. R. E. Cramer, K. T. Higa, S. L. Pruskin, and J. W. Gilje, *J. Am. Chem. Soc.*, **105**, 6749 (1983).

18. R. E. Cramer, U. Engelhardt, K. T. Higa, and J. W. Gilje, *Organometallics*, **6**, 41 (1987).

Chapter Five

TRANSITION METAL CLUSTER COMPLEXES

36. TETRAIRON CARBIDO CARBONYL CLUSTERS

Submitted by ERNESTINE W. HILL* and JOHN S. BRADLEY*
Checked by JUANITA CASSIDY† and KENTON H. WHITMIRE†

The reactivity of the μ_4-carbido ligand in butterfly shaped tetrairon clusters has given rise to an intriguing class of organometallic clusters in which the organic fragment is bound to an open fourfold cluster site.[1] The utility of this chemistry for syntheses is determined by the ease of access to the Fe_4C starting materials. During our investigations into the chemistry of μ_4-carbido clusters, we have developed a preparative procedure for these materials which allows the isolation of the clusters in reproducibly high yields, from readily available starting materials.

The common precursor to $Fe_4(CO)_{13}C$, $[Fe_4(CO)_{12}C]^{2-}$, $[HFe_4(CO)_{12}C]^-$, and $HFe_4(CO)_{12}CH$ is $[Fe_4(CO)_{12}C(CO_2CH_3)]^-$, prepared in good yield from $[Fe_6(CO)_{16}C]^{2-}$. The latter is readily available by published procedures from $Fe(CO)_5$.[2] The syntheses of both the precursor clusters are included here, with those minor modifications to the published methods that have proved useful in our hands.

■ **Caution.** *All manipulations involving metal carbonyls must be carried out in a well-ventilated fume hood. All of the procedures described here should be performed using standard Schlenk glassware under an atmosphere of*

*Exxon Research and Engineering Company, Annandale, NJ 08801.
†Department of Chemistry, Rice University, Houston, TX 77251.

nitrogen,[3] *although the clusters, once isolated, can be handled for short periods in air. However, it is not uncommon in the manipulation of reduced iron cluster compounds for minute amounts of pyrophoric iron to be formed, and to preclude the loss by fire of carefully prepared material, inert handling techniques*[3] *should be employed whenever possible.*

A.   BIS(TETRAETHYLAMMONIUM) CARBIDOHEXADECA-CARBONYLHEXAFERRATE(2−), $(Et_4N)_2[Fe_6(CO)_{16}C]$

$$6Fe(CO)_5 + 2NaMn(CO)_5$$

$$\longrightarrow Na_2Fe_6(CO)_{16}C + Mn_2(CO)_{10} + 12CO + CO_2$$

$$Na_2Fe_6(CO)_{16}C + 2Et_4NBr \longrightarrow (Et_4N)_2[Fe_6(CO)_{16}C] + 2NaBr$$

Procedure

A solution of 9.0 g (0.041 mol) of $Na[Mn(CO)]_5$ (ref. 4) in 200 mL of diglyme* (dried by refluxing over sodium for 24 h and then distilling) is prepared under nitrogen in a dry 500-mL three-necked round-bottomed flask fitted with a reflux condenser, a nitrogen inlet, and a rubber septum. To the magnetically stirred solution is added 16.0 mL (23.8 g, 0.122 mol) of $Fe(CO)_5$ from a hypodermic syringe, and the solution is heated to reflux for 4 h. During this time the color of the reaction mixture changes to dark brown (after 10 min), deep blue (characteristic of the intermediate $[MnFe_2(CO)_{12}]^-$), and finally deep red-purple. The solution is cooled to room temperature, added to 800 mL of water, and filtered through a 2-cm deep bed of Celite® (Ace Scientific). The Celite is washed with water until the washings are colorless, and a saturated solution of 10 g of Et_4NBr in water is added to the filtrate. The resulting purple precipitate is collected by filtration on a bed of Celite as above, and washed with 100 mL of water. After drying under reduced pressure (0.1 torr) in a vacuum dessicator the solids are extracted into methanol (1 L), and the product is crystallized by reducing the volume of the resulting solution to 250 mL. The black crystalline product is filtered, washed with 15 mL of cold methanol, and dried under reduced pressure (0.1 torr). Yield: 18.5 g (86% based on $Fe(CO)_5$). The initial filtration and washing procedures may be carried out in air for convenience; the small amount of orange oxidation product, which forms on the solid product, is easily removed by washing with 0.1 N HCl and water prior to drying and recrystallization. This procedure yields a product of adequate purity for use in the next step.

*Diglyme = 1, 1′-oxybis(2-methoxyethane).

Anal. Calcd. for $C_{33}H_{40}O_{16}N_2Fe_6$: C, 37.53; H, 3.82; N, 2.65. Found: C, 38.16; H, 3.82; N, 2.65.

Salts of the cluster dianion with other cations may be isolated by use of the appropriate water soluble salt in place of Et_4NBr.

Properties

The compound $(Et_4N)_2[Fe_6(CO)_{16}C]$ is a black slightly air sensitive crystalline solid, soluble in polar organic solvents to give deep purple solutions. Infrared (CH_2Cl_2): 2032.3 (w), 1966.7 (s), 1929.6 (sh), and 1767.0 (w) cm^{-1}. ^{13}C NMR $[(CD_3)_2CO]$: 228 ppm (carbonyl), and 485 ppm (carbide). (All ^{13}C spectra are at 25 °C, 25 MHz.) The structure of the dianion has been determined by single crystal X-ray diffraction.[5] It comprises a distorted octahedron of iron atoms surrounding a central carbon atom, with 12 terminal and 4 bridging carbonyl groups.

B. TETRAETHYLAMMONIUM DODECACARBONYL-[μ₄-(METHOXYCARBONYL)METHYLIDYNE]-TETRAFERRATE(1−), $Et_4N[Fe_4(CO)_{12}C(CO_2Me)]$

$$(Et_4N)_2[Fe_6(CO)_{16}C] + 6Fe^{3+} + MeOH$$

$$\longrightarrow Et_4N[Fe_4(CO)_{12}C(CO_2Me)] + 8Fe^{2+} + H^+ + Et_4N^+ + 3CO$$

Procedure

Various oxidizing agents are effective in this reaction, but the presence of bromide ion is essential to ensure fragmentation of the cluster to the Fe_4 stage.[6] A solution of $(Et_4N)_2[Fe_6(CO)_{16}C]$ (10.0 g, 9.5 mmol) in 700 mL methanol is prepared in a 2-L two-necked flask fitted with a nitrogen inlet. Tetraethylammonium bromide (10 g, excess) is added, followed by anhydrous ferric chloride (10.8 g, 7 equiv) in portions of ~ 1 g over a period of 15 min. The purple solution of the starting material becomes green-black, and after a further 15 min the reaction mixture is evaporated to dryness under reduced pressure (~ 10 torr) and washed with distilled water. The black residue is dried under reduced pressure (0.1 torr) for 1 h, washed with diethyl ether until the washings are colorless, and recrystallized from dichloromethane–hexane (50:50) by slow evaporation in a stream of nitrogen. Yield: 5.2 g (70%).

Anal. Calcd. for $C_{23}H_{23}O_{14}Fe_4N$: C, 36.31; H, 3.04; N, 1.84. Found: C, 36.50; H, 3.04; N, 1.84.

Properties

$Et_4N[Fe_4(CO)_{12}C(CO_2Me)]$ is a black, slightly air-sensitive crystalline solid, soluble in polar organic solvents to give dark green solutions. Infrared (CH_2Cl_2): 2068 (w), 2025 (s), 1993 (s), 1986 (sh), 1921 (m), and 1650 (w) cm^{-1}. ^{13}C NMR (CD_2Cl_2): 237 ppm (methylidyne), 214 ppm (carbonyl), and 175 ppm (carbomethoxy carbonyl). The crystal structure of the anion has been determined by single crystal X-ray diffraction[7] to comprise a butterfly shaped array of iron atoms with the (methoxycarbonyl)methylidyne group located between the wingtips of the butterfly, in the plane of the backbone of the butterfly, the methylidyne carbon atom being essentially equidistant from the four iron atoms. The 12 carbonyl ligands are all terminal.

C. CARBIDOTRIDECACARBONYLTETRAIRON, $Fe_4(CO)_{13}C$

$$Et_4N[Fe_4(CO)_{12}C(CO_2Me)] + CF_3SO_3H$$
$$\longrightarrow Fe_4(CO)_{13}C + MeOH + (Et_4N)(O_3SCF_3)$$

Procedure

■ **Caution.** *Trifluoromethanesulfonic acid causes severe skin and eye burns; its vapor is an eye and respiratory irritant. Wear gloves and safety goggles, and work in a well-ventilated fume hood when handling this reagent.*

Dichloromethane (100 mL) is added to $Et_4N[Fe_4(CO)_{12}C(CO_2Me)]$ (1.0 g, 1.3 mmol) in a 250-mL flask fitted with a nitrogen inlet and a rubber septum. Trifluoromethanesulfonic acid (0.25 mL) is added, and the dark brown mixture is stirred for 5 min. Distilled water (100 mL) is added, and the two-phase mixture is stirred rapidly for several minutes. The organic layer is collected and evaporated to dryness. The residue is extracted with warm hexane (50 °C, 3 × 50 mL) and the brown hexane solution is cooled to −20 °C for several h in a refrigerator. The product crystallizes as black prisms that are collected by filtration, washed with cold hexane (10 mL), and dried. Yield: 0.75 g (95%).

Anal. Calcd. for $C_{14}O_{13}Fe_4$: C, 28.04; Fe, 37.26. Found: C, 27.56; Fe, 37.28.

Properties

The compound $Fe_4(CO)_{13}C$ is a black, slightly air-sensitive solid, soluble in organic solvents to give brown solutions, which react readily with alcohols[7] to form anionic clusters of the type $[Fe_4(CO)_{12}C(CO_2R)]^-$. Infrared

(CH_2Cl_2): 2060 (sh), 2049 (s), 2032 (s), 2009 (sh), 1983 (w), and 1895 (m) cm^{-1}; (hexane) 2060 (s), 2048 (s), 2038 (s), 2031 (s), 2012 (w), 1998 (m), 1987 (m), 1954 (w), and 1898 (m) cm^{-1}. ^{13}C NMR (CD_2Cl_2): 208.4 ppm (carbonyl), and 468.9 ppm (carbide). The structure of the cluster has been determined by single crystal X-ray diffraction[7] to comprise a butterfly core of four iron atoms with the carbide carbon atom situated between the wingtips of the butterfly, within bonding distance of all four iron atoms. Twelve carbonyls are terminal, 3 on each iron atom, and the 13th bridges the 2 iron atoms on the backbone of the butterfly.

D.　TETRAETHYLAMMONIUM CARBIDODODECACARBONYL-HYDRIDOTETRAFERRATE(1 −), $Et_4N[HFe_4(CO)_{12}C]$

$$(Et_4N)[Fe_4(CO)_{12}C(CO_2Me)] \xrightarrow{BH_3 \cdot THF} Et_4N[HFe_4(CO)_{12}C]$$

Procedure

■　**Caution.**　*$BH_3 \cdot THF$ reacts with water to liberate flammable gases. It should be stored in a refrigerator, protected from atmospheric moisture by a rubber septum, and handled in a well-ventilated hood.*

To a solution of $Et_4N[Fe_4(CO)_{12}C(CO_2Me)]$ (1.0 g, 1.3 mmol) in dry (distilled from sodium benzophenone under dry nitrogen) tetrahydrofuran (THF) (50 mL) is prepared in a 100-mL flask fitted with a nitrogen inlet and a rubber septum. To the solution is added 1.5 mL of a 1.0 M THF solution of $BH_3 \cdot THF$ (Aldrich). The solution is stirred at room temperature for 30 min, during which time the initially green solution becomes dark red. After removal of the solvent under reduced pressure the gummy residue is extracted into CH_2Cl_2 (25 mL) and hexane (10 mL) is added. The volume of the solution is slowly reduced to 15 mL in a steam of nitrogen, and crystallization is completed by cooling to − 20 °C in a refrigerator for several h. Yield: 0.85 g (93%).

Anal. Calcd. for $C_{21}H_{21}O_{12}Fe_4N$: C, 35.88; H, 3.00; N, 1.99. Found: C, 35.38; H, 3.00; N, 1.95.

Properties

The compound $Et_4N[HFe_4(CO)_{12}C]$ is a deep red slightly air sensitive solid, soluble in polar organic solvents to give red quite air sensitive solutions. Infrared (THF): 2015 (s), 2007 (s), 1988 (s), 1977 (sh), and 1933 (m) cm^{-1}. ^{1}H NMR (CD_2Cl_2): − 26.5 ppm (hydride). ^{13}C NMR (CD_2Cl_2): 464.2 ppm (carbide), 216.0, and 214.9 ppm (carbonyl). The molecular structure of the anion has been determined by X-ray diffraction[8] to comprise a butterfly core of four iron atoms each bonded to the carbide carbon atom that sits between the wingtips of the cluster. Three terminal carbonyl groups are bonded to each iron atom, and the hydride ligand bridges the backbone of the butterfly.

E. BIS(TETRAETHYLAMMONIUM) CARBIDODODECA-CARBONYLTETRAFERRATE(2−), $(Et_4N)_2[Fe_4(CO)_{12}C]$

$$Et_4N[HFe_4(CO)_{12}C] + KOH + Et_4NBr$$

$$\longrightarrow (Et_4N)_2[Fe_4(CO)_{12}C] + KBr + H_2O$$

Procedure

A solution of $Et_4N[HFe_4(CO)_{12}C]$ (1.0 g, 1.4 mmol) in methanol (100 mL), containing excess Et_4NBr (0.60 g, 2.9 mmol) is prepared in a 250-mL flask fitted with a nitrogen inlet and a rubber septum. To the solution is added a solution of potassium hydroxide (0.08 g, 1.4 mmol) in the minimum amount of methanol. After stirring for 15 min, the red-brown crystalline precipitate of $(Et_4N)_2[Fe_4(CO)_{12}C]$ is collected by filtration, washed with cold methanol (2 × 10 mL) and diethyl ether (2 × 10 mL), and dried under reduced pressure (0.01 torr). A second crop of the dianion salt can be obtained by cooling the filtrate to −20 °C. Yield: 0.90 g (77%). The air sensitive crystals may be recrystallized from acetone–2-propanol (60:40) by slow evaporation.

Anal. Calcd. for $C_{29}H_{40}O_{12}Fe_4N_2$: C, 41.87; H, 4.85; N, 3.36. Found: C, 40.54; H, 4.85; N, 3.44.

Properties

The compound $(Et_4N)_2[Fe_4(CO)_{12}C]$ is a black, somewhat air-sensitive, crystalline solid, soluble in polar organic solvents to give brown solutions. Infrared (THF): 2024(w), 1968(s), 1945(s), 1917(w), and 1892(w) cm^{-1}. ^{13}C NMR (CD_2Cl_2): 478.0 ppm (carbide), 222.8, and 220.8 ppm (carbonyl). The dianion has characteristic butterfly geometry, with three terminal carbonyl ligands on each iron atom and the carbide carbon midway between the wingtip iron atoms. Reaction with iodomethane yields $[Fe_4(CO)_{12}CC(O)CH_3]^-$.[9]

References

1. J. S. Bradley, *Adv. Organomet. Chem.*, **22**, 1 (1983).
2. M. R. Churchill, J. Wormald, J. Knight, and M. J. Mays, *J. Am. Chem. Soc.*, **93**, 3073 (1971).

3. D. F. Shriver, *The Manipulation of Air-Sensitive Compounds*, McGraw-Hill, New York, 1969.
4. R. B. King and F. G. A. Stone, *Inorg. Syn.*, **7**, 198 (1963).
5. M. R. Churchill and J. Wormald, *J. Chem. Soc. Dalton Trans.*, **1974**, 2410.
6. J. S. Bradley, E. W. Hill, G. B. Ansell, and M. A. Modrick, *Organometallics*, **1**, 1634 (1982).
7. J. S. Bradley, G. B. Ansell, M. E. Leonowicz, and E. W. Hill, *J. Am. Chem. Soc.*, **103**, 4968 (1981).
8. E. M. Holt, K. H. Whitmire, and D. F. Shriver, *J. Organomet. Chem.*, **213**, 125 (1981).
9. J. H. Davis, M. A. Beno, J. M. Williams, J. Zimmie, M. Tachikawa, and E. L. Muetterties, *Proc. Natl. Acad. Sci. USA*, **78**, 668 (1981).

37. TETRAETHYLAMMONIUM DODECARBONYL-TRICOBALTFERRATE(1−) AND (TRIPHENYLPHOSPHINE)-GOLD(1+) DODECARBONYLTRICOBALTFERRATE(1−)

Submitted by ARTHUR A. LOW* and JOSEPH W. LAUHER*
Checked by PIERRE BRAUNSTEIN,† ANNE DEGREMONT,† and JACKY ROÉ†

$$2Fe(CO)_5 + \tfrac{7}{2}Co_2(CO)_8 \xrightarrow[\text{acetone}]{\Delta} [Co(solvent)_x][FeCo_3(CO)_{12}]_2 + 14CO$$

$$[Co(solvent)_x][FeCo_3(CO)_{12}]_2 + 2Et_4NI$$

$$\xrightarrow[H_2O]{} 2[Et_4N][FeCo_3(CO)_{12}] + [Co(solvent)_x]I_2$$

$$[Et_4N][FeCo_3(CO)_{12}] + Au(NO_3)(PPh_3)$$

$$\xrightarrow[CH_2Cl_2]{} Ph_3PAuFeCo_3(CO)_{12} + Et_4NNO_3$$

A. [Et$_4$N][FeCo$_3$(CO)$_{12}$]

The anion [FeCo$_3$(CO)$_{12}$]$^-$ was first prepared by Chini et al. in 1960[1] and was one of the first mixed metal clusters reported. The procedure outlined here is based on Chini's original method although the quantities involved have been cut 10-fold and less temperature control is specified.

*Department of Chemistry, State University of New York at Stony Brook, Stony Brook NY 11794.
†Laboratoire de Chimie de Coordination, UA 416 CNRS, Institut Le Bel, Université Louis Pasteur, 4 rue Blaise Pascal, F-67070 Strasbourg Cedex, France.

Procedure

■ **Caution.** *Reactions and transfer of iron pentacarbonyl are done in a well-ventilated fume hood because of the high toxicity of iron pentacarbonyl and carbon monoxide. The compound $Co_2(CO)_8$ is pyrophoric and is weighed out into the reaction flask in a dry nitrogen filled glove box.*

A 3.00 g sample of $Co_2(CO)_8$ (Alfa) (8.77 mmol) is placed, with a magnetic stirring bar, into a 250-mL round-bottomed side-arm flask inside a nitrogen filled glove box. The flask is then capped with a septum, brought out of the glove box, and attached to a standard dual manifold Schlenk vacuum–nitrogen line. A 1.04 g sample of $Fe(CO)_5$ (Alfa) (0.73 mL, 5.31 mmol) is added to the flask via a cannula. Then, 95 mL of deareated acetone is added dropwise through the cannula while the reaction mixture is stirred. After the acetone has been added, the septum is removed and is quickly replaced with a condenser. The solution is heated at reflux for 14 h, after which it is red-brown. The solvent is then removed by vacuum at room temperature, and the residue is pumped on for at least an additional 1.5 h to remove any unreacted iron pentacarbonyl. The remaining black powder is dissolved in the minimum amount of acetone (40–50 mL) and filtered through a Schlenk filter to remove any possible contaminents. A solution of 1.25 g of Et_4NI (Aldrich) in 75.0 mL of deareated distilled H_2O is added dropwise via a syringe. A red-brown precipitate of $[Et_4N][FeCo_3(CO)_{12}]$ forms immediately. It is filtered out of the solution through a Schlenk filter and dried *in vacuo*. The color of the filtrate should be light pink. If there is any dark color, then more Et_4NI is added to precipitate additional $[Et_4N][FeCo_3(CO)_{12}]$. The precipitate is washed with water and dried in vacuum for at least 5 h. Yield: 3.33 g (95%, based on $Co_2(CO)_8$). The compound may be recrystallized by dissolving a sample in warm dichloromethane followed by addition of an equal volume of hexane and cooling.

Anal. Calcd. for $C_{20}H_{20}FeCo_3O_{12}N$: C, 34.37; H, 2.88. Found: C, 34.47; H, 2.60.

Properties

The compound $[Et_4N][FeCo_3(CO)_{12}]$ is a reddish-brown powder. It is air stable in the solid state and in solution. It is soluble in tetrahydrofuran (THF) and dichloromethane and very soluble in acetone. It is sparingly soluble in chloroform. The IR spectrum of this compound contains the following C—O stretching vibrations (CH_2Cl_2): 2063 (w), 2007 (s), 1976 (m), 1929 (m), and 1812 (m) cm^{-1}. Acidification of solutions of the $[FeCo_3(CO)_{12}]^-$ anion produces the hydride compound $[HFeCo_3(CO)_{12}]$ as reported by Chini.[1]

B. Ph$_3$PAuFeCo$_3$(CO)$_{12}$

The synthesis of Ph$_3$PAuFeCo$_3$(CO)$_{12}$ was reported by Lauher and Wald in 1981.[2] The reported synthesis involved mixing Au(NO$_3$)(PPh$_3$) and [Et$_4$N][FeCo$_3$(CO)$_{12}$] in acetone. It was subsequently found that the reaction occurs as easily in dichloromethane. Since the product eventually dissociates in acetone, dichloromethene is the solvent of choice. The silver analog of this compound has been synthesized in the same manner.[3] The compound Ph$_3$PRuCo$_3$(CO)$_{12}$ was reported by Braunstein in 1982.[4] It was synthesized by adding a diethyl ether solution of AuCl(PPh$_3$) to a toluene suspension of Na[RuCo$_3$(CO)$_{12}$]. The same procedure can be used with K[FeCo$_3$(CO)$_{12}$] to produce the title compound.[3]

The Au(NO$_3$)(PPh$_3$) starting material is prepared by a simple AgNO$_3$ metathesis reaction with commercial AuCl(PPh$_3$) using a procedure first reported by Malatesta, et al.[5]

■ **Caution.** *Benzene is a human carcinogen, and chloroform is a suspected carcinogen. Protective gloves should be worn, and all manipulations should be carried out in a well-ventilated fume hood.*

In a open beaker, 0.72-g of AuCl(PPh$_3$) (Alfa) (1.46 mmol) is dissolved in 10 mL of dichloromethane, an ethanol solution of AgNO$_3$ (0.25 g, 1.47 mmol) is added and the resulting silver chloride precipitate is filtered off. The filtrate is evaporated to dryness by vacuum, and the residual product, Au(NO$_3$)(PPh$_3$),[5] is dissolved in 30.0 mL of dichloromethane. Under a nitrogen atmosphere using standard Schlenk techniques, the solution of Au(NO$_3$)(PPh$_3$) is added via cannula to a solution of 1.00 g of [Et$_4$N][FeCo$_3$(CO)$_{12}$] (1.43 mmol) dissolved in 10 mL of CH$_2$Cl$_2$ in a 100-ml round-bottomed sidearm flask containing a magnetic stirring bar. The color of the solution immediately changes from red-brown to a deep purple. After 5 min of stirring the solvent is removed at room temperature in vacuum leaving a dark residue. This residue is extracted under a nitrogen atmosphere with 10-mL aliquots of a 7:3 benzene–chloroform solution until the extracts are colorless (50–80 mL total should be required). The extracts are combined, and the solvent is removed in vacuum, leaving a dark black crystalline solid. Yield: 1.35 g (91%). The compound may be recrystallized by dissolving a sample in warm toluene followed by the addition of an equal quantity of hexane and cooling.

Anal. Calcd. for C$_{30}$H$_{15}$AuFeCo$_3$PO$_{12}$: C, 35.05; H, 1.47. Found: C, 35.26; H, 1.53.

Properties

The compound Ph$_3$PAuFeCo$_3$(CO)$_{12}$ is a deep purple-black color in the solid state. In solution, the color is a very intense purple. It is soluble in acetone, dichloromethane, chloroform, benzene, diethyl ether, and toluene. It is slightly soluble in hexane. In polar solvents, such a acetone, the compound dissociates into $[Au(PPh_3)(solvent)]^+$ cations and $[FeCo_3(CO)_{12}]^-$ anions. This process may be reversed by decreasing the polarity of the solution. The compound is air stable in the solid state and moderately air stable in solution. The IR spectrum of the compound in dichloromethane contains the following C—O stretching vibrations: 2072(m), 2016(vs), 1990(m), and 1860(s) cm^{-1}.

The molecular structure of this compound consists of a trigonal bipyramid of the metal atoms as determined by X-ray analysis. Solutions of the compound instantly dissociate upon addition of phosphines producing $[Ph_3PAuPR_3]^+$ cations and $[FeCo_3(CO)_{12}]^-$ anions. Similar chemistry has been observed by Braunstein et al. for $Ph_3PAuRuCo_3(CO)_{12}$.[4]

References

1. P. Chini, I., Colli, and M. Peraldo, *Gazz. Chim. Ital.*, **90**, 1005 (1960).

2. J. W. Lauher and K. Wald, *J. Am. Chem. Soc.*, **103**, 7648 (1981).

3. S. J. Sherlock, "Synthesis and Characterization of Some Novel Group IB and Mercury Mixed Metal Clusters," Ph.D. thesis, State University of New York, Stony Brook, New York, Chapter 2, 1985.

4. P. Braunstein, J. Rose, Y. Dusausoy, and J. P. Mangeot, *C. R. Acad. Sci. Ser. C*, **294**, 967 (1982). P. Braunstein, J. Rose, A. Dedieu, Y. Dusausoy, J. P. Mangeot, A. Tiripicchio, and an M. Tiripicchio-Camellini, *J. Chem. Soc. Dalton Trans.*, **1986**, 225.

5. L. Malatesta, L. Naldini, A. Simonetta, and F. Cariati, *Coord. Chem. Rev.*, **1**, 255 (1966).

38. CHIRAL TRINUCLEAR NiCoMo AND MoRuCo CLUSTERS

Submitted by H. VAHRENKAMP*
Checked by DA QUIANG XU AND H. D. KAESZ (PART A)[†]
M. L. WILLIAMS and F. G. A. STONE (PART B)[‡]

The reactivity of organometallic clusters has been studied to a point where systematization becomes possible,[1,2] and details of mechanisms and stereo-

*Institut für Anorganische Chemie der Universität, Albertstr. 21, D-7800 Freiburg, Federal Republic of Germany. Based on thesis work by H. Beurich, R. Blumhofer, W. Bernhardt, and H. Bantel.
†Department of Chemistry, University of California, Los Angeles, CA 90024.
‡Department of Inorganic Chemistry, University of Bristol, Cantock's Close, Bristol, B58 1TS, United Kingdom.

chemical pathways can be investigated. The bases for these developments are pioneering studies of organic interconversions in the ligand sphere[3] and of elementary cluster framework reactions.[4] Further developments are to be expected using functional cluster frameworks and ligands with multicenter attachment. Described herein are the syntheses of two clusters combining these two properties. Both have chiral tetrahedral frameworks consisting of an organic μ_3-ligand and three different metal atoms, and in both cases the organic μ_3 unit (alkylidine or alkyne) lends itself to further interconversions. The approach in these syntheses is based on metal exchange,[5] which allows the formation of chiral clusters with different transition metals and different μ_3 units.

General Procedures and Techniques

All materials are handled under an inert gas atmosphere in Schlenk-type vessels and in well-ventilated hoods. All solvents must be dried by Na–K alloy, degassed, and distilled. The silica gel for chromatography (Macherey-Nagel 0.06–0.2 mm) is dried for 12 h at 160 °C in high vacuum. For thin layer chromatography (TLC) tests, silica gel on aluminum (Merck-60 F_{254}) is used. The term "hexane" is used for the petroleum ether fraction boiling between 60 and 70 °C.

The syntheses of the starting clusters $(\mu_3\text{-MeC})Co_3(CO)_9$ (ref. 6) and $RuCo_2(CO)_{11}$ (ref. 7) are well documented. The reagents $Na[Mo(C_5H_5)(CO)_3]$ (ref. 8) and $[(C_5H_5)Ni(CO)]_2$ (ref. 9) can be prepared according to published procedures. 2-Butyne is commercially available. All intermediate and product clusters have been fully characterized including elemental analyses.[10-12]

▪ **Caution.** *The complex $[(C_5H_5)Ni(CO)]_2$ is volatile and should be handled with the caution associated with $Ni(CO)_4$, from which it is made. Synthesis and handling of $[(C_5H_5)Ni(CO)]_2$ should be carried out in a well-ventilated hood. Benzene is a human carcinogen. It should be handled only in a well-ventilated hood and gloves should be worn. Replacing benzene by toluene in these preparations requires longer reaction times and longer columns for chromatography.*

A. CYCLO-μ_3-ETHYLIDYNE-1:2:3-$\kappa^3 C$-PENTACARBONYL-1$\kappa^2 C$, 2$\kappa^3 C$-BIS[1, 3(η^5)-CYCLOPENTADIENYL]COBALT-MOLYBDENUMNICKEL(*Co—Mo*)(*Co—Ni*)(*Mo—Ni*), (μ_3-$CH_3 C$)NiCoMo($C_5 H_5$)$_2$(CO)$_5$

This chiral cluster[10] is obtained in a two-step metal-exchange sequence with an average yield of 25% based on $(\mu_3\text{-}CH_3 C)Co_3(CO)_9$:

$$(\mu_3\text{-}CH_3C)Co_3(CO)_9 + Na[Mo(C_5H_5)(CO)_3]$$

$$\longrightarrow (\mu_3\text{-}CH_3C)Co_2Mo(C_5H_5)(CO)_8 + Na[Co(CO)_4]$$

$$(\mu\text{-}CH_3C)Co_2Mo(C_5H_5)(CO)_8 + \tfrac{1}{2}[C_5H_5Ni(CO)]_2$$

$$\longrightarrow (\mu_3\text{-}CH_3C)NiCoMo(C_5H_5)_2(CO)_5 + \tfrac{1}{2}Co_2(CO)_8$$

Procedure

1. $(\mu_3\text{-}CH_3C)Co_2Mo(C_5H_5)(CO)_8$

A 100-mL Schlenk flask fitted with a magnetic stirring bar is charged with 2.00 g (4.38 mmol) of $(\mu_3\text{-}CH_3C)Co_3(CO)_9$,[6] 1.90 g (7.09 mmol) of $Na[Mo(C_5H_5)(CO)_3]$, and 50 mL of tetrahydrofuran (THF), and closed with a pressure equalizing bubbler. The mixture is stirred for ~ 2 h after which time TLC spot tests with hexane indicate that no further product is formed. The solvent is removed in an oil pump vacuum, the residue is dissolved in 50 mL of benzene, and the solution is filtered through a D3 frit filled with a 3-cm layer of silica gel. After washing with three 20-mL portions of benzene, the dark green combined filtrate is reduced in volume to 5 mL by a rotary evaporator. A 50-mL volume of hexane is added, and the solution is cooled to $-35\,°C$ in a refrigerator. Quick filtration through a medium porosity frit and washing with 5 mL of cold hexane yield 1.58 to 1.74 g (68–75%) of black-green, crystalline, air sensitive $(\mu_3\text{-}CH_3C)Co_2Mo(C_5H_5)(CO)_8$.

The compound can be identified by its IR bands at 2086 (vw), 2072 (m), 2043 (sh), 2032 (vs), 2016 (m), 2003 (vs), 1990 (m), 1948 (m), and 1890 (vw, br) cm^{-1} in C_6H_{12} and by its NMR signals at 3.56 (CH_3) and 4.50 (C_5H_5) ppm versus int TMS in C_6D_6.

2. $(\mu_3\text{-}CH_3C)NiCoMo(C_5H_5)_2(CO)_5$

A 100-mL Schlenk flask fitted with a magnetic stirring bar is charged with 0.76 g (1.43 mmol) of $(\mu_3\text{-}CH_3C)Co_2Mo(C_5H_5)(CO)_8$, 1.30 g (4.28 mmol) of $[C_5H_5Ni(CO)]_2$, and 40 mL of benzene. A reflux condenser closed with a pressure equalizing bubbler is attached, and the mixture is heated with stirring in an oil bath kept at 60 °C for 5 days, after which time TLC spot tests with hexane indicate that the amount of starting cluster is not becoming any less. The solvent is removed in an oil pump vacuum. The residue is suspended in 30 mL of hexane and chromatographed with benzene–hexane (1:5) on a 3.5 × 50-cm column filled with silica gel. In the first, dark green, fraction ~ 0.3 g ($\sim 40\%$) of the starting cluster is recovered. The second, brown, fraction, after evaporation to dryness in a rotary evaporator, leaves behind 0.21–0.30 g (28–41%) of brown $(\mu_3\text{-CH}_3\text{C})\text{NiCoMo}(\text{C}_5\text{H}_5)_2(\text{CO})_5$. This material is spectroscopically pure. Recrystallization from hexane at $-35\,^\circ\text{C}$ to obtain analytically pure material reduces the yield to 0.17–0.24 g (21–33%). Taking the recovery of the starting cluster into account, the isolated amount of 0.30 g corresponds to a yield of 67%.

Properties

Air sensitive $(\mu_3\text{-CH}_3\text{C})\text{NiCoMo}(\text{C}_5\text{H}_5)_2(\text{CO})_5$ forms dark brown crystals. It is reasonably soluble in all nonpolar solvents except aliphatic hydrocarbons. It shows IR bands at 2053 (m), 2034 (s), 2005 (m), 1990 (vs), 1976 (s), 1968(s), 1947 (vw), 1934 (m), 1916 (w), 1898 (vw), and 1870 (vw) cm^{-1} in C_6H_{12} and NMR signals at 3.70 (CH$_3$), 4.68 (C$_5$H$_5$), and 4.84 (C$_5$H$_5$) ppm versus int TMS in C_6D_6. It is the basis compound for a series of $(\mu_3\text{-RC})\text{NiCoMo}$ clusters the properties of which, including enantiomer separation, have been investigated.[11]

B.   *CYCLO*-[μ_3-1(η^2):2(η^2):3(η^2)-2-BUTYNE] OCTACARBONYL-1κ^2C, 2κ^3C, 3κ^3C-[1(η^5)-CYCLOPENTADIENYL]COBALT-MOLYBDENUMRUTHENIUM(*Co—Mo*)(*Co—Ru*)(*Mo—Ru*), (μ_3-C$_2$(CH$_3$)$_2$)MoRuCo(C$_5$H$_5$)(CO)$_8$

This cluster,[12] which is a chiral example of the large class of alkyne bridged trinuclear clusters, is obtained from the ternary metal carbonyl $\text{RuCo}_2(\text{CO})_{11}$ by a two-step sequence involving a capping reaction and a metal exchange, with an overall average yield of 42%:

$$\text{RuCo}_2(\text{CO})_{11} + \text{C}_2(\text{CH}_3)_2 \longrightarrow (\mu_3\text{-C}_2(\text{CH}_3)_2)\text{RuCo}_2(\text{CO})_9 + 2\text{CO}$$

$$\{\mu_3\text{-C}_2(\text{CH}_3)_2\}\text{RuCo}_2(\text{CO})_9 + \text{Na}[\text{Mo}(\text{C}_5\text{H}_5)(\text{CO})_3]$$

$$\longrightarrow \{\mu_3\text{-C}_2(\text{CH}_3)_2\}\text{MoRuCo}(\text{C}_5\text{H}_5)(\text{CO})_8 + \text{Na}[\text{Co}(\text{CO})_4]$$

Procedure

1.   (μ_3-C$_2$(CH$_3$)$_2$)RuCo$_2$(CO)$_9$

A 0.80-g quantity (1.52 mmol) of $\text{RuCo}_2(\text{CO})_{11}$ (ref. 7) is dissolved in 100 mL of hexane in a 250-mL Schlenk flask containing a magnetic stirring bar and cooled to 5 °C in an ice bath. To this solution 0.12 g (2.22 mmol) of 2-butyne is added, and the flask is closed with a pressure equalizing bubbler. (Because of the volatility of 2-butyne it is convenient to use it as a 0.5 M solution in hexane prepared by condensing the appropriate quantity of 2-butyne into a Schlenk flask at $-78\,°C$ and then diluting it with the appropriate amount of hexane.) The mixture is stirred for 6 h with cooling. Then the solution is filtered and reduced in volume to 5–10 mL in a ratory evaporator. After cooling to $-35\,°C$ in a refrigerator, quick filtration through a medium porosity frit, and washing with 5 mL of cold hexane, 0.60–0.68 g (75–85%) of red, crystalline, air sensitive, $(\mu_3\text{-}C_2(CH_3)_2)RuCo_2(CO)_9$ is obtained.

The compound can be identified by its IR bands at 2091 (w), 2056 (vs), 2041 (vs), 2029 (s), 2020 (w), 2006 (w), and 1900 (w, br) cm^{-1} in C_6H_{12} and its single NMR resonance at 2.49 ppm versus int TMS in CDCl$_3$.

2. $(\mu_3\text{-}C_2(CH_3)_2)MoRuCo(C_5H_5)(CO)_8$

A 100-mL Schlenk flask fitted with a magnetic stirring bar is charged with 0.68 g (1.30 mmol) of $(\mu_3\text{-}C_2(CH_3)_2)RuCo_2(CO)_9$, 0.45 g (2.22 mmol) of Na[Mo(C$_5$H$_5$)(CO)$_3$], and 30 mL of THF. A reflux condenser, closed with a pressure equalizing bubbler, is attached, and the mixture is stirred for 15 h at room temperature and then for 2 h in an oil bath kept at 40 °C. The mixture is reduced in volume to 10 mL in an oil-pump vacuum and transferred to a 2.5 × 30-cm column filled with silica gel, which was introduced as a slurry in benzene–hexane (1:4). Chromatography with benzene–hexane (1:4) elutes first a red-brown band containing some remaining $(\mu_3\text{-}C_2(CH_3)_2)RuCo_2(CO)_9$ and then a red band containing a small amount of [(C$_5$H$_5$)Mo(CO)$_3$]$_2$. Benzene–hexane (1:4) then elutes the product cluster as a red band. The eluate is evaporated to dryness in a rotary evaporator, leaving behind 0.40–0.55 g (51–70%) of red $(\mu_3\text{-}C_2(CH_3)_2)MoRuCo(C_5H_5)(CO)_8$, which is spectroscopically pure. Recrystallization from hexane at $-35\,°C$ to obtain analytically pure material reduces the yield to 0.35–0.47 g (45–60%).

Properties

The compound $(\mu_3\text{-}C_2(CH_3)_2)MoRuCo(C_5H_5)(CO)_8$ forms air sensitive red crystals. It is reasonably soluble in all nonpolar solvents except aliphatic hydrocarbons. It shows IR bands at 2079 (s), 2046 (vs), 2029 (sh), 2020 (s), 2011 (sh), 1989 (w), 1977 (m), 1898 (w, br), and 1874 (w, br) cm^{-1} in C_6H_{12} and NMR signals at 2.38 (CH$_3$), 2.47 (CH$_3$), and 5.24 (C$_5$H$_5$) ppm versus int TMS in CDCl$_3$. In solution it decomposes very slowly, producing a precipitate and [(C$_5$H$_5$)Mo(CO)$_3$]$_2$. It is the basis compound for a series of chiral alkyne and vinylidene bridged trimetal clusters for which the chemical and stereochemical features of hydrogenations and dehydrogenations of the organic ligands have been investigated.[12]

References

1. H. Vahrenkamp, *Adv. Organomet. Chem.*, **22**, 169 (1983).
2. R. D. Adams and I. T. Horvath, *Prog. Inorg. Chem.*, **33**, 127 (1985).
3. D. Seyferth, *Adv. Organomet. Chem.*, **14**, 97 (1976).
4. B. F. G. Johnson and J. Lewis, *Adv. Inorg. Chem. Radiochem.*, **24**, 225 (1981); J. Lewis and B. F. G. Johnson, *Pure Appl. Chem.*, **54**, 97 (1982).
5. H. Vahrenkamp, *Comments Inorg. Chem.*, **4**, 253 (1985).
6. D. Seyferth, J. E. Hallgren, and P. L. K. Hung, *J. Organomet. Chem.*, **50**, 265 (1973).
7. H. Vahrenkamp, *Inorg. Synth.*, **26**, 351 (1989).
8. U. Behrens and F. Edelmann, *J. Organomet. Chem.*, **263**, 179 (1984).
9. R. B. King, *Organometallic Syntheses*, Vol. 1, Academic Press, New York, 1965, p. 119.
10. H. Beurich, R. Blumhofer, and H. Vahrenkamp, *Chem. Ber.*, **115**, 2409 (1982).
11. R. Blumhofer and H. Vahrenkamp, *Chem. Ber.*, **119**, 683 (1986).
12. T. Albiez, W. Bernhardt, C. von Schnering, E. Roland, H. Banttel, and H. Vahrenkamp, *Chem. Ber.*, **120**, 141 (1987).

39. NONACARBONYL-TRI-μ-HYDRIDO-μ₃-METHYLIDYNE-TRIRUTHENIUM AND -TRIOSMIUM

Submitted by JEROME B. KEISTER,* JOHN R. SHAPLEY,† and DEBRA A. STRICKLAND†
Checked by LING-SHWU HWANG and YUN CHI‡

Much of the early work in metal cluster chemistry concerned syntheses and reactions of alkylidynetricobalt clusters of the general formula [$Co_3(\mu_3$-CX)(CO)$_9$], where X = H, alkyl, aryl, halo, carboalkoxy, and others.[1] Until the last few years alkylidyne clusters of other metals were limited to a few examples with X = H or CH_3; no other series of functionalized methylidyne clusters was available for physical and chemical studies.

The first alkylidynetriruthenium cluster, [$Ru_3H_3(\mu$-CCH$_3$)(CO)$_9$], was prepared in 1972 in 12% yield by the reaction of [$Ru_4H_4(CO)_{12}$] with ethylene.[2] The structure was shown by a variety of techniques to be comprised of a triangle of ruthenium atoms capped by an μ_3-ethylidyne moiety; each of the three hydride ligands bridges one Ru—Ru edge.[3] The osmium analog has been prepared by a two-step synthesis from $Os_3(CO)_{12}$, first heating

*Department of Chemistry, University at Buffalo, State University of New York, Buffalo, NY 14214.
†Department of Chemistry, University of Illinois, Urbana, IL 61801.
‡Department of Chemistry, National Tsing Hua University, P.O. Box 2-46 Hsinchu, Taiwan 30043.

with ethylene to give $[Os_3H_2(CCH_2)(CO)_9]$ and then hydrogenating to give the ethylidyne product.[4] The complex $[Os_3H_3(\mu_3\text{-}CH)(CO)_9]$ has been prepared by pyrolysis of the tautomeric mixture $[Os_3H(CH_3)(CO)_{10}]/$-$[Os_3H_2(CH_2)(CO)_{10}]$,[5] and has also been isolated from the reaction of $[Os_3(CO)_{12}]$ with acetaldehyde.[6] The ruthenium analog, $[Ru_3H_3(\mu\text{-}CH)(CO)_9]$, has been identified as a minor component of the product mixture formed by reduction of $[Ru_3(CO)_{12}]$ with sodium tetrahydroborate$(1-)$.[7]

We describe here syntheses of $[Ru_3H_3(\mu_3\text{-}CX)(CO)_9]$ $(X = OCH_3, Cl,$ and Br) and $[Os_3H_3(\mu_3\text{-}CX)(CO)_9]$ $(X = OCH_3, CO_2CH_3, Cl,$ and Br).[8] Also reported are syntheses for the precursors $[Ru_3H(\mu\text{-}COMe)(CO)_{10}]$, $[Os_3H(\mu\text{-}COMe)(CO)_{10}]$, and the tautomeric mixture $[Os_3H_2(\mu\text{-}CH_2)$-$(CO)_{10}]/[Os_3H(\mu\text{-}CH_3)(CO)_{10}]$. These compounds have been used as precursors for the syntheses of $[M_3H(CR)(CO)_{10}]$ $(M = Os, X = H$ (ref. 9) or Ph (ref. 10); $M = Ru, X = NR_2$ (ref. 11) or H (ref. 12)) and $[M_3H_3(\mu_3\text{-}CX)(CO)_9]$ $(M = Ru, X = Ph,^{11b}CO_2Me,^{11b}OEt,^{11b}H,^{11b}$ or SEt (ref. 13); $M = Os, X = Ph, H,^5$ or F (ref. 14). Related clusters prepared by other methods include $[Os_3H(CX)(CO)_{10}]$, $X = NR_2$ (ref. 15) or $CH_2CHMe_2,^{16}$ $[Os_3H_3(\mu\text{-}CCO_2H)(CO)_9]$,[17] and $[\{Os_3H_3(CO)_9(\mu_3\text{-}CO)\}_3(B_3O_3)]$.[18] These compounds are of interest because of the comparisons and contrasts that can be made with the physical and chemical properties of the analogous cobalt clusters and because of the additional properties associated with the three hydride ligands. A review of the chemistry of these compounds has appeared.[19]

Most of the syntheses of the $[M_3H_3(\mu_3\text{-}CX)(CO)_9]$ clusters given here begin with the preparation of the appropriate $[M_3H(\mu\text{-}COCH_3)(CO)_{10}]$ cluster by methylation with methyl fluorosulfonate* of the corresponding $[M_3H(CO)_{11}]$ anion, which is derived from $[M_3(CO)_{12}]$. The synthesis of $[Ru_3H(\mu\text{-}COCH_3)(CO)_{10}]$ has also been accomplished by methylation with trimethyloxonium tetrafluoroborate of the $[Ru_3H(CO)_{11}]$ anion, which was prepared by reduction of $[Ru_3(CO)_{12}]$ with sodium tetrahydroborate$(1-)$.[20] The osmium cluster has been prepared by reaction between methyl fluorosulfonate and the $[Os_3H(CO)_{11}]^{1-}$ ion, derived by reduction of $[Os_3(CO)_{12}]$ with potassium hydroxide in methanol.[21] These procedures are fully comparable in yield and convenience to the syntheses given here. All procedures are best carried out using a vacuum–nitrogen double manifold.[22] Although these syntheses are performed under a nitrogen atmosphere unless otherwise indicated, the work-ups may be done without the exclusion of air.

■ **Caution.** *All manipulations involving metal carbonyls must be carried out in a well-ventilated fume hood.*

---

*Methyl fluorosulfonate is no longer commercially available. The checkers substituted methyl trifluoromethanesulfonate and obtained satisfactory results.

A. DECACARBONYL-μ-HYDRIDO-μ-(METHOXY-METHYLIDYNE)-*TRIANGULO*-TRIRUTHENIUM

$$[Ru_3(CO)_{12}] + N(C_2H_5)_3 + H_2O \longrightarrow [NH(C_2H_5)_3][Ru_3H(CO)_{11}] + CO_2$$

$$[NH(C_2H_5)_3][Ru_3H(CO)_{11}] + CH_3SO_3F \longrightarrow [Ru_3H(\mu\text{-}COCH_3)(CO)_{10}]$$
$$+ [NH(C_2H_5)_3][SO_3F]$$

Procedure

A 500-mL three-necked round-bottomed flask is equipped with a magnetic stirring bar, a reflux condenser topped with a stopcock connected to a nitrogen line, a Vigreux column topped with a stopcock connected to a vacuum trap, and a 50-mL, pressure equalizing addition funnel. In the flask under dry nitrogen is placed dodecacarbonyltriruthenium (1.045 g, 1.64 mmol)[23] and tetrahydrofuran (THF) 100 mL, (Fisher reagent grade, used without purification, previously deoxygenated by saturation with a stream of nitrogen gas). Then triethylamine (20 mL) and water (20 mL), both deoxygenated by nitrogen saturation, are added from the addition funnel. The resulting mixture is stirred under nitrogen and heated with an oil bath at 60 to 70 °C. The color of the solution rapidly changes from orange to dark red, characteristic of the $[Ru_3H(CO)_{11}]^{1-}$ ion. After 1 h the solution is cooled to room temperature, and the triethylamine and THF are removed by closing the stopcock for the nitrogen inlet and opening the stopcock to the vacuum trap, which is cooled with liquid nitrogen or with a Dry Ice–acetone bath. When only water and a brick red precipitate of $[NH(C_2H_5)_3][Ru_3H(CO)_{11}]$ remain, most of the water is removed with a pipet or filter stick under a nitrogen blanket, and the remainder of the water is evaporated with vacuum. The product must be thoroughly dry before proceeding to the next step; residual water will react with methyl fluorosulfonate to produce, ultimately, $[Ru_3(CO)_{12}]$ and triethylammonium fluorosulfonate.

■ **Caution.** *The methylating agents, methyl fluorosulfonate or methyl trifluoromethanesulfonate, used in this procedure are highly toxic and volatile. Because methyl fluorosulfonate, used in the original procedure, is no longer commercially available, the checkers have successfully replaced this reagent with methyl trifluoromethanesulfonate. Although less hazardous than methyl fluorosulfonate, methyl trifluoromethanesulfonate is an extremely dangerous chemical: Inhaling its vapors or absorption through the skin can cause a potentially fatal pulmonary edema if not promptly treated. All work with either of these reagents should be done in a well-ventilated fume hood, and suitable* *protective clothing should be worn. Small amounts of unreacted methyl trifluoromethanesulfonate can be destroyed by hydrolysis with alcoholic KOH (4 g KOH in 100 mL methanol or ethanol). The hydrolyzing mixture should be allowed to stand for at least 1 day. Excess methyl fluorosulfonate can be destroyed in a similar manner by treatment with a suspension of potassium carbonate in methanol.*

Next a solution of methyl fluorosulfonate (500 µL, $\sim$ 7 mmol)* in dry dichloromethane (50 mL) is added under nitrogen from the addition funnel. The resulting solution is stirred under nitrogen for 24 h. Finally, the dichloromethane and unreacted methyl fluorosulfonate are vacuum transferred to a trap cooled with liquid nitrogen, and the methyl fluorosulfonate is destroyed by treatment with a suspension of potassium carbonate in methanol ($\sim$ 100 mL, allowed to stand for 24 h). Methanol (25 mL) is added to the product residue, and after stirring for 1 h under nitrogen the $[Ru_3H(\mu\text{-}COCH_3)(CO)_{10}]$ is extracted from $[Ru_3(CO)_{12}]$ with a minimum amount of methanol ($\sim$ 25 mL). The solution is cooled to $-16\,°C$ to give orange-yellow crystals of $[Ru_3H(\mu\text{-}COCH_3)(CO)_{10}]$ (461 mg). The mother liquor is evaporated to dryness by using a rotary evaporator at room temperature and the residue is purified by preparative thin layer chromatography (TLC) on silica gel, eluting with hexanes or cyclohexane.[†] Additional product is obtained by extraction of the second, orange band with dichloromethane and evaporation of this solution to dryness. The product may be contaminated with a small amount of dark red $[Ru_4H_2(CO)_{13}]$, but it is pure enough for use in the following syntheses. Very pure material is obtained by careful preparative TLC or by recrystallization from methanol.[‡] Total yield is 749 mg (73%).

Anal. Calcd. for $Ru_3C_{12}H_4O_{11}$: C, 22.95; H, 0.64; Ru, 48.35. Found: C, 22.95; H, 0.71; Ru, 48.49.

The major side product is $Ru_3(CO)_{12}$. Yield: 144 mg (14%).

---

*The checkers substituted an equivalent amount of methyl trifluoromethanesulfonate (Aldrich). Yield: 67%.

[†]Thin layer chromatography is carried out by using 20 × 20-cm preparative TLC plates made with silica [Kieselgel 60G (EM Science) or Silica Gel GF (10 µm, Analtech)] to an approximate thickness of 0.5 mm. The mixtures are applied as dichloromethane solutions using a drawn-out Pasteur pipet. Development of the plates is conducted in air. The capacity of each plate is $\sim$ 100 mg.

[‡]All recrystallizations are conducted by saturating a methanol solution of the compound at room temperature and cooling the solution to $-16\,°C$ in a freezer.

Properties

The complex $[Ru_3H(\mu\text{-COCH}_3)(CO)_{10}]$ is an orange-yellow, crystalline solid that is slightly air sensitive and is very soluble in common organic solvents. The IR spectrum of $[Ru_3H(\mu\text{-COCH}_3)(CO)_{10}]$ in cyclohexane exhibits sharp bands at 2104(w), 2064(vs), 2054(s), 2030(vs), 2018(m), 1990(w), and 1968(w) cm^{-1}. The ^1H NMR spectrum of $[Ru_3H(\mu\text{-COCH}_3)(CO)_{10}]$ in chloroform-d solution consists of singlets at δ 4.56 and -14.85 in a 3:1 ratio.

B.   NONACARBONYL-TRI-μ-HYDRIDO-μ$_3$-(METHOXY-METHYLIDYNE)-*TRIANGULO*-TRIRUTHENIUM

$$[Ru_3H(\mu\text{-COCH}_3)(CO)_{10}] + H_2 \longrightarrow [Ru_3H_3(\mu_3\text{-COCH}_3)(CO)_9] + CO$$

■   **Caution.**   *Hydrogen gas is highly flammable. The following procedure should be carried out in a well-ventilated hood and well away from any potential source of ignition.*

Procedure

Hydrogenation of $[Ru_3H(\mu\text{-COCH}_3)(CO)_{10}]$ is carried out in a 250-mL three-necked round-bottomed flask equipped with a condenser topped with a gas outlet to an oil bubbler, a magnetic stirring bar, and an inlet tube for introduction of hydrogen gas into the solution. A solution of $[Ru_3H(\mu\text{-}COCH_3)(CO)_{10}]$ (209 mg, 0.13 mmol) in hexane (100 mL) is heated at reflux for 2 h with hydrogen bubbling through the solution. The solvent is then removed with a rotary evaporator, and the red-orange residue is recrystallized from methanol (see the footnote in the procedure in Section A) to give bright orange crystals. The mother liquor is evaporated to dryness, and the residue is purified by preparative TLC on silica gel eluting with cyclohexane (see the footnote in the procedure in Section A). The second, orange to red band is extracted with dichloromethane to give additional product. Yield: 186 mg (93%).

Anal. Calcd. for $Ru_3C_{11}H_6O_{10}$: C, 21.95; H, 1.01; Ru, 50.43. Found: C, 22.19; H, 1.11; Ru, 50.26.

Properties

The compound $[Ru_3H_3(\mu_3\text{-COCH}_3)(CO)_9]$ forms bright orange crystals that are mildly air sensitive. It dissolves readily in common organic solvents to give somewhat more air-sensitive solutions. Its IR spectrum in cyclohexane is 2106(vw), 2078(s), 2075(s), 2036(vs), 2028(m), 2018(sh), 2014(m), and 2000(w)cm^{-1}; its ^{1}H NMR spectrum in chloroform-d solution consists of two singlets of equal intensities at δ 3.77 and -17.53. The compound decomposes slowly in air to form initially a dark red, insoluble compound of unknown formulation. This impurity may be removed by extraction of $H_3Ru_3(\mu_3\text{-COCH}_3)(CO)_9$ with methanol or cyclohexane or by chromatography.

C. μ₃-(BROMOMETHYLIDYNE)-NONACARBONYL-TRI-μ-HYDRIDO-*TRIANGULO*-TRIRUTHENIUM

$$[Ru_3H_3(\mu_3\text{-COCH}_3)(CO)_9] + BBr_3 \longrightarrow [Ru_3H_3(\mu_3\text{-CBr})(CO)_9]$$
$$+ B(OCH_3)Br_2$$

Procedure

To a stirred solution of $[Ru_3H_3(\mu_3\text{-COCH}_3)(CO)_9]$ (128 mg, 0.208 mmol) in dichloromethane (100 mL) is added boron tribromide (50 μL, 0.54 mmol) under nitrogen.

■ **Caution.** *Boron tribromide reacts violently with water, is very toxic by inhalation, and causes severe burns. A well-ventilated fume hood must be used and suitable protective clothing should be worn.*

After 30 min the solution is evaporated to dryness under vacuum, and methanol (10 mL) is added to the residue to destroy any remaining boron tribromide. The product mixture is purified by preparative TLC (see the footnote in the procedure in Section A), eluting with dichloromethane/hexanes (1:5 v/v). A single, bright yellow band yields $[Ru_3H_3(\mu\text{-CBr})(CO)_9]$ (116 mg, 86%) after extraction with dichloromethane.

Anal. Calcd. for $Ru_3C_{10}H_3O_9Br$: C, 18.46; H, 0.46; Br, 12.31. Found: C, 18.56; H, 0.43; Br, 11.97.

Properties

The complex $[Ru_3H_3(\mu_3\text{-CBr})(CO)_9]$ is a bright yellow, air-stable, crystalline solid, which is very soluble in dichloromethane and soluble in most other common organic solvents. Its IR spectrum in cyclohexane solution displays sharp bands at 2087(s), 2080(w), 2040(s), 2030(m), 2000(vw), and 1992(vw)cm^{-1}; its ^{1}H NMR spectrum in chloroform-d consists of a singlet at $\delta -17.80$.

D. DECACARBONYL-μ-HYDRIDO-μ-(METHOXY-METHYLIDYNE)-*TRIANGULO*-TRIOSMIUM

$$[Os_3(CO)_{12}] + N(C_2H_5)_3 + H_2O + [N(C_2H_5)_4]Br$$
$$\longrightarrow [N(C_2H_5)_4][Os_3H(CO)_{11}] + CO_2 + [NH(C_2H_5)_3]Br$$
$$[N(C_2H_5)_4][Os_3H(CO)_{11}] + CH_3SO_3F$$
$$\longrightarrow [Os_3H(\mu\text{-}COCH_3)(CO)_{10}] + [N(C_2H_5)_4][SO_3F]$$

Procedure

A 1000-mL three-necked round-bottomed flask is equipped with a magnetic stirring bar, a stopcock connected to a nitrogen line, a reflux condenser topped with a stopcock connected to an oil bubbler, and a 500-mL pressure-equalizing addition funnel. Dodecacarbonyltriosmium (1.513 g, 1.67 mmol)[24] is placed in the flask and THF (320 mL, distilled from sodium benzophenone under nitrogen) is added under a stream of nitrogen. Triethylamine (80 mL) and distilled water (80 mL), both deoxygenated by nitrogen saturation, are added from the addition funnel. The resulting mixture is heated at 60 °C with an oil bath while stirring under nitrogen, causing the color of the solution to change from yellow to dark red within 10 min. After 2 h the solution is allowed to cool, and tetraethylammonium chloride (1.0 g, 6.06 mmol) in deoxygenated water (20 mL) is added. The triethylamine and THF are removed by closing the stopcock for the nitrogen inlet and connecting the remaining stopcock to a liquid nitrogen-cooled vacuum trap. The cloudy aqueous layer that remains is removed from the brick red precipitate of $[N(C_2H_5)_4][Os_3H(CO)_{11}]$ with a pipet, and the residue is washed with two 10-mL portions of deoxygenated water, again removing the water with a pipet. The flask is placed under vacuum overnight to remove all traces of water.

■ **Caution.** *The methylating agents, methyl fluorosulfonate or methyl trifluoromethanesulfonate, used in this procedure are highly toxic and volatile. Because methyl fluorosulfonate, used in the original procedure, is no longer commercially available, the checkers have successfully replaced this reagent with methyl trifluoromethanesulfonate. Although less hazardous than methyl fluorosulfonate, methyl trifluoromethanesulfonate is an extremely dangerous chemical: Inhaling its vapors or absorption through the skin can cause a potentially fatal pulmonary edema if not promptly treated. All work with either of these reagents should be done in a well-ventilated fume hood, and suitable protective clothing should be worn. Small amounts of unreacted methyl trifluoromethanesulfonate can be destroyed by hydrolysis with alcoholic KOH (4 g KOH in 100 mL methanol or ethanol). The hydrolyzing mixture should* *be allowed to stand for at least 1 day. Excess methyl fluorosulfonate can be destroyed by treatment with a suspension of potassium carbonate in methanol.*

Next, a solution of methyl fluorosulfonate (400 μL, ~ 5.6 mmol, stored over anhydrous sodium carbonate)* in dry dichloromethane (400 mL) is placed in the addition funnel and added to the dry solid. The resulting solution is stirred under nitrogen for 1 h, during which time the solution changes from red to bright yellow. The dichloromethane and unreacted methyl fluorosulfonate are vacuum transferred to a liquid nitrogen-cooled trap, and the methyl fluorosulfonate is destroyed by treatment with a suspension of potassium carbonate in methanol (20 mL, allowed to stand for 24 h). Most of the $[Os_3H(\mu\text{-}COCH_3)(CO)_{10}]$ can be collected after crystallization from methanol (see the footnote in the procedure in Section A). The mother liquor is then evaporated to dryness and purified by preparative scale TLC (see the footnote in the procedure in Section A) eluting with petroleum ether (bp 30–60 °C). Additional product is obtained after the first yellow band, trailing a trace of purple $[Os_3H_2(CO)_{10}]$, is extracted with dichloromethane. Yield: 1.108 g (74%).

Anal. Calcd. for $Os_3C_{12}H_4O_{11}$: C, 16.10; H, 0.45. Found: C, 16.16, H, 0.41.

Variable amounts of the impurity $[Os_3H(\mu\text{-}C(O)OCH_3)(CO)_{10}]$ (ref. 25) (^{1}H NMR spectrum (CDCl₃): δ 3.65 (s, 3H) and -14.30 (s, 1H)) may also be formed. This impurity does not affect the following procedure and is easily removed by TLC after hydrogenation of $[Os_3H(\mu\text{-}COMe)(CO)_{10}]$ to $[Os_3H_3(\mu_3\text{-}COMe)(CO)_9]$; however, $[Os_3H(\mu\text{-}C(O)OCH_3)(CO)_{10}]$ may be removed from $[Os_3H(\mu\text{-}COMe)(CO)_{10}]$ with difficulty by very careful TLC.

Properties

$[Os_3H(\mu\text{-}COCH_3)(CO)_{10}]$ is an air-stable, bright yellow, crystalline solid. The compound is very soluble in common organic solvents. The IR spectrum in cyclohexane includes sharp bands at 2109 (w), 2064 (s), 2057 (m), 2025 (s), 2012 (m), 1998 (m), and 1982 (w, sh) cm^{-1}. The ^{1}H NMR spectrum in chloroform-*d* consists of singlets at δ 4.59 and -16.41 in a 3:1 ratio.

E. NONACARBONYL-TRI-μ-HYDRIDO-μ₃-(METHOXY-METHYLIDYNE)-*TRIANGULO*-TRIOSMIUM

$$[Os_3H(\mu\text{-}COCH_3)(CO)_{10}] + H_2 \longrightarrow [Os_3H_3(\mu_3\text{-}COCH_3)(CO)_9] + CO$$

---

*The checkers substituted an equivalent amount of methyl trifluoromethanesulfonate (Aldrich). Yield: 52%.

■ **Caution.** *Hydrogen gas is highly flammable. The following procedure should be carried out in a well-ventilated hood and well away from any potential source of ignition.*

Procedure

Hydrogenation of $[Os_3H(\mu\text{-}COCH_3)(CO)_{10}]$ is carried out in a 500-mL three-necked round-bottomed flask equipped with a magnetic stirring bar, a reflux condenser topped with a stopcock connected to an oil bubbler, and a fixed inlet tube for introduction of hydrogen gas into the solution. A solution of $[Os_3H(\mu\text{-}COCH_3)(CO)_{10}]$ (315 mg, 0.35 mmol) in decane (270 mL) is heated at 120 °C for 1 h, with hydrogen bubbling vigorously through the solution, to produce a somewhat paler yellow solution. The solvent is removed by vacuum into a Dry Ice–2-propanol cooled trap.* Purification of the the residue is accomplished by preparative scale TLC (see the footnote in the procedure in Section A) eluting with petroleum ether (bp 30–60 °C). Extraction of the second major yellow band trailing purple $[Os_3H_2(CO)_{10}]$ provides $[Os_3H_3(\mu_3\text{-}COCH_3)(CO)_9]$. Yield from $[Os_3H(\mu\text{-}COCH_3)(CO)_{10}]$ is 264 mg (86%). The compound can be recrystallized from methanol (see the footnote in the procedure in Section A).

Anal. Calcd. for $Os_3C_{11}H_6O_{10}$: C, 15.20; H, 0.70. Found: C, 15.08; H, 0.67.

Properties

The compound $[Os_3H_3(\mu_3\text{-}COCH_3)(CO)_9]$ is a very light yellow, air-stable, crystalline solid that is soluble in most organic solvents. The IR spectrum in cyclohexane displays sharp bands at 2107 (w), 2077 (s), 2074 (s), 2022 (vs), 2013 (m), 2008 (m), and 1995 (w, v) cm^{-1}. Its ^1H NMR spectrum consists of singlets of equal intensities at δ 3.80 and -18.58.

F.   NONACARBONYL-TRI-μ-HYDRIDO-μ₃-(METHOXY-CARBONYLMETHYLIDYNE)-*TRIANGULO*-TRIOSMIUM

$$[Os_3H_3(\mu_3\text{-}CBr)(CO)_9] + AlCl_3 + CO + 4CH_3OH$$

$$\longrightarrow [Os_3H_3(\mu_3\text{-}CCO_2CH_3)(CO)_9] + Al(OCH_3)_3 + 3HCl + HBr$$

---

*The checkers report that evaporation of decane is excessively slow. They found that the synthesis works equally well using xylenes as solvent, and the evaporation is more efficient.

Procedure

In a nitrogen-purged glove bag [$Os_3H_3(\mu_3$-$CBr)(CO)_9$] (336 mg, 0.37 mmol) and aluminum trichloride (510 mg, 3.8 mmol) are added to a 250-mL three-necked round-bottomed flask equipped with a magnetic stirring bar, a gas outlet with a stopcock, and two septum stoppers. Dry dichloromethane (5 mL) is added by cannula, producing a cloudy orange solution. The gas outlet is then connected to an oil bubbler and a septum stopper is replaced by a gas inlet tube. Dichloromethane is added to total volume of 150 mL while carbon monoxide is bubbled vigorously through the stirred solution at room temperature. After 55 min, methanol (5 mL; distilled from potassium carbonate) is added, causing an immediate color change to bright yellow. The solution is poured into 5% aqueous HCl (200 mL), the organic layer is separated by using a separatory funnel, and the aqueous layer is extracted with two 10-mL portions of dichloromethane. The combined organic fractions are dried over magnesium sulfate, filtered, and evaporated to dryness with a rotary evaporator. The yellow residue is applied to a fluorescent preparative scale thin layer chromatography plate (see the footnote in the procedure in Section A). Elution with petroleum ether (bp 30–60 °C)–dichloromethane (3:2) produces a bright yellow band followed by a colorless band observed with an UV lamp). Extraction of this second band provides 198 mg [$Os_3H_3(\mu_3$-$CCO_2CH_3)(CO)_9$] (60%). The compound can be recrystallized from methanol (see the footnote in the procedure in Section A).

Anal. Calcd. for $Os_3C_{12}H_6O_{11}$: C, 16.07; H, 0.67; Os, 63.63. Found: C, 16.30; H, 0.60; Os, 64.65.

Properties

The compound [$Os_3H_3(\mu_3$-$CCO_2CH_3)(CO)_9$] is a very pale yellow, air-stable, crystalline solid that is soluble in common organic solvents, particularly dichloromethane. Its IR spectrum in cyclohexane displays bands at 2116(vs), 2089(vs), 2081(w, sh), 2030(s), 2025(vs), 2015(w), and 1688(vw) cm^{-1}. Its ^{1}H NMR spectrum in chloroform-*d* consists of singlets of equal intensity at δ 3.83 and − 19.39.

G. NONACARBONYL-μ₃-(HALOMETHYLIDYNE)-TRI-μ-HYDRIDO-*TRIANGULO*-TRIOSMIUM

$$[Os_3H_3(\mu_3\text{-}COCH_3)(CO)_9] + BX_3 \longrightarrow [Os_3H_3(\mu_3\text{-}CX)(CO)_9] + B(OCH_3)X_2$$

$$X = Br \text{ and } Cl$$

Procedure

To a stirred solution of $[Os_3H_3(\mu_3\text{-}COCH_3)(CO)_9]$ (112 mg, 0.13 mmol) in dry dichloromethane (50 mL) under nitrogen is added boron tribromide (200 μL of a 2 M solution in pentane; 0.4 mmol) by syringe.

■ **Caution.** *Boron tribromide reacts violently with water, is very toxic by inhalation, and causes severe burns. A well-ventilated fume hood must be used and suitable protective clothing should be worn.*

An immediate color change to a very pale gold is observed. After 5 min the solution is evaporated to dryness with a rotary evaporator, and methanol (10 mL) is added to the residue to destroy any remaining boron tribromide. Purification of the product is accomplished by preparative scale TLC (see the footnote in the procedure in Section A) on fluorescent silica gel, eluting with petroleum ether (bp 30–60 °C). The complex $[Os_3H_3(\mu_3\text{-}CBr)(CO)_9]$ (100 mg, 90%) is obtained after extraction of the single colorless band with dichloromethane.

The chloride derivative of the osmium compound is prepared in a similar fashion. The complex $[Os_3H_3(\mu_3\text{-}CCl)(CO)_9]$ is obtained in 73% yield after a solution of $[Os_3H_3(\mu_3\text{-}COCH_3)(CO)_9]$ and boron trichloride (obtained as a 1.0 M solution in CH_2Cl_2) is stirred for 40 min, followed by the work-up described previously.

Properties

The compounds $[Os_3H_3(\mu_3\text{-}CBr)(CO)_9]$ and $[Os_3H_3(\mu_3\text{-}CCl)(CO)_9]$ are very pale yellow air-stable crystalline solids that are soluble in common organic solvents. Their IR spectra in cyclohexane consist of bands at 2087 (vs), 2079 (m, sh), 2027 (vs), 2015 (m), and 1986 (w) cm^{-1}. The ^{1}H NMR spectra in chloroform-d consist of sharp singlets at $\delta = -18.90$ for $H_3Os_3(\mu_3\text{-}CBr)(CO)_9$ and at $\delta -18.80$ for the chloro derivative.

H. DECACARBONYL-DI-μ-HYDRIDO-μ-METHYLENE-*TRIANGULO*-TRIOSMIUM AND ITS HYDRIDOMETHYL TAUTOMER

$$[Os_3H_2(CO)_{10}] + CH_2N_2$$

$$\longrightarrow [Os_3H(CH_3)(CO)_{10}]/[Os_3H_2(CH_2)(CO)_{10}] + N_2$$

■ **Caution.** *Diazomethane is highly toxic, extremely volatile and potentially explosive. This synthesis should be done in a well-ventilated hood and suitable protective clothing should be worn. Diazomethane should be used only* *in dilute solution and contact with ground glass joints should be avoided. Excess can be destroyed by treatment with dilute acetic acid.*

Procedure

A solution of ethereal diazomethane is prepared in a minidistillation apparatus consisting of a dropping funnel, Teflon stopcock, 10-mL flask, and Clear-Seal 14/20 joints (Wheaton). A solution of 2-(2-ethoxyethoxy)ethanol (cabitol, 3.5 mL), diethyl ether (2.0 mL), and potassium hydroxide (0.6 g) in water (1.0 mL) is added to the distillation flask. N-methyl-N-nitroso-p-toluenesulfonamide [Diazald (Aldrich), 2.15 g, 10 mmol] is dissolved in diethyl ether (20 mL) and the solution is placed in the dropping funnel. The receiving flask, containing diethyl ether (4.0 mL), is immersed in an ice bath. When the distillation flask is set in a water bath preheated to 68 °C, the ether begins to distill. The Diazald solution is added through the dropping funnel over a 20-min period. Ether and diazomethane codistill at this temperature to give a bright yellow distillate that can be stored in a Clear-Seal stoppered flask for periods of several weeks at -10 °C in the dark.

The compound $[Os_3H_2(CO)_{10}]$ (40 mg, 0.047 mmol)[26] is dissolved in dichloromethane (5 mL) in a 50-mL three-necked round-bottomed flask equipped with a magnetic stirring bar, a gas inlet–outlet tube connected to a nitrogen line, and a 25-mL pressure equalizing addition funnel. Ethereal diazomethane is added dropwise from the addition funnel to the stirred solution of $[Os_3H_2(CO)_{10}]$ until the color change from purple to yellow is complete (~ 5 min). Samples of "$[Os_3(CH_4)(CO)_{10}]$" prepared in this way contain some polymethylene as an impurity, the amount of which can be minimized by not adding an excess of diazomethane and by keeping the rate of addition slow. The product is purified by preparative scale TLC (see the footnote in the procedure in Section A) eluting with hexanes. A single yellow band is observed, which is extracted with dichloromethane. Contact time with the silica should be kept to a minimum. Yield: 31.3 mg (77%).

Anal. Calcd. for $Os_3C_{11}H_4O_{10}$: Os, 64.83; C, 15.24. Found: Os, 65.13; C, 15.25.

Properties

The crystalline compound $[Os_3H_2(CH_2)(CO)_{10}]$, in solution exists as a mixture of methyl and methylene tautomers. A fully equilibrated sample in cyclohexane displays IR bands at 2126 (w), 2108 (vw), 2089 (s), 2067 (sh), 2059 (vs), 2021 (vs), 2005 (vs), 2005 (sh), 1985 (m), and 1978 (m) cm^{-1}. The ^{1}H NMR spectrum in chloroform-d exhibits four multiplets of equal intensities at δ 5.12, 4.32, $-$ 15.38, and $-$ 20.71 attributable to the methylene isomer, and weaker signals at $\delta -$ 3.65 (d) and $-$ 15.02 (q) in a 3:1 ratio due to the methyl isomer.

References

1. D. Seyferth, *Adv. Organometal. Chem.*, **14**, 97 (1976).

2. A. J. Canty, B. F. G. Johnson, J. Lewis, and J. R. Norton, *J. Chem. Soc. Chem. Commun.*, **1972**, 1331.

3. (a) G. M. Sheldrick and J. P. Yesinowski, *J. Chem. Soc. Dalton Trans.*, **1975**, 873; (b) A. D. Buckingham, J. P. Yesinowski, A. J. Canty, and A. J. Rest, *J. Am. Chem. Soc.*, **95**, 2732 (1973).

4. (a) A. J. Deeming and M. Underhill, *J. Organometal. Chem.*, **42**, C60 (1972); (b) A. J. Deeming and M. Underhill, *J. Chem. Soc. Chem. Commun.*, **1973**, 277.

5. R. B. Calvert and J. R. Shapley, *J. Am. Chem. Soc.*, **99**, 5225 (1977).

6. K. A. Azam and A. J. Deeming, *J. Chem. Soc. Chem. Commun.*, **1977**, 472.

7. C. R. Eady, B. F. G. Johnson, and J. Lewis, *J. Chem. Soc. Dalton Trans.*, **1977**, 477.

8. These procedures are modifications of previously published syntheses (ref. 9).

9. J. R. Shapley, M. E. Cree-Uchiyama, G. M. St. George, M. R. Churchill, and C. Bueno, *J. Am. Chem. Soc.*, **105**, 140 (1983).

10. W.-Y. Yeh, J. R. Shapley, Y. Li, and M. R. Churchill, *Organometallics*, **4**, 767 (1985).

11. (a) J. B. Keister, M. W. Payne, and M. J. Muscatella, *Organometallics*, **2**, 219 (1983); (b) J. B. Keister and T. L. Horling, *Inorg. Chem.*, **19**, 2304 (1980).

12. J. S. Holmgren and J. R. Shapley, *Organometallics*, **3**, 1322 (1984).

13. M. R. Churchill, J. W. Ziller, D. M. Dalton, and J. B. Keister, *Organometallics*, **6**, 806 (1987).

14. H.-J. Kneuper, D. S. Strickland, and J. R. Shapley, *Inorg. Chem.*, **27**, 1110 (1988).

15. R. D. Adams and N. M. Golembeski, *J. Am. Chem. Soc.*, **101**, 2579 (1979).

16. M. Green, A. G. Orpen, C. J. J. Schaverien, *J. Chem. Soc. Chem. Commun.*, **1984**, 37.

17. J. Krause, D.-Y. Jan, and S. G. Shore, *J. Am. Chem. Soc.*, **109**, 4416 (1987).

18. S. G. Shore, D.-Y. Jan, W.-L. Hsu, L.-Y. Hsu, S. Kennedy, J. C. Huffman, T.-C. L. Wang, and A. G. Marshall, *J. Chem. Soc. Chem. Commun.*, **1984**, 392.

19. J. B. Keister, *Polyhedron*, **7**, 846 (1988).

20. B. F. G. Johnson, J. Lewis, A. G. Orpen, P. R. Raithby, and G. Suss, *J. Organometal. Chem.*, **173**, 187 (1979).

21. P. D. Gavens and M. J. Mays, *J. Organometal. Chem.*, **162**, 389 (1978).

22. D. F. Shriver, *The Manipulation of Air-Sensitive Compounds*, McGraw-Hill, New York, 1969, pp. 145–146.

23. A. Montorani and S. Cenini, *Inorg. Synth.*, **16**, 47 (1975).

24. B. F. G. Johnson and J. Lewis, *Inorg. Synth.*, **13**, 92 (1972).

25. M. Tachikawa, Ph. D. thesis, University of Illinois, IL 1977.

26. S. A. R. Knox, J. W. Koepke, M. A. Andrews, and H. D. Kaesz, *J. Am. Chem. Soc.*, **97**, 3942 (1975).

40. ANIONIC AND HETERONUCLEAR TRIOSMIUM CLUSTERS

Submitted by KEVIN BURGESS* and ROSS P. WHITE
Checked by RICHARD D. ADAMS† and JAMES E. BABIN

It has been suggested[1] that electronic similarities between hydride and gold(I)phosphine "ligands" are reflected in the structures of cluster compounds related by interchange of these fragments. An ideal compound for the study of such isolobal relationships is the dihydride $Os_3(\mu\text{-}H)_2(CO)_{10}$.[2] Coordinative unsaturation in this molecule causes it to be activated towards a variety of substrates hence it is a useful starting material for syntheses of many other clusters.[3] Mono- and di-gold homologs of $Os_3(\mu\text{-}H)_2(CO)_{10}$, that is, $Os_3(\mu\text{-}H)AuPR_3(CO)_{10}$, and $Os_3(AuPR_3)_2(CO)_{10}$ (R = Ph and Et), have been prepared from the anions $[Os_3(\mu\text{-}H)(\mu\text{-}CO)(CO)_{10}]^-$ and $[Os_3(CO)_{11}]^{2-}$, respectively.[4,5] Similarities between the hydridotriosmium cluster and its gold analog were confirmed. The syntheses of these gold–triosmium clusters have been outlined previously.[4,5] detailed procedures appear here.

The relatively stable polymetallic $Os_3(CO)_{12}$ (ref. 6) may be activated by conversion to anionic derivatives. Reduction of $Os_3(CO)_{12}$ with potassium benzophenone gives the reactive, air-sensitive triosmium dianion $[Os_3(CO)_{11}]^{2-}$[7]. Generally, it is convenient to generate and use this complex immediately, without isolation; Section B describes an example of this procedure

■ **Caution.** *Carbon monooxide and osmium compounds are highly toxic. These reactions should be carried out in well-ventilated fume hood. Potassium reacts violently with water; care should be taken to ensure anhydrous conditions. Furthermore, potassium oxides can be explosive, so that only fresh samples with little or no surface oxide should be cut as described here.*

A. DECACARBONYL-μ-HYDRIDO-μ-[(TRIPHENYL-PHOSPHINE)GOLD]TRIOSMIUM AND DECACARBONYL-μ-HYDRIDO-μ-[(TRIETHYLPHOSPHINE)GOLD]TRIOSMIUM

$$[(PPh_3)_2N][Os_3(\mu\text{-}H)(\mu\text{-}CO)(CO)_{10}] + [AuPR_3][BF_4]$$

$$\longrightarrow Os_3(\mu\text{-}H)(AuPR_3)(CO)_{10} + CO + (PPh_3)_2NBF_4$$

R = Ph or Et

*Department of Chemistry, Rice University, P.O. Box 1892, Houston, TX 77251.
†Department of Chemistry, University of South Carolina, Columbia, SC 29208.

Procedure

A 0.0024-g (0.12 mmol) sample of Ag[BF$_4$] (Aldrich) is weighed, in a glove bag filled filled with nitrogen, and placed in a vial. The vial is removed from the glove bag, and the silver salt is introduced, against a moderate flow of nitrogen, into a 50-mL, oven-dried Schlenk tube containing a magnetic stirrer. A 0.060-g (0.12 mmol) sample of Au(PPh$_3$)Cl (weighed in air)[8] is introduced in the same way. The Schlenk tube is capped with a septum and alternately evacuated and flooded with nitrogen three times. Dichloromethane (0.7 mL, distilled from CaH$_2$ and taken directly from the still head under nitrogen) is introduced by syringe, and the mixture is stirred for 15 min at 20 °C, during which time AgCl forms as a lilac-white precipitate. Meanwhile, 0.142 g (0.10 mmol) of [(PPh$_3$)$_2$N][Os$_3$(μ-CO)(CO)$_{10}$] (weighed in the air)[9] is introduced into another 50-mL, oven-dried Schlenk tube equipped with a magnetic stirrer. This vessel is capped with a septum and successively evacuated and flooded with nitrogen three times. Dichloromethane (0.7 mL) is added by syringe, and the mixture is stirred, giving a red solution. When the suspension of silver salts has been stirred for 15 min, it is cannulated into the red solution via a needle with a small circle of filter paper wired onto the end immersed in the silver salt suspension. Any solution remaining with the precipitate is washed through with two 1-mL portions of dichloromethane. The Schlenk tube containing the mixture of silver and osmium complexes is then immersed in an oil bath at 45 °C, and maintained at that temperature with stirring for 30 min. A dark brown-green solution forms and removal of the solvent under vacuum gives a brown-green residue. The residue is dissolved in a minimum volume (~ 0.5 mL) of dichloromethane (in air). This solution is then taken up into a drawn-out pipette and loaded as thin lines onto six 20-cm^2 silica thin layer (~ 2 mm thick) chromatography (TLC) plates* and eluted with ~ 150 mL of 1:4 dichloromethane–hexane (in air). An intense green band, $rf = 0.4$, is isolated and extracted from the silica with 25 mL of dichloromethane. Removal of the solvent under vacuum affords 0.0881 g (63% based on the triosmium anion) of decacarbonyl-μ-hydrido-μ-[(triphenylphosphine)gold]triosmium. Recrystallization from warm hexane gives 0.068 g (51% overall yield) of this compound as well-formed dark green crystals.

Anal. Calcd. for C$_{28}$H$_{16}$AuO$_{10}$Os$_3$P:C, 25.65; H, 123; P, 2.37. Found: C, 25.35; H, 1.14; P, 2.49.

Decacarbonyl-μ-hydrido-μ-[(triethylphosphine)gold]triosmium is prepared in the same way from 0.042 g (0.12 mmol) of Au(PEt$_3$)Cl,[8] giving a green band $rf = 0.4$ (solvent as above); 0.059 g (51% yield) before

---

*These plates were prepared in the laboratory; 20-cm^2, silica-coated glass plates from Merck can also be used and these give better results.

recrystallization from warm hexane, 0.037 g (32% overall yield) after crystallization.

Anal. Calcd. for $C_{16}H_{16}AuO_{10}Os_3P$:C, 16.46; H, 1.37. Found: C, 16.70; H, 1.54.

Properties

Solid decacarbonyl-μ-hydrido-μ-[triphenylphosphine)gold]triosmium is stable in air for several weeks; in solution it is stable for several hours. It is soluble in tetrahydrofuran (THF), diethyl ether, and dichloromethane, but only slightly soluble in hydrocarbons. This complex is less reactive than $Os_3(\mu\text{-H})_2(CO)_{10}$; for instance, $Os_3(\mu\text{-H})(AuPPh_3)(CO)_{10}$ will react slowly with carbon monoxide under 2–3 atm of pressure over a period of ~ 1 h, whereas the dihydride adds carbon monoxide instantaneously at room temperature and atmospheric pressure. The IR spectrum of $Os_3(\mu\text{-H})(AuPPh_3)(CO)_{10}$ contains several bands corresponding to terminal carbonyl groups (hexane cm^{-1}):2090(m), 2047(s), 2040(m), 2008(m), 1997(s), and 1944(vw). The ^1H NMR spectrum (250 MHz, dichloromethane-d_2, δ in parts per million downfield from TMS, ambient temperature) consists of a broad multiplet at 7.5 due to aromatic protons and a sharp metal hydride singlet at -11.9. The ^{13}C NMR spectrum (250 MHz, chloroform-d_1, δ in parts per million downfield from TMS, ambient temperature) contains signals at 186.2, 185.3, 183.6, 178.7, 178.6, 134.0, 133.7, 131.4, 130.5, 129.8, 129.3, 129.2. Low resolution mass spectra of this compound exhibit a molecular ion peak at $m/z = 1316$ (^{196}Os and ^{197}Au).

Decacarbonyl-μ-hydrido-μ-[(triethylphosphine)gold]triosmium has similar properties and spectral characteristics. IR (hexane cm^{-1}):2089(m), 2045(s), 2039(m), 2008(s), 1995(s), 1978(s), and 1943(vw). The ^1H NMR spectrum (250 MHz, chloroform-d_1, δ parts per million downfield from TMS, ambient temperature) shows a doublet of triplets ($J = 19$ and 9 Hz) at 1.15 resulting from the methyl hydrogen atoms, a multiplet at 1.85 for the methylene protons, and a sharp hydride singlet at -11.84. Low resolution mass spectra of this compound contain a peak at $m/z = 1172$ corresponding to the molecular ion (^{192}Os, ^{197}Au).

B. DECACARBONYLBIS{μ-[(TRIPHENYLPHOSPHINE)GOLD]}-TRIOSMIUM AND DECACARBONYLBIS{μ-[(TRIETHYL-PHOSPHINE)GOLD]}TRIOSMIUM

$$Os_3(CO)_{12} + 2K \longrightarrow Os_3(CO)_{11}K_2 + CO$$

$$Os_3(CO)_{11}K_2 + 2Au(PR_3)Cl \longrightarrow Os_3(AuPR_3)_2(CO)_{10} + 2KCl + CO$$

Procedure

A moderate flow of nitrogen is passed through an oven-dried Schlenk tube containing a magnetic stirrer and 0.91 g (5 mmol) of benzophenone (Aldrich); 0.39 g (10 mmol) of potassium (cut into small pieces and weighed under dry heptane) is introduced. The Schlenk vessel is capped with a septum and alternately evacuated and flooded with nitrogen three times. Tetrahydrofuran (10 mL, distilled from sodium benzophenone and taken directly from the still head under nitrogen) is added via syringe. A purple-blue color develops, and intensifies as the suspension is stirred for 1 h at 20 °C. During that interval a second 50-mL, oven-dried Schlenk tube containing a magnetic stirrer and 0.091 g (0.1 mmol) of finely ground $Os_3(CO)_{12}$ (ref. 6) is capped with a septum and successively evacuated and flooded with nitrogen three times. Tetrahydrofuran (5 mL) is then added via syringe, and the mixture is stirred vigorously at 60 °C to dissolve as much of the cluster as possible. The resulting solution [with some undissolved $Os_3(CO)_{12}$] is allowed to cool to 20 °C. Approximately 0.45 mL of the purple-blue benzophenone radical anion solution is introduced via syringe, dropwise over 10 min, into the stirred solution of $Os_3(CO)_{12}$ until an "end point" is reached. This is judged as follows: the solution of $Os_3(CO)_{12}$ initially is yellow; as drops of the radical anion solution are added it becomes more orange until, at the end point, a green solution forms and remains that color on stirring for 5 min. It is important that the radical anion should be added slowly. If it is introduced too fast diminished yields result. This solution is then stirred for an additional hour, during which time the deep orange color reforms, indicative of $[Os_3(CO)_{11}]^{2-}$. The septum is then removed, and 0.108 g (0.22 mmol) of $Au(PPh_3)Cl$ (ref. 8) is introduced rapidly against a moderate flow of nitrogen; the septum is immediately replaced. The reaction vessel is then immersed in an oil bath at 65 °C, and maintained at that temperature, with stirring, for 3 h. During this time the solution darkens to brown-green. Removal of the solvent gives a brown-green residue, which is then dissolved in a minimum volume (~ 0.5 mL) of dichloromethane (in air). This solution is introduced as thin lines on six 20 × 20-cm^2 silica TLC plates (~ 2 mm thickness) and eluted with ~ 150 mL of 11:9 dichloromethane–hexane. The intense green band that develops ($rf \sim 0.5$) is isolated and extracted with 25 mL of dichloromethane. Removal of the solvent under vacuum gives 0.132 g [75% yield based on $Os_3(CO)_{12}$] of decacarbonylbis{μ-[(triphenylphosphine)-gold]}triosmium. Recrystallization from warm hexane affords 0.080 g (46% yield overall) of well-formed dark green crystals.

Anal. Calcd. for $C_{46}H_{30}Au_2O_{10}Os_3P_2$: C, 31.23; H, 1.71; P, 3.50. Found: C, 31.23; H, 1.76; P, 4.04.

Decacarbonyl bis{μ-[(triethylphosphine)gold]}triosmium is prepared in the same way from 0.077 g (0.22 mmol) of Au(PEt$_3$)Cl, giving a green band, $rf = 0.6$ (solvent as before), from which 0.082 g (55% yield) of the product is extracted and affords 0.061 g (41% overall yield) of green crystals after recrystallization from hexane.

Anal. Calcd. for $C_{22}H_{30}Au_2O_{10}Os_3P_2$:C, 17.84; H, 2.04. Found: C, 18.11; H, 2.04.

Properties

Decacarbonylbis{μ-[(triphenylphosphine)gold]}triosmium is air stable in the solid state for several weeks and in solution for several hours. It is soluble in THF, diethyl ether, and dichloromethane and sparingly soluble in hydrocarbons. This cluster is less reactive than $Os_3(\mu\text{-}H)_2(CO)_{10}$ or $Os_3(\mu\text{-}H)(AuPPh_3)(CO)_{10}$. For example, dichloromethane solutions of $Os_3(AuPPh_3)(CO)_{10}$ remain unchanged after prolonged periods at 45 °C under 2 to 3 atm of carbon monoxide. The IR spectrum (hexane, cm^{-1}) shows five terminal carbonyl stretching bands: 2071(m), 1029(s), 1985(s), 1968(w), and 1946(m). The ^{1}H NMR spectrum (250 MHz, dichloromethane-d_2, δ in parts per million downfield from TMS, ambient temperature) consists of a multiplet at 7.5 resulting from aromatic protons. The ^{13}C NMR spectrum (250 MHz, chloroform-d_1, δ in parts per million downfield from TMS, ambient temperature) shows signals at 196.6, 189.1, 186.0, 183.0, 182.8, 134.1, 134.0, 133.9, 132.2, 132.1, 131.3, 131.0, 130.6, 129.6, 128.6, 128.4. Low-solution mass spectra of this complex have a peak at $m/z = 1774$, corresponding to the molecular ion (^{192}Os, ^{197}Au).

Decacarbonylbis{μ-[(triethylphosphine)gold]}triosmium has similar properties and spectral characteristics. IR (hexane, cm^{-1}):2068(m), 2015(s), 1980(s), 1968(w), and 1943(m). The ^{1}H NMR spectrum (250 MHz, chloroform-d_1, δ in parts per million downfield from TMS, ambient temperature) shows a doublet of triplets at 1.1 ($J = 19$ and 7 Hz) for the methyl protons and a multiplet at 1.77 for the methylene protons. Low-resolution mass spectra of this complex contain a molecular ion peak a $m/z = 1486$ (^{192}Os, ^{197}Au).

References

1. J. W. Lauher and K. Wald, *J. Am. Chem. Soc.*, **103**, 7648 (1981).
2. A. J. Demming, *Transition Metal Clusters*, (Ed.), B. F. G. Johnson, Wiley, New York, 1980, p. 391.
3. K. Burgess, *Polyhedron*, **3**, 1175 (1984).
4. B. F. G. Johnson, D. A. Kaner, J. Lewis, and P. R. Raithby, *J. Organomet. Chem.*, **215**, C33 (1981).

5. K. Burgess, B. F. G. Johnson, D. A. Kaner, J. Lewis, P. R. Raithby, and S. N. A. B. Syed-Mustaffa *J. Chem. Soc. Chem. Commun.*, **1983**, 455.

6. B. F. G. Johnson and J. Lewis, *Inorg. Synth.*, **13**, 93 (1972).

7. C. C. Nagel, J. C. Bricker, A. G. Alway, and S. G. Shore, *J. Organomet. Chem.*, **219**, C9 (1981)

8. B. J. Gregory and C. K. Ingold, *J. Chem. Soc.* (*B*), **1969**, 276. M. F. Bruce, B. K. Nicholson, and O. bin Shawkataly, *Inorg. Synth.*, **26**, 324 (1989).

9. K. Burgess and R. P. White, *Inorg. Synth.*, **25**, 193 (1989).

10. C. R. Eady, B. F. G. Johnson, J. Lewis, and M. C. Malatesta, *J. Chem. Soc. Dalton Trans.*, **1978**, 1358.

41. HEXACHLORODODECAKIS(TRIPHENYL-PHOSPHINE)PENTAPENTACONTAGOLD, $Au_{55}[P(C_6H_5)_3]_{12}Cl_6$

Submitted by GÜNTER SCHMID*
Checked by BRUCE D. ALEXANDER, JURGEN BARTHELMES, ANN M. MUETING, and LOUIS H. PIGNOLET†

The compound $Au_{55}[P(C_6H_5)_3]_{12}Cl_6$ (ref. 1) was the first full-shell cluster with the magic number of 55 metal atoms to be described. A few examples with the magic number 13 have been described before.[2] Full-shell clusters consist of cubic or hexagonal close packed structures; less dense icosahedral structures are observed in Au_{13} units. Two-shell clusters with 55 atoms (13 + 42 atoms, corresponding to $10 n^2 + 2$ atoms for the *n*th shell) have fewer ligands than surface atoms, since atoms in planes cannot each be coordinated by a ligand. In $Au_{55}[P(C_6H_5)_3]_{12}Cl_6$ and other M_{55} clusters only the vertices bind large ligands, normally phosphines, gaps are occupied more or less by Cl atoms.

Conditions for the successful synthesis of full-shell clusters are first, the possibility of unhindered collisions of the metal atoms, and second, the prevention of a local ligand excess, leading to smaller clusters and complexes. The only method of fulfilling these conditions hitherto described is the use of diborane(6) as a reducing agent for metal halides and additionally as a Lewis acid to bind phosphine to such an extent that only very small amounts of free ligand molecules are present in solution.

*Institut für Anorganische Chemie, Universität Essen, Universitätsstr. 5–7, 4300 Essen 1, Federal Republic of Germany.
†Department of Chemistry, University of Minnesota Kolthoff and Smith Halls, 207 Pleasant Street, S. E., Minneapolis, MN 55455.

Procedure

■ **Caution.** *Diborane(6) is a very toxic, flammable, and ill-smelling gas. It has an autoignition tempeature of 38 to 52 °C. Some people cannot smell diborane(6); for those individuals, the handling of this compound may be especially dangerous. All work with diborane(6) must be carried out in a well-ventilated fume hood.*

The solvents used are previously dried by reflux over potassium (benzene) or Li[AlH$_4$] (dichloromethane, 1, 2-dimethoxyethane, pentane) and distilled under nitrogen. All operations are carried out in an atomsphere of pure nitrogen, unless diborane(6) fills the flasks.

A. DIBORANE(6)

$$3NaBH_4 + 4BF_3 \cdot O(C_2H_5)_2 \longrightarrow 3NaBF_4 + 2B_2H_6 + 4(C_2H_5)_2O$$

Diborane(6), used for the synthesis of Au$_{55}$[P(C$_6$H$_5$)$_3$]$_{12}$Cl$_6$, is generated according to the equation,[3] using an apparatus as shown in Fig. 1. The 500-mL three-necked round-bottomed flask A is equipped with a pressure-equalizing dropping funnel (D), a Dry Ice condenser (E), a vacuum

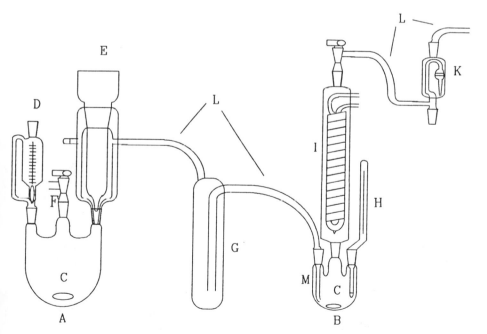

Fig. 1. Laboratory apparatus for the preparation of Au$_{55}$[P(C$_6$H$_5$)$_3$]$_{12}$Cl$_6$.

stopcock (F), and a Teflon-coated magnetic stirring bar (C). The condenser E is fitted with a gas exit tube. A flexible polyethylene tube (L) leads to a trap (G), cooled with Dry Ice. The trap G is connected with a 250-mL three-necked flask (B) by another flexible polyethylene tube ending in a gas inlet tube (M). In addition, B is fitted with a reflux condenser (I), which is connected with a mercury safety valve (K) to avoid any air contact, a thermometer (H), and a magnetic stirrer (C). Before charging, the whole apparatus is evacuated and backfilled with dry nitrogen gas. Then, under a stream of nitrogen, D is filled with 100 mL of BF_3–ether, while A is charged with 20 g of $Na[BH_4]$ and 100 mL of 1, 2-dimethoxyethane. The low temperature reflux condenser E is filled with methanol–Dry Ice. To generate diborane(6), the BF_3–ether is dropped into the stirred mixture of $Na[BH_4]$ and 1, 2-dimethoxyethane at such a speed that the 100-mL volume is used after ~ 40 min for the reaction with $AuCl[P(C_6H_5)_3]$ (see below).

To avoid venting excess of diborane(6) into the fume hood, it may be absorbed by passing it into a trap containing triethylamine.

B. $Au_{55}[P(C_6H_5)_3]_{12}Cl_6$

This compound is prepared by reaction of $AuCl[P(C_6H_5)_3]$ with B_2H_6 in warm benzene. Besides the cluster itself, only $(C_6H_5)_3P$—BH_3 has been identified as a product. The formation of chlorinated boranes is assumed but not proven. A stoichiometrically correct equation for the reaction cannot be given. The three-necked flask B in Fig. 1 is charged with 3.94 g of $AuCl[P(C_6H_5)_3]$ (see Section. 42.A)[4] and 150 mL of benzene.

■ **Caution.** *Benzene is a human carcinogen. Wear gloves and work in a well-ventilated fume hood!*

A stream of diborane(6), generated as described previously, is passed through the solution and the temperature is increased to 50 °C. After ~ 40 min the diborane evolution should be finished. During that time the originally colorless solution becomes dark brown. After cooling to room temperature, a dark precipitate is obtained and the solution is almost colorless. Now, the atmosphere of diborane in the apparatus is driven out by a stream of nitrogen. After that, the precipitate in flask B is removed by filtration and is then treated with 100 mL of dichloromethane. The red-brown solution is filtered to separate small amounts of insoluble matter, and then filtered a second time, through a 5-cm column of Celite® to remove any traces of colloidal gold. The compound $Au_{55}[P(C_6H_5)_3]_{12}Cl_6$ is precipitated by adding 250 mL of pentane to the stirred solution over a period of ~ 2 h. This slow rate of pentane addition is very important, since the triphenylphosphine ligands are partially dissociated from the cluster in solution. Rapid precipitation therefore yields products with fewer ligands. The precipitate is collected and vacuum dried. Yield: 0.6 g (29%).

Anal. Calcd. for $C_{216}H_{180}Au_{55}Cl_6P_{12}$:C, 18.28; H, 1.28; Au, 76.33; Cl, 1.50; P, 2.62. Found: C, 17.66; H, 1.28; Au, 76.10; Cl, 1.70; P, 2.60.

The reaction mixture in flask A can be hydrolyzed with moist 1, 2-dimethoxyethane under an atmosphere of nitrogen in a fume hood.

Properties

The compound $Au_{55}[P(C_6H_5)_3]_{12}Cl_6$ is a dark brown powder soluble in dichloromethane and pyridine, insoluble in petroleum ether, benzene, and alcohols. It is air stable, but on warming it decomposes in the solid state as well as in solution into $AuCl[P(C_5H_5)_3]$ and metallic gold. The IR spectrum of $Au_{55}[P(C_6H_5)_3]_{12}Cl_6$ shows the Au—Cl vibration at $280 \, cm^{-1}$ (Nujol mull). A ^{31}P NMR signal is observed as a singlet at 32 ± 0.5 ppm versus H_3PO_4 (dichloromethane). The Mössbauer data agree with the presence of 4 different kinds of gold atoms[5]: 13 central Au atoms: IS $= -1.4 \, mm \, s^{-1}$; 24 uncoordinated peripheral Au atoms: IS $= 0.3 \, mm \, s^{-1}$; QS $= 1.4 \, mm \, s^{-1}$; 12 Au atoms coordinated to triphenylphosphine: IS $= 0.6 \, mm \, s^{-1}$; QS $= 7.1 \, mm \, s^{-1}$; 6Au atoms coordinated to chlorine: IS $= 0.1 \, mm \, s^{-1}$; QS $= 4.4 \, mm \, s^{-1}$ (4.2 K, ^{197}Pt). The cluster size is proven by molecular weight determinations with an ultracentrifuge and by high resolution transmission electron microscopy.[6]

The 12 triphenylphosphine ligands in $Au_{55}[P(C_6H_5)_3]_{12}Cl_6$ can be completely exchanged by 12 of the *m*-sulfonated derivative molecules $P(C_6H_5)_2C_6H_4SO_3Na$,[7] by stirring a dichloromethane solution of Au_{55}-$[P(C_6H_5)_3]_{12}Cl_6$ with an aqueous solution of $P(C_6H_5)_2C_6H_4SO_3Na$.[7] The cluster moves from the organic phase into the water layer to give $Au_{55}[P(C_6H_5)_2C_6H_4SO_3Na]_{12}Cl_6$, which is completely dissociated into $12Na^+$ cations and the 12 fold negative anion $Au_{55}[P(C_6H_5)_2C_6H_4SO_3^-]_{12}$-$Cl_6$.

Analogous Complexes

The $M_{55}L_{12}Cl_6$ and $M_{55}L_{12}Cl_{20}$ clusters with M = Rh,[8,9] Ru,[9] and Pt (ref. 9) can be prepared by similar procedures. The ligands vary from $P(C_6H_5)_3$, and $P(t\text{-}C_4H_9)_3$ to $As(t\text{-}C_4H_9)_3$; the number of Cl ligands is 6 when L is $P(C_6H_5)_3$ and is 20 for L = $P(t\text{-}C_4H_9)_3$ and $As(t\text{-}C_4H_9)_3$.

References

1. G. Schmid, R. Pfeil, R. Boese, F. Bandermann, S. Meyer, G. H. M. Calis, and J. W. A. van der Velden, *Chem. Ber.*, **114**, 3634 (1981).
2. G. Schmid, *Struct. Bonding*, **62**, 51 (1985).
3. R. Köster and P. Binger, *Inorg. Synth.*, **15**, 141 (1974).

4. C. A. Reed and W. R. Roper, *J. Chem. Soc.* (*A*), **1970**, 506. P. Braunstein, H. Lehner, and D. Matt, *Inorg. Synth.*, **27**, 218 (1990).
5. H. H. A. Smit, R. C. Thiel, L. J. de Jongh, G. Schmid, and N. Klein, *Solid State Commun.*, **65**, 915 (1988).
6. L. R. Wallenberg, J.-O. Bovin, and G. Schmid, *Surface. Sci.*, **156**, 256 (1985).
7. G. Schmid, N. Klein, L. Korste, U. Kreibig, and D. Schönauer, *Polyhedron*, **7**, 605 (1988).
8. G. Schmid, U. Giebel, W. Huster, and A. Schwenk, *Inorg. Chim. Acta*, **85**, 97 (1984).
9. G. Schmid and W. Huster, *Z. Naturforsch. Teil B*, **41**, 1028 (1986).

42. A PLATINUM-GOLD CLUSTER: CHLORO-1κ*Cl*-BIS(TRIETHYLPHOSPHINE-1κ*P*)BIS(TRIPHENYL-PHOSPHINE)-2κ*P*, 3κ*P*-*TRIANGULO*- DIGOLD-PLATINUM(1 +) TRIFLUOROMETHANESULFONATE

Submitted by PIERRE BRAUNSTEIN,* HANS LEHNER,*
and DOMINIQUE MATT*
Checked by KEVIN BURGESS,† and MICHAEL J. OHLMEYER†

There is a considerable interest in heterometallic molecular clusters containing gold. This includes the development of methods of synthesis, the study of their bonding, and their chemical and catalytic reactivities.[1-4] Furthermore, metal–metal bonded compounds of gold represent an important aspect to the chemistry of this element.

The high-yield synthesis of the first platinum–gold cluster[5] is reported here and involves the reaction of chlorohydridobis(triethylphosphine)platinum(II) with the cation $[Au(PPh_3)]^+$, itself generated from chloro(triphenylphosphine)gold(I).

A. AuCl(PPh₃)

Procedure‡

$$HAuCl_4 + 2PPh_3 + H_2O \xrightarrow{EtOH} AuCl(PPh_3) + 3HCl + O{=}PPh_3$$

This synthesis is performed under an atmosphere of argon and derives from the original preparation.[6a] The solvents are degassed prior to use by bubbling

*Laboratoire de Chimie de Coordination, UA 416 CNRS, Institut Le Bel, Université Louis Pasteur, 4 rue Blaise Pascal, F-67070 Strasbourg Cédex, France.
†Department of Chemistry, Box 1892, Rice University, Houston, TX 77251.
‡The following procedure appears to give consistently better yields of AuCl(PPh₃) than the procedure in *Inorg. Synth.*, **26**, 324 (1989).

argon for 2 min. Hydrogen tetrachloroaurate [$HAuCl_4 \cdot xH_2O$, $x \sim 3$ (Johnson Matthey), 1.000 g, 2.54 mmol of Au, or prepared from gold metal according to ref. 7] is introduced into a 250-mL round-bottomed Schlenk flask equipped with a stopcock side arm and containing a magnetic stirring bar. The flask is then evacuated and backfilled twice with argon. Degassed 95% EtOH (technical grade, 35 mL) is added to dissolve the hydrogen tetrachloroaurate. To this solution is added under argon a solution of PPh_3 (1.364 g, 5.20 mmol) in 50 mL of degassed 96% EtOH, using a glass pipette.* The reaction mixture immediately becomes colorless, and a white precipitate appears after a few seconds. The mixture is stirred for 2 min, and the product is then removed by filtering through a glass frit (medium porosity), washed with Et_2O (technical grade, three 15 mL portions), and dried *in vacuo*. The solid on the frit is then directly dissolved into a 250-mL Schlenk flask with CH_2Cl_2 (10 mL). Slow addition of pentane (60 mL) and cooling to $-25\,°C$ result within a few minutes in the formation of white needles of $AuCl(PPh_3)$. Yield: 1.156 g (92%).

Anal. Calcd. for $C_{18}H_{15}AuClP$:C, 43.70; H, 3.06. Found: C, 43.75; H, 3.00. Mp 236–237 °C (lit[6] 242 °C).[†]

Properties

The gold(I) center in chloro(triphenylphosphine)gold(I) has linear two coordination. It is stable in air and soluble in organic solvents such as dichloromethane, chloroform, or tetrahydrofuran (THF). The $^{31}P\{^1H\}$ NMR ($CDCl_3$) spectrum displays a peak at 33.7 ppm (external reference H_3PO_4).

B. $[PtCl\{AuP(C_6H_5)_3\}_2\{P(C_2H_5)_3\}_2](CF_3SO_3)$

Procedure

$$AuCl(PPh_3) + Ag(CF_3SO_3) \xrightarrow{THF} [Au(thf)\{P(C_6H_5)_3\}](CF_3SO_3) + AgCl$$

$$trans\text{-}[PtHCl\{P(C_2H_5)_3\}_2] + 2[Au(thf)\{P(C_6H_5)_3\}](CF_3SO_3)$$

$$\longrightarrow [PtCl\{AuP(C_6H_5)_3\}_2\{P(C_2H_5)_3\}_2](CF_3SO_3) + CF_3SO_3H$$

tetrahydrofuran = thf(ligand) and THF(solvent)

---

*Checkers used an aluminum cannula.
[†]Checkers obtained a mp 242 to 243 °C.

All operations are performed in an atmosphere of argon dried over molecular sieves (4 Å) (Merck 5708). Tetrahydrofuran and diethyl ether are freshly distilled from sodium benzophenone under argon. Pentane is freshly distilled over sodium wire.

The chlorotriphenylphosphine gold complex (0.990 g, 2.00 mmol) is introduced into a 250-mL round-bottomed Schlenk flask equipped with a stopcock side arm and fitted with a Teflon-encased magnetic stirring bar. The round-bottomed flask is evacuated and backfilled twice with argon. Freshly distilled THF (30 mL) is added to dissolve the $AuCl(PPh_3)$ completely. To this solution is added under argon a solution of $AgCF_3SO_3$ (Aldrich) (0.501 g, 1.95 mmol) in THF (20 mL) using a pipette or Teflon cannula. After being stirred at room temperature for 2 min, this mixture is filtered under argon through a Celite®-padded glass frit (medium porosity) into a 500-mL round-bottomed Schlenk flask equipped with a stopcock side arm and a Teflon-encased magnetic stirring bar, and containing a solution of *trans*-$PtHCl(PEt_3)_2$ (prepared according to ref. 8) (0.468 g, 1.00 mmol) in THF (20 mL), which has been precooled to $-78\,°C$ in a Dry Ice–acetone bath. Upon stirring the reaction mixture at $-78\,°C$, a white precipitate is rapidly formed. After 15 min of stirring at $-78\,°C$, cold ($-78\,°C$) pentane (100 mL) is added, and the resulting white suspension is rapidly filtered under argon through a glass frit (porosity 4). The precipitate is sometimes contaminated with a yellow-cream oil. In this case, it should be treated with a small portion of a cold ($-20\,°C$) 1:1 THF–pentane mixture (~ 15 mL), which dissolves the yellow impurity while the oil transforms into a white powder. After filtration, the white solid is washed on the frit with distilled diethyl ether (2 portions of 10 mL) and dried *in vacuo*. This solid is then washed directly into a 250-mL Schlenk flask using THF (30 mL). Addition of distilled pentane (100 mL) and cooling to $-25\,°C$ result in the formation of a white microcrystalline powder of $[PtClAu_2(PPh_3)_2(PEt_3)_2](CF_3SO_3)$ (1.380 g, 90% based on Pt). Spectroscopically pure samples may have a slight cream color. If desired, a second recrystallization from THF (minimum amount)–pentane can be performed. This complex appears to be air stable for several days and can be stored under inert atmosphere at $0\,°C$ in the dark for months.

Anal. Calcd. for $C_{49}H_{60}Au_2ClF_3O_3P_4PtS$: C, 38.35; H, 3.94. Found: C, 38.2; H, 3.98.

Mp $149\,°C$ (dec.)*

---

*Checkers obtained a 27% yield, mp 138 °C. They believe that their low yield is a result of the loss of product when the yellow-cream oil is removed by treatment with THF.

Properties

This platinum–gold cluster, of core composition *triangulo*-$PtAu_2$, is a white product, which is soluble in THF, CH_2Cl_2, and $CHCl_3$. The compound is best characterized by its $^{31}P\{^1H\}$ NMR spectrum [in CH_2Cl_2 distilled from P_4O_{10}, external reference H_3PO_4 and external D_2O lock, $\partial\,35.7$ (t with ^{195}Pt satellites, PEt_3, $^3J_{(PP)} \sim 4\,Hz$, $^1J_{(PPt)} = 2197\,Hz$), 42.0 (t with ^{195}Pt satellites, PPh_3, $^3J_{(PP)} \sim 4\,Hz$, $^2J_{(PPt)} = 792\,Hz$)]. The crystal structure of this cluster has been reported.[5] The core of the molecule is that of an almost isosceles triangle, with Au—Pt distances of 2.600(3) and 2.601(4) Å, and an Au—Au distance of 2.737(3) Å, much shorter than in gold metal (2.884 Å). The structure has been described as involving two-electron three-center bonding, and a detailed molecular orbital calculation has been performed on this and related molecules.[9]

References

1. K. P. Hall and D. M. P. Mingos, *Prog. Inorg. Chem.*, **32**, 237 (1984).

2. J. Schwank, *Gold Bull.*, **18**, 1 (1985).

3. P. Braunstein and J. Rosé, *Gold Bull.*, **18**, 17 (1985).

4. H. Lehner, D. Matt, P. S. Pregosin, L. M. Venanzi, and A. Albinati, *J. Am. Chem. Soc.*, **104**, 6825 (1982).

5. P. Braunstein, H. Lehner, D. Matt, A. Tiripicchio, and M. Tiripicchio Camellini, *Angew. Chem. Int. Ed. Engl.*, **23**, 304 (1984).

6. (a). A. Levi-Malvano, *Atti R. Accad. Naz. Lincei, Mem. Cl. Sci. Fis. Mat. Nat.* **17**, 857 (1908).
   (b) L. Malatesta, L. Naldini, G. Simonetta, and F. Cariati, *Coord. Chem. Rev.*, **1**, 255 (1966).

7. B. P. Block, *Inorg. Synth.*, **4**, 14 (1953).

8. G. W. Parshall, *Inorg. Synth.*, **12**, 26 (1970).

9. D. I. Gilmour and D. M. P. Mingos, *J. Organomet. Chem.*, **302**, 127 (1986).

Chapter Six

PHOSPHORUS COMPLEXES AND COMPOUNDS

43. CHLORO(η^2-*TETRAHEDRO*-TETRAPHOSPHORUS)-BIS(TRIPHENYLPHOSPHINE)RHODIUM(I)

$$RhCl(PPh_3)_3 + P_4 \xrightarrow[CH_2Cl_2]{-78\,^\circ C} RhCl(\eta^2\text{-}P_4)(PPh_3)_2 + PPh_3$$

Submitted by W. E. LINDSELL* and A. P. GINSBERG[†]

Checked by A. L. BALCH,[‡] S. ROWLEY,[‡] and S. REIMER[‡]

The reactive, tetrahedral allotrope of elemental phosphorus, P_4, has long interested theoreticians and is an important intermediate in the commercial preparation of phosphorus compounds.[1] Its strained, unsaturated bonding enables P_4 to coordinate to transition metal centers either as a simple *monohapto* ligand[2] or, as in the complex described here, in a *dihapto* mode.[3] Such coordination to rhodium in $[RhCl(\eta^2\text{-}P_4)(PPh_3)_2]$ causes significant stabilization of P_4 to oxidation in the solid state, although the P_4 molecule is readily displaced by other ligands, including CO and 1,2-bis(diphenylphosphino)ethane, in solution.[3] The following preparative method can also be employed, under identical conditions, to prepare bromo and iodo

*Department of Chemistry, Heriot-Watt University, Riccarton, Edinburg, United Kingdom.
[†]Work done AT & T Bell Laboratories; present address: P.O. Box 986, New Providence, NJ 07974.
[‡]Department of Chemistry, University of California, Davis, CA 95616.

derivatives $[RhX(\eta^2\text{-}P_4)(PPh_3)_2]$ (X = Br and I) (ref. 3) by using the corresponding molar quantities of $[RhX(PPh_3)_3]$ (X = Br and I*) (ref. 4) as precursors.

Procedure

All operations are performed under dry deoxygenated nitrogen, using conventional Schlenk techniques.[5] Solvents are dried by distillation under nitrogen after being heated at reflux over appropriate drying reagents: P_4O_{10} for CH_2Cl_2; $Li[AlH_4]$ for diethyl ether. White (yellow) phosphorus, freshly cut from the center of a stick (Alfa), is washed well with distilled water and dried under vacuum (10^{-3} torr) for 20 min before use.

■ **Caution.** *White phosphorus is extremely toxic and causes severe burns. All operations should be carried out in a well-ventilated hood, and gloves should be worn when handling the solid or solutions. Disposal of small amounts of white phosphorus, kept safely under water, can be carried out by drying and controlled burning in an efficient hood or in an open, isolated area.*

Chlorotris(triphenylphosphine)rhodium(I)[4] (Strem) (0.50 g, 0.54 mmol) dissolved in CH_2Cl_2 (15 mL) is stirred in a 250-mL Schlenk tube cooled to $-78\,°C$ by a Dry Ice–acetone bath. A solution of white phosphorus (0.070 g, 0.56 mmol) dissolved in CH_2Cl_2 (20 mL) is added dropwise to the cold solution from a pressure equalizing dropping funnel over 15 min. Stirring is continued for a further 45 min at $-78\,°C$, during which time the solution changes from deep red to yellow. Dropwise addition of diethyl ether (200 mL) from the dropping funnel to the cold solution precipitates the yellow microcrystalline product. This solid is collected by Schlenk filtration of the cold reaction mixture on a fritted funnel and washed well at ambient temperature with diethyl ether. The product tends to retain solvents, which can be removed by drying under vacuum (10^{-3} torr) for 16 h in a drying pistol at $82\,°C$ (bp of 2-propanol). Yield: 0.305 g (72%)[†]; mp 171 to 173 °C (with dec).

Anal. Calcd. for $C_{36}H_{30}ClP_6Rh$: C, 54.95; H, 3.8; P, 23.6; Cl, 4.5. Found: C, 54.8; H, 4.0; P, 23.5; Cl, 4.7.

Properties

Chloro(η^2-*tetrahedro*-tetraphosphorus)bis(triphenylphosphine)rhodium(I) is a yellow solid that is stable under nitrogen or vacuum at ambient temperature.

---

*Small amounts of $[Rh_2I_2(PPh_3)_4]$ in $[RhI(PPh_3)_3]$ do not appear to affect the preparation of the iodo complex.

[†]The checkers obtained a yield of 20%.

It is recovered unchanged after short exposures to air, although slow aerial oxidation does occur. It forms a 1:2 solvate, as amber prismatic crystals from CH_2Cl_2 at $-78\,°C$, which slowly desolvates at ambient temperature. It is soluble in CH_2Cl_2 and $CHCl_3$, buy solutions darken and decompose with ligand dissociation at temperatures $> -20\,°C$, even under inert atmospheres. The IR spectrum (Nujol mull) shows, in addition to typical coordinated PPh_3 bands, $\nu_{(Rh—Cl)}$ at 276(s) cm^{-1} and bands associated with the coordinated P_4 at 569(s), 433(s), 387(m), ~ 376(sh), and 349(w) cm^{-1}; oxidized samples show broad absorptions in the region 900 to 1200 cm^{-1}. $^{31}P\{^1H\}$ NMR ($-22\,°C$, CD_2Cl_2, δ relative to 85% H_3PO_4): $\delta + 43.2$, complex doublet, $^1J_{(Rh—P)} = 115.3$ Hz (PPh_3 ligands); multiplet around $\delta - 280$ [A_2B_2 resonance of η^2-$P_2^A P_2^B$ with $\delta_{(PA)} - 279.4$, $\delta_{(PB)} - 284.0$, $^1J_{(Rh—PA)} = 33.9$ Hz, $^1J_{(PA—PB)} = 175$ Hz). A study by X-ray diffraction has established η^2-P_4 coordination in this complex.[3]

References

1. D. E. C. Corbridge, *Phosphorus. An Outline of its Chemistry, Biochemistry and Technology*, 3rd ed., Elsevier, Amsterdam, 1985.
2. P. Dapporto, S. Midollini, and L. Sacconi, *Angew. Chem. Int. Ed. Engl.*, **18**, 469 (1979); P. Dapporto, L. Sacconi, P. Stoppioni, and F. Zanobini, *Inorg. Chem.*, **20**, 3834 (1981).
3. A. P. Ginsberg, W. E. Lindsell, K. J. McCullough, C. R. Sprinkle, and A. J. Welch, *J. Am. Chem. Soc.*, **108**, 403 (1986).
4. J. A. Osborn and G. Wilkinson, *Inorg. Synth.*, **10**, 67 (1967).
5. D. F. Shriver, The Manipulation of Air-Sensitive Compounds, McGraw-Hill, New York, 1969.

44. TETRACARBONYLBIS(η^5-CYCLOPENTADIENYL)-(μ-η^2-DIPHOSPHORUS)DIMOLYBDENUM(I) AND DICARBONYL(η^5-CYCLOPENTADIENYL)-(η^3-*cyclo*-TRIPHOSPHORUS)MOLYBDENUM(I)

Submitted by O. J. SCHERER,* J. SCHWALB,* and H. SITZMANN*
Checked by W. E. LINDSELL†

$$[Mo(\eta^5\text{-}C_5H_5)(CO)_3]_2 + P_4 \xrightarrow[\Delta]{\text{toluene}}$$

$$\cdot[Mo_2(\eta^5\text{-}C_5H_5)_2(CO)_4(\mu\text{-}\eta^2\text{-}P_2)]$$

$$+ [Mo(\eta^5\text{-}C_5H_5)(CO)_2(\eta^3\text{-}P_3)] + \text{polymer}$$

*Fachbereich Chemie, Universität Kaiserslautern, D-6750 Kaiserslautern, Federal Republic of Germany.
†Department of Chemistry, Heriot-Watt University, Edinburgh EH14 4AS, United Kingdom.

The concepts of diagonal relationship and "isoelectronic families" point to a manifold analogy between carbon and phosphorus compounds.[1c] Besides P_2, the diphosphorus molecule, *cyclo*-P_3, *cyclo*-P_5^-, and *cyclo*-P_6, the all-phosphorus analogs of the carbocyclic π systems $C_3H_3^+$, $C_5H_5^-$, and C_6H_6, as well as white phosphorus, P_4, can be stabilized in the coordination spheres of different transition metal complex fragments.[1]

Procedure

The reaction must be performed under argon (or N_2). Prior to use, toluene, pentane, and heptane are freshly distilled from sodium; dichloromethane is distilled from P_4O_{10}. Basic aluminum oxide (Brockman activity I) is heated for 6 h at 200 °C under an oil pump vacuum; after cooling, 2% H_2O is added. White phosphorus (P_4) is dried in an oil pump vacuum.

■ **Caution.** *White phosphorus is extremely toxic and causes severe burns. All operations should be carried out in a well-ventilated hood, and gloves should be worn when handling the solid or solution.*

In a 100-mL two-necked round-bottomed flask equipped with a magnetic stirring bar, an argon inlet, and a reflux condenser connected to a mineral oil bubbler, $[Mo(\eta^5 C_5H_5)(CO)_3]_2$ (Strem) (1.0 g, 2.0 mmol)[2] is dissolved in dry toluene (60 mL). The mixture is heated with stirring to gentle reflux (static Ar atmosphere) for 12 h {formation of $[Mo(\eta^5\text{-}C_5H_5)(CO)_2]_2$ (Mo$\equiv$Mo)[3]}. Freshly dried white phosphorus (0.5 g, 4.0 mmol) is added at room temperature through the second neck of the flask (argon input through the reflux condenser), which is then closed with the argon inlet. The mixture is stirred and heated to maintain gentle reflux for 10 h. Then this mixture is poured, under argon, into a centrifuge tube. After centrifugation* to remove the brown precipitate (0.56 g, insoluble in all usual solvents), the solution is evaporated to $\sim$ 10 mL, stirred with aluminum oxide ($\sim$ 10 mL, basic, 2% H_2O), and dried under an oil pump vacuum until a freely flowing powder is obtained. The powder is applied under argon to a column (25 $\times$ 2.0 cm), which is filled with alumina (basic, 2% H_2O) and pentane. Elution with pentane–toluene (10:1) affords a faint yellow fraction[†], which after evaporation to dryness and recrystallization from heptane ($-$ 16 °C), affords flat yellow crystals of $[Mo(\eta^5\text{-}C_5H_5)(CO)_2(\eta^3\text{-}P_3)]$. Yield: 65 mg (5%) based on $[Mo(\eta^5\text{-}C_5H_5)(CO)_3]_2$.

Anal. Calcd. for $C_7H_5MoO_2P_3$:C, 27.12; H, 1.63. Found: C, 26.80; H, 1.63.

---

*The checker separated the brown precipitate by decantation and filtration through a frit.
[†]The checker notes that the faint yellow band of $[Mo(\eta^5\text{-}C_5H_5)(CO)_2(\eta^3\text{-}P_3)]$ is almost impossible to see on the column during chromatography, but the onset of elution of yellow solution is quite clear.

Dichloromethane elutes an orange-red fraction, which, after evaporation and drying under an oil pump vacuum gives NMR spectroscopically pure $[(Mo_2(\eta^5\text{-}C_5H_5)_2(CO)_4(\mu\text{-}\eta^2\text{-}P_2)]$. Recrystallization from heptane ($-16\,°C$) affords orange-red to red crystals. Yield: 250 mg (25%) based on $[Mo(\eta^5\text{-}C_5H_5)(CO)_3]_2$.

Anal. Calcd. for $C_{14}H_{10}Mo_2O_4P_2$: C, 33.90; H, 2.03. Found: C, 33.10; H, 2.04.

Properties

Dicarbonyl(η^5-cyclopentadienyl) (η^3-*cyclo*-triphosphorus)molybdenum(I) is a yellow crystalline solid that may be handled briefly in air, but it should be stored in an inert atmosphere. It dissolves readily in benzene, toluene, dichloromethane, diethyl ether, and acetonitrile. The crystals are moderately soluble in pentane and hexane. The ^1H NMR spectrum (200 MHz, C_6D_6) shows a quartet at $\delta_{TMS}4.21$ ($^3J_{(PH)} = 0.5$ Hz). The $^{31}P\{^1H\}$ NMR spectrum (C_6H_6, 85% H_3PO_4 as reference) shows a singlet at $\delta - 351.1.$* The IR spectrum (C_6H_6) shows $\nu_{CO} = 1995$ and $1938\,cm^{-1}$. The molecular structure has been determined by X-ray diffraction studies.[4]

Tetracarbonylbis(η^5-cyclopentadienyl) ($\mu\text{-}\eta^2$-diphosphorus)dimolybdenum(I) is an orange-red to red solid that may be handled briefly in air but should be stored in an inert atmosphere. It dissolves readily in benzene, toluene, and dichloromethane; it is sparingly soluble in pentane and hexane at room temperature. The ^1H NMR spectrum (200 MHz, C_6D_6) shows a singlet at $\delta_{TMS}4.54$. The $^{31}P\{H\}$ NMR spectrum (C_7D_8, 85% H_3PO_4 as reference) shows a singlet at $\delta - 42.9$. The IR spectrum (CH_2Cl_2) shows $\nu_{CO} = 1990$ (small shoulder), 1965 and $1913\,cm^{-1}$. The complex, the molecular structure of which has been determined by X-ray diffraction studies,[5] can be used as a ligand.[6]

References

1. (a) O. J. Scherer, *Angew. Chem.*, **97**, 905 (1985); (b) *Angew. Chem. Int. Ed, Engl.*, **24**, 924 (1985). (c) O. J. Scherer, *Comments Inorg. Chem.*, **6**, 1(1987); (d) M. Di Vaira, P. Stoppioni, and M. Peruzzini, *Polyhedron* **6**, 351 (1987).
2. (a) R. B. King, *Organomet. Synth.*, **1** 109 (1965); (b) R. Birdswhistell, P. Hackett, and A. R. Manning, *J. Organomet. Chem.*, **157**, 239 (1978).
3. (a) M. D. Curtis and R. J. Klingler, *J. Organomet. Chem.*, **161**, 23 (1978); (b) M. D. Curtis, N. A. Fotinos, L. Messerle, and A. P. Sattelberger, *Inorg. Chem.*, **22**, 1559 (1983).

*The checker reports $\delta - 347.4$. He suggests that the variation from the submitters result may be due to a concentration effect.

4. O. J. Scherer, H. Sitzmann, and G. Wolmershäuser, *Acta Cryst.*, **C41**, 1761 (1985).
5. O. J. Scherer, H. Sitzmann, and G. Wolmershäuser, *J. Organomet. Chem.*, **268**, C9 (1984).
6. (a) O. J. Scherer, H. Sitzmann, and G. Wolmershäuser, *Angew. Chem.*, **96**, 979 (1984); (b) *Angew. Chem. Int. Ed. Engl.*, **23**, 968 (1984).

45. TRILITHIUM HEPTAPHOSPHIDE, DILITHIUM HEXADECAPHOSPHIDE, AND TRISODIUM HENICOSAPHOSPHIDE

Submitted by M. BAUDLER* and K. GLINKA*
Checked by RICHARD A. JONES[†]

Trilithium heptaphosphide, dilithium hexadecaphosphide, and trisodium henicosaphosphide are polyphosphides with isolated $P_n{}^{m-}$ anions in the solid state as well as in solution. These metalated polyphosphorus compounds are of interest because of their structural relationship to the phosphorus hydrides (phosphanes) P_nH_m and organophosphanes P_nR_m and particularly because of their potential use as building blocks for the syntheses of polyphosphorus compounds.

The structural peculiarity of trilithium heptaphosphide is its fluctuating anion.[1,2] On account of the structural similarity to the hydrocarbon bullvalene,[3] the $P_7{}^{3-}$ ion is capable of a degenerate Cope rearrangement and at room temperature has to be described by 1680 valence-tautomeric forms of equal structure.

The ions $P_{16}{}^{2-}$ and $P_{21}{}^{3-}$ exhibit a striking resemblance to structural units of Hittorfs phosphorus.[4]

$P_7{}^{3-}$ $P_{16}{}^{2-}$ $P_{21}{}^{3-}$

The increasing complexity of these homoatomic anions is well reflected by the following systematic names:

*Institut für Anorganische Chemie der Universität Köln, Greinstr. 6, D-5000 Köln 41, Federal Republic of Germany.
†Department of Chemistry, The University of Texas at Austin, Austin, TX 78712.

Tricyclo[2.2.1.0^{2,6}]heptaphosphide(3 −)

Octacyclo[7.7.0.0^{2,6}·0^{3,8}·0^{5,7}·0^{10,14}·0^{11,16}·0^{13,15}]hexadecaphosphide(2 −)

Decacyclo[9.9.1.0^{2,10}·0^{3,7}·0^{4,9}·0^{6,8}·0^{12,20}·0^{13,17}·0^{14,19}·0^{16,18}]henicosaphosphide(3 −)

The compound Li_3P_7 was first prepared by the action of butyllithium on diphosphane.[1] The route given below, however, is a much better approach; it is based on a nucleophilic cleavage of white phosphorus with lithium dihydrogen phosphide.[5] The latter is readily available by the reaction of butyllithium and phosphine.[6]

A high yield of Li_2P_{16} is obtained in the reaction of P_4 and $LiPH_2$ if the stoichiometric ratio of reactants and the reaction time are significantly modified.[7] The formation of the $P_{16}{}^{2-}$ ion has also been observed in the reaction of Na_3P_7 with $(Ph_4P)Cl$ (ref. 8) and in the decomposition of Li_2HP_7 and LiH_2P_7.[9]

The compound Na_3P_{21} is formed when sodium sand is allowed to react with white phosphorus at a Na:P ratio of 1:2 (not 1:7!).[10]

The three polyphosphides precipitate in the form of definite solvent adducts. This is also observed for the starting compound $LiPH_2$.[6] On drying, the adducts lose a part of the solvent with concomitant decomposition of the salts.

A. TRILITHIUM HEPTAPHOSPHIDE

1. Lithium Dihydrogen Phosphide

$$n\text{-}C_4H_9Li + PH_3 \xrightarrow{\text{monoglyme}} n\text{-}C_4H_{10} + LiPH_2 \cdot 1 \text{ monoglyme}$$

Procedure

■ **Caution.** *All operations involving the use of butyllithium and phosphine gas must be carried out in an inert atmosphere and inside a well-ventilated hood. Phosphine has a foul odor and is very toxic (threshold limit value in air is 0.3 ppm). A gas mask approved for use with phosphine should be immediately available in case of emergency.*

Through one of the side arms of a clean, dry, 1-L three-necked round-bottomed flask (F1, Fig. 1) a gas inlet device (D) is inserted consisting of an inlet tube, which is fitted with a three-way stopcock (T-bore) and a septum.

The septum will allow the introduction of a steel wire reaching through the plug to the bottom of the tube. The inlet device is connected with a phosphine

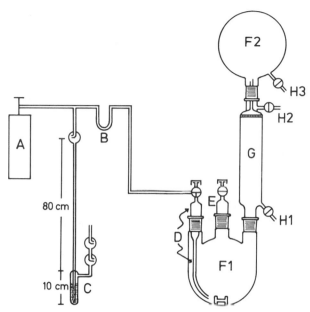

Fig. 1. Apparatus for synthesis of LiPH₂: A, PH₃ cylinder; B, bubble counter; C, mercury safety valve; D, inlet tube with three-way stopcock and septum; E, adapter with stopcock and septum; F1, flask with two parallel sidenecks; F2, flask with lateral stopcock; G, sintered glass funnel with lateral stopcocks; H1–H3, stopcocks.

gas cylinder (A) via a bubble counter (B, paraffin oil) and a mercury safety valve (C) using polyethylene tubing. A magnetic stirring bar is inserted through the main neck of the flask, which is then equipped with a stopcock (E) topped by a septum. The other side arm is connected with a filtering device consisting of a sintered-glass funnel (G, 500 mL, medium porosity) and a 1-L round-bottomed flask (F2), which are both equipped with lateral stopcocks (H1, H3). A vacuum line, including another mercury valve and a supply of purified nitrogen or argon, is connected to the stopcocks H1, H2, and H3. The exits of both mercury valves are combined by polyethylene tubing and terminate in a short glass tube (I.D. = 3 mm). The tube is mounted horizontally within a 2-cm distance of the flame of a Bunsen burner, which must be burning throughout the following steps of the procedure to ensure the complete combustion of any phosphine gas leaving the safety valves.

The apparatus, including the connection to the gas cylinder and the volumes between plug and septum in D and E, is evacuated and refilled with inert gas five times. Under reduced pressure 300 mL of monoglyme (1, 2-dimethoxyethane; dried over potassium) is sucked into the reaction flask

via a cannula bridge[11] through the septum inlet of the main neck. The flask is cooled to $-78\,°C$ (Dry Ice–acetone) and commercial 1.6 M butyllithium solution in hexane (300 mL, 0.48 mol) is slowly added in the same manner.
■ **Caution.** *Exothermic reaction!* When the addition is complete, the main neck is connected with the vacuum line (under inert gas flow).

While maintaining the low temperature, a moderate stream of phosphine gas is passed through the solution with vigorous stirring. The precipitation of white $LiPH_2$ monoglyme begins immediately. If the inlet tube should be plugged by this salt, the tube can be opened easily by moving the steel wire up and down. The reaction is complete as soon as the phosphine gas is no longer absorbed by the solution but is flowing through the mercury valves. After an additional 15 min the phosphine supply is cut off. The reaction mixture is stirred and allowed to slowly warm to room temperature over ~ 2 h. As it warms, a considerable amount of excess phosphine gas evolves and is burned.

The monoglyme suspension is poured into the filter by turning the entire apparatus upside down (H1 closed), while its ground glass joints are held together by appropriate clips. Filtration proceeds by evacuating the 1-L flask via the lateral stopcock H3 and is followed by three washings with 20-mL portions of cold monoglyme. Cautious drying of the white $LiPH_2\cdot$monoglyme is effected at 10 torr by passing a stream of inert gas through the filter cake for 1 min. In a typical experiment a yield of 56.5 g (90.5%)* of lithium dihydrogen phosphide monoglyme adduct is obtained. The purity of the product is easily checked by its $^{31}P\{^1H\}$ NMR spectrum (singlet at $\delta - 283$). Although the salt can be stored in the pure state, it is more convenient to store and use it either in the form of a 1 to 2 M solution in tetrahydrofuran (THF) or as an ~ 0.3 M solution in monoglyme. The solutions are stable at room temperature for some weeks but should be estimated prior to use by titration with 0.1 N HCl.

2.   Trilithium Heptaphosphide

$$3P_4 + 6LiPH_2 \xrightarrow{\text{monoglyme}} 2Li_3P_7\cdot 3\text{ monoglyme} + 4PH_3$$

Procedure

■ **Caution.** *White phosphorus must be handled with care because it ignites spontaneously on contact with oxygen, causes severe burns on the skin, and is*

---

*The checker obtained a yield of 82%. He notes that $LiPH_2\cdot$monoglyme can be recrystallized from monoglyme at $-20\,°C$ and that pumping on $LiPH_2\cdot$monoglyme results in loss of PH_3 and formation of Li_2PH.[6]

toxic. All operations must be carried out in a well-ventilated hood under a flow of inert gas.

A clean, dry, 500-mL three-necked round-bottomed flask is equipped with a reflux condenser and a pressure-equalizing dropping funnel topped by a septum adapter. The third neck is connected via a selector valve with a source of inert gas (including a mercury bubbler) and an oil pump, respectively. The apparatus is repeatedly evacuated and filled with inert gas. A suspension of 4.1 g of freshly distilled white phosphorus[12] (33.1 mmol based on P_4) in 50 mL of dry monoglyme is transferred to the flask while maintaining an inert atmosphere. The dropping funnel is charged with a solution of 9.6 g (73.8 mmol) $LiPH_2 \cdot$ monoglyme in 270 mL of dry monoglyme via a cannula bridge.[11]

With vigorous magnetic stirring the suspension is heated so that the solvent refluxes. The $LiPH_2$ solution is added dropwise over a 2-h period during which time the generated PH_3 gas leaves the apparatus via the mercury bubbler and burns in the flame of a Bunsen burner (see Section A.1). As the addition proceeds, the color of the reaction mixture changes from red to black (maximum viscosity) then to red again. The final mixture is allowed to reflux for a further 30 min. After this period a clear, orange-yellow solution is present. Lowering the temperature to $0\,°C$ for a 40-min period and to $-20\,°C$ for 30 min leads to precipitation of $Li_3P_7 \cdot 3$ monoglyme.*

At $-20\,°C$, the reflux condenser and the dropping funnel are replaced under a countercurrent inert gas flow with a stopper and a sintered glass funnel (250 mL, medium porosity), which is connected to a 1-L round-bottomed flask. The funnel and flask each bear a lateral stopcock (see Fig. 1) and have previously been evacuated and filled with inert gas five times. The filtration as well as the washings (3×15-mL portions of cold monoglyme) are carried out at $-78\,°C$ under reduced pressure and as rapidly as possible. Both flasks at the ends of the frit are replaced with socket caps under inert gas flow via the stopcocks of the frit (corresponding to H1 and H2 in Fig. 1). The frit is evacuated at room temperature for 5 min and refilled with inert gas three times. This drying operation is interrupted as soon as the slightest reddish color of the substance becomes visible. Yield: 8.4 g (75%).

Anal.[1] Calcd. for $C_{12}H_{30}Li_3O_6P_7$: C, 28.37; H, 5.95; Li, 4.10; P, 42.68. Found: C, 28.36; H, 5.72; Li, 3.96; P, 42.65.

---

*The checker reports that after the reaction mixture was refluxed for 30 min an unidentified red oil had formed. This was removed by cooling the solution to $0\,°C$ and decanting the supernatant solution into another flask. Further cooling of the solution to $-30\,°C$ gave crystalline needles of $Li_3P_7 \cdot 3$monoglyme. Yield: 58%.

Properties

Trilithium heptaphosphide is formed as an adduct with 3 mol of monoglyme. It is a bright beige solid, which is stable under inert gas at room temperature. On contact with air the compound ignites spontaneously. It is readily soluble in THF, monoglyme, and diglyme. Its ^{31}P NMR spectrum exhibits a singlet at $\delta - 122$ ($+60\,°C$). At $-60\,°C$ this coalescence signal splits into three separate groups of signals corresponding to the three different types of phosphorus atoms: In the apex ($\delta - 57$), in the bridges ($\delta - 103$), and in the three-membered ring ($\delta - 162$) of the $P_7{}^{3-}$ ion. The Raman and the IR spectra of Li_3P_7 have also been described.[1]

The reaction of trilithium heptaphosphide with bromomethane in THF at $-60\,°C$ leads to trimethylheptaphosphane, $P_7(CH_3)_3$. Yield: 93%.[1,13] The condensation of Li_3P_7 with the diphosphane $Cl(t\text{-}C_4H_9)P\!\!-\!\!P(t\text{-}C_4H_9)Cl$ yields the tetracyclic nonaphosphide $LiP_9(t\text{-}C_4H_9)_2$, which has a skeleton analogous to that of the hydrocarbon deltacyclane.[14]

B. DILITHIUM HEXADECAPHOSPHIDE

$$23P_4 + 12LiPH_2 \xrightarrow{\text{THF}} 6Li_2P_{16}\cdot8THF + 8PH_3$$

Procedure

Following the precautions and using equipment analogous to that employed in the preparation of Li_3P_7 (see Section A.2), a clean, dry, 250-mL round-bottomed flask is charged with 2.82 g (22.76 mmol) of freshly distilled white phosphorus and 150 mL of THF (dried over potassium). Under reflux and with vigorous magnetic stirring, a solution of 1.54 g (11.84 mmol) of $LiPH_2\cdot$monoglyme in 10.6 mL of dry THF is injected by means of a syringe within 3 s. The reaction mixture immediately turns dark brown to black; simultaneously a violent evolution of phosphine gas (~ 175 mL) takes place. All ground glass joints are carefully secured by appropriate clips so that PH_3 leaves the apparatus via the mercury safety valve and is burned as described in Section A. The reaction mixture is allowed to reflux for another 3 min with stirring. The hot mixture is filtered, and the precipitate is washed with four 20-mL portions of THF. After drying under oil-pump vacuum for some minutes a yield of 6.2 g (96%) of dark red-brown $Li_2P_{16}\cdot8$ THF is obtained.

Anal.[7] Calcd. for $C_{32}H_{64}Li_2O_8P_{16}$:C, 35.38; H, 5.94; Li, 1.28; P, 45.62. Found: C, 35.10; H, 5.52; Li, 1.23; P, 44.95.

Extension of the reaction time favors the formation of by-products, although under these conditions a small amount of red octahedral crystals of the title compound can be isolated from the filtrate after standing for some hours.

Properties

Dilithium hexadecaphosphide is formed as an adduct with 8 mol of THF and is stable under inert gas at room temperature. It dissolves in N,N-dimethylformamide (DMF) and dimethyl sulfoxide (DMSO), but shows slight decomposition in both solvents. The compound is insoluble in ethers as well as in aliphatic and aromatic hydrocarbons. On heating, the crystals lose a part of the solvent of crystallization with concomitant deepening of their color. The Raman spectrum of Li_2P_{16} is similar to that of Li_3P_7 and exhibits two additional bands of medium intensity at 340 and 401 cm^{-1}.[7]

A solution of Li_2P_{16} in DMF shows a ^{31}P NMR spectrum with six signal groups at $\delta + 60$, $+ 38$, $+ 6$, $- 34$, $- 134$, and $- 172$ (intensity ratio 2:1:1:1:1:2).[9,15]

The reaction of a solution of Li_2P_{16} with bromomethane and bromoethane, respectively, yields the diorgano-hexadecaphosphanes $P_{16}R_2$ (R = CH_3 and C_2H_5).[16]

C. TRISODIUM HENICOSAPHOSPHIDE

$$21P_4 + 12Na \xrightarrow{\text{THF}} 4Na_3P_{21} \cdot 15THF$$

Procedure

■ **Caution.** *Careful handling of white phosphorus and powdered sodium under an atmosphere of dry, deoxygenated inert gas at all times is strongly recommended because of their spontaneous inflammability.*

Sodium powder is prepared by melting 5 g of carefully cleaned cut sodium metal in 250 mL of boiling xylenes. After removal of the heating bath a slow stream of nitrogen is passed over the solvent, and a high speed stirrer (20,000 rpm, e.g., Ultra-Turrax, model T 18/10)* is introduced and run for 15 s. The finely divided metal is filtered and washed (three 15-mL portions of pentane) under inert gas and sucked dry. It can be stored in a dry box for months without loss of activity.[†]

Trisodium henicosaphosphide is obtained using equipment similar to that

---

*Janke & Kunkel GmbH, D-7813 Staufen, Federal Republic of Germany.
[†]The checker prepared sodium powder by washing Na dispersion in light oil (Aldrich 21, 712–3) with hexane until the oil had been removed.

employed in the preparation of dilithium hexadecaphosphide (Section B). A 3.0 g (24.2 mmol) quantity of freshly distilled white phosphorus, and 1.1 g (47.8 mmol) of powdered sodium in 150 mL of dry THF are stirred slowly and heated under reflux for 2.5 h under inert gas. The solution turns red, and a fine black precipitate separates. After removal of the solid by filtration (slightly reduced pressure) at room temperature, the filtrate is cooled to $-20\,°C$ for ~ 12 h. The crude product is removed by filtering in the cold (without reduced pressure!) and washed five times with 10-mL portions of cold ($-20\,°C$) THF. Dissolution in 20 mL of THF at 0 °C and repeated crystallization at $-20\,°C$ yield pure $Na_3P_{21} \cdot 15THF$, which is filtered cold and dried under oil pump vacuum for 15 s. Yield: 3.0 g $Na_3P_{21} \cdot 15THF$ (36%). A further amount of the compound can be obtained from the mother liquor.

Anal. Calcd. for $C_{60}H_{120}Na_3O_{15}P_{21}$:C, 40.01; H, 6.72; Na, 3.83; P, 36.12. Found: C, 41.22; H, 6.57; Na, 3.82; P, 37.28.

Properties

Trisodium henicosaphosphide is obtained as an adduct with 15 mol of THF and forms orange needle-shaped crystals, which liquefy above $-10\,°C$. At room temperature the melt undergoes a slow disproportionation. The crystals are slightly sensitive to oxidation but strongly sensitive to hydrolysis. A solution of Na_3P_{21} in THF–DMF (~ 2:1) exhibits a ^{31}P NMR spectrum with seven signal groups at $\delta + 72, + 61, - 15, - 108, - 118, - 146$, and $- 169$ (intensity ratio 2:8:2:1:2:2:4).[10]

In the alkylation or silylation of trisodium henicosaphosphide the monometalated henicosaphosphanes(3) $NaP_{21}R_2$ [$R = CH_3$, C_2H_5, i-C_3H_7, and $(CH_3)_3Si$] have been obtained as the final products. A substitution on the central one-atom bridge of the P_{21}^{3-} ion does not occur for steric reasons.[17]

References

1. M. Baudler, H. Ternberger, W. Faber, and J. Hahn, *Z. Naturforsch.*, **B34**, 1690 (1979).

2. M. Baudler, Th. Pontzen, J. Hahn, H. Ternberger, and W. Faber, *Z. Naturforsch.*, **B35**, 517, (1980).

3. W. v. E. Doering and W. R. Roth, *Angew. Chem.*, **75**, 27 (1963); *Angew. Chem. Int. Ed. Engl.*, **2**, 115 (1963). G. Schröder, *Angew Chem.*, **75**, 722 (1963); *Angew. Chem. Int. Ed. Engl.*, **2**, 481 (1963).

4. H. Thurn and H. Krebs, *Angew. Chem.*, **78**, 1101 (1966); *Angew. Chem. Int. Ed. Engl.*, **5**, 1047 (1966); *Acta Crystallogr.*, **B25**, 125 (1969).

5. M. Baudler and W. Faber, *Chem. Ber.*, **113**, 3394 (1980).

6. H. Schäfer, G. Fritz, and W. Hölderich, *Z. Anorg. Allg. Chem.*, **428**, 222 (1977).

7. M. Baudler and O. Exner, *Chem. Ber.*, **116**, 1268 (1983).

8. H. G. von Schnering, V. Manriquez, and W. Hönle, *Angew. Chem.*, **93**, 606 (1981); *Angew. Chem. Int. Ed. Engl.*, **20**, 594 (1981).

9. M. Baudler, *Angew. Chem.*, **94**, 520 (1982); *Angew. Chem. Int. Ed. Engl.*, **21**, 492 (1982).

10. M. Baudler, D. Düster, K. Langerbeins, and J. Germeshausen, *Angew. Chem.*, **96**, 309 (1984); *Angew. Chem. Int. Ed. Engl.*, **23**, 317 (1984).

11. F. Fehér, G. Kuhlbörsch, and H. Luhleich, *Z. Naturforsch.*, **B14**, 466 (1959); *Z. Anorg. Allg. Chem.*, **303**, 294 (1960).

12. T. W. DeWitt and S. Skolnik, *J. Am. Chem. Soc.*, **68**, 2305 (1946).

13. M. Baudler, W. Faber, and J. Hahn, *Z. Anorg. Allg. Chem.*, **469**, 15 (1980).

14. M. Baudler and W. Göldner, *Chem. Ber.*, **118**, 3268 (1985).

15. M. Baudler, R. Heumüller, and J. Hahn, *Z. Anorg. Allg. Chem.*, **529**, 7 (1985).

16. M. Baudler and R. Becher, *Z. Naturforsch.*, **B40**, 1090 (1985).

17. M. Baudler, R. Becher, and J. Germeshausen, *Chem. Ber.*, **119**, 2510 (1986).

46. PHOSPHORUS COMPOUNDS CONTAINING STERICALLY DEMANDING GROUPS

**Submitted by ALAN H. COWLEY,* NICHOLAS C. NORMAN,†
and MAREK PAKULSKI***
Checked by G. BECKER,‡ M. LAYH,‡ E. KIRCHNER,‡ and M. SCHMIDT‡

Stable compounds containing unsupported double bonds between phosphorus and/or arsenic atoms have been isolated recently. This has been achieved by the use of sterically demanding substituents that offer a high degree of kinetic and thermodynamic stability.[1] The synthesis of representative examples of these compounds is described in Section 47.[2] Herein are described the preparations of the relevant starting materials.

■ **Caution.** *Phosphines, and their chloro-substituted derivatives, are toxic and air-sensitive. Silyllithium phosphides, (naphthalene)sodium, and alkyllithium reagents are very sensitive to oxygen and moisture. All operations involving these materials should be carried out under an inert atmosphere and in a well-ventilated fume hood.*

---

*Department of Chemistry, The University of Texas at Austin, Austin, TX 78712.
†Department of Inorganic Chemistry, University of Newcastle upon Tyne Newcastle upon Tyne, NE1 7RU, United Kingdom.
‡Institut für Anorganische Chemie der Universität Stuttgart Pfaffenwaldring 55, D-7000 Stuttgart 80, Federal Republic of Germany.

A.   1-BROMO-2, 4, 6,-TRI-*TERT*-BUTYLBENZENE

$$2, 4, 6\text{-}t\text{-}Bu_3C_6H_3 + Br_2 \longrightarrow 1\text{-}Br\text{-}2, 4, 6\text{-}t\text{-}Bu_3C_6H_2 + HBr$$

$$(MeO)_3P{=}O + HBr \longrightarrow MeBr + (MeO)_2POOH$$

Procedure

This procedure is a modification of that published by Pearson et al.[3] A 1.5-L two-necked flask equipped with a reflux condenser, magnetic stirrer bar, and surrounded by aluminum foil (to minimize exposure to light) is charged with trimethyl phosphate (Aldrich) (1 L) and dessicator dried (P_2O_5) tri-*tert*-butyl-benzene (Aldrich) (76 g, 0.3 mol). The mixture is then stirred and warmed to 85 °C by means of an oil bath until the hydrocarbon dissolves. The temperature is lowered to 70 °C and bromine (56 g, 18 mL, 0.35 mol) is added rapidly.

■   **Caution.**   *Bromine is toxic by inhalation and causes severe burns. It must be handled in a well-ventilated fume hood, and rubber gloves and safety goggles should be worn.*

The temperature is then maintained at 65–70 °C and stirring is continued for 30 h. The exclusion of air is not vital; however, the reaction must be carried out in moisture-free conditions and a nitrogen atmosphere is advised. After completion of the reaction, the solution is allowed to cool to room temperature, and it is then added to 1 L of water at 0 °C (no further precautions to exclude air are required), causing immediate precipitation of the product. Extraction with 5 × 200-mL portions of hexane, followed by evaporation of the combined fractions affords the product as a white solid, crude yield, 92 g (95%). Further purification can be achieved by double recrystallization from ethanol–2-propanol mixtures at − 20 °C, first from 1 L of 5:1, then from 500 mL of 1:5, affording white crystals. Yield: 54 g (55%); mp 172–174 °C. ^1H NMR (C_6D_6): δ 1.35 (s, 9H, *p-t*-Bu), 1.58 (s, 18H, *o-t*-Bu), 7.42 (s, 2H, *m*-H).

B.   (2, 4, 6-TRI-*TERT*-BUTYLPHENYL)PHOSPHONOUS DICHLORIDE

$$1\text{-}Br\text{-}2, 4, 6\text{-}t\text{-}Bu_3C_6H_2 + n\text{-}BuLi \longrightarrow 1\text{-}Li\text{-}2, 4, 6\text{-}t\text{-}Bu_3C_6H_2 + BuBr$$

$$1\text{-}Li\text{-}2,4,6\text{-}t\text{-}Bu_3C_6H_2 + PCl_3 \longrightarrow (2, 4, 6\text{-}t\text{-}Bu_3C_6H_2)PCl_2 + LiCl$$

Procedure

This procedure is a modification of that published by Yoshifuji et al.[4] A solution of 1-Br-2, 4, 6-*t*-Bu$_3$C$_6$H$_2$ (16.27 g, 0.05 mol) in dry (potassium benzophenone), degassed THF (200 mL) is added to a 250-mL Schlenk flask equipped with a rubber septum and magnetic stirrer bar. The flask and contents are cooled to and maintained at $-78\,°C$ by means of a Dry ice–ethanol bath. A hexane solution of BuLi (Aldrich) (34.4 mL of 1.6 M, 0.055 mol) is then added via syringe over a period of a few minutes and the resulting mixture is stirred for 2 h and maintained at low temperature. After this time the solution, containing $1\text{-Li-2,4,6-}t\text{-Bu}_3C_6H_2$, is transferred via a cannula to a second 250-mL Schlenk flask containing a stirred solution of PCl_3 (7.6 g, 5 mL, 0.055 mol) in THF (10 mL) over a period of 30 min. The temperature of both flasks is maintained at $-78\,°C$ throughout this process. After complete addition, the flask is allowed to warm to room temperature and the mixture is stirred for 2 h. All volatile substances, including excess PCl_3, BuBr, and traces of $BuPCl_2$, are removed *in vacuo* ($30\,°C/10^{-4}$ torr), resulting in an essentially quantitative yield of white crystalline $(2,4,6\text{-}t\text{-Bu}_3C_6H_2)PCl_2$ of suitable purity for further use. Extraction of the solid with Et_2O (200 mL), followed by filtration through Celite®, affords the product free of LiCl. Further purification may be achieved by recrystallization from saturated Et_2O solution at $-78\,°C$. ^{31}P NMR δ(THF) 155.

C. (2, 4, 6-TRI-*TERT*-BUTYLPHENYL)PHOSPHINE

$$(2,4,6\text{-}t\text{-Bu}_3C_6H_2)PCl_2 \xrightarrow{\text{Li[AlH}_4]} (2,4,6\text{-}t\text{-Bu}_3C_6H_2)PH_2$$

Procedure

The phosphonous dichloride $(2,4,6\text{-}t\text{-Bu}_3C_6H_2)PCl_2$(13.9 g, 40 mmol) in THF (170 mL), contained in a 250-mL Schlenk flask, is slowly added via a cannula to a second similar flask of 500-mL capacity containing a stirred slurry of $LiAlH_4$ (3 g, 80 mmol) in THF (180 mL) at $-5\,°C$. The resulting suspension is allowed to warm to room temperature and is then refluxed for 2 h. After cooling to $0\,°C$, 120 mL of 20% HCl is added carefully with stirring.

■ **Caution.** *Addition must be slow to avoid excessive gas evolution.*

After complete addition, the mixture is allowed to warm to room temperature. When stirring is stopped, an upper yellow organic layer separates from the lower aqueous layer, and the former is then removed *via* a cannula to a separate flask.* This solution is then dried by addition of anhydrous $MgSO_4$, filtered through Celite,® and finally all volatile substances

---

*If good separation of the organic and aqueous layers does not occur, the addition of 3×50-mL portions of pentane is recommended.

are removed *in vacuo*, affording a crude yellow-white product. Recrystalliza-tion from *i*-PrOH affords pure $2,4,6$-*t*-$Bu_3C_6H_2PH_2$. Yield: 8.9 g (80%); mp 144–146 °C. ^{31}P NMR (CH_2Cl_2): $\delta - 132$ (t' $J_{PH} = 209$ Hz).

D. (2,4,6-TRI-*TERT*-BUTYLPHENYL) (TRIMETHYLSILYL)PHOSPHINE

$$RPH_2 + n\text{-}BuLi \longrightarrow RPHLi + C_4H_{10}$$

$$RPHLi + ClSiMe_3 \longrightarrow RP(SiMe_3)H + LiCl$$

$$R = 2,4,6\text{-}t\text{-}Bu_3C_6H_2$$

Procedure

A solution of $2,4,6$-*t*-$Bu_3C_6H_2PH_2$ (5.57 g, 20.0 mmol) in 250 mL of THF is placed in a 500-mL Schlenk flask equipped with a rubber septum and a magnetic stirrer bar. A solution of BuLi (13.8 mL, 1.6 *M*) is added dropwise with stirring at 0 °C. After warming to room temperature and stirring for 30 min, the reaction mixture is recooled to 0 °C and Me_3SiCl (2.39 g. 22.0 mmol) is added by means of a syringe. After stirring the reaction mixture for 1 h at room temperature, the solvent and volatile substances are removed *in vacuo*. The white solid residue is extracted with 150 mL of hexane and filtered to remove LiCl. Evacuation of the hexane leaves colorless, oily $(2,4,6$-*t*-$Bu_3C_6H_2)P(SiMe_3)H$, which crystallizes on cooling to 0 °C. ^{31}P NMR: (hexane) $\delta - 127.5$, $^1J_{PH} = 212$ Hz. Yield: 6.7 g(95%).

E. TRIS(TRIMETHYLSILYL)METHANE

$$Me_3SiCl + 2Li \longrightarrow (Me_3Si)Li + LiCl$$

$$3(Me_3Si)Li + CHCl_3 \longrightarrow (Me_3Si)_3CH + 3LiCl$$

Procedure

This procedure is a modification of that published by Merker and Scott.[5] A carefully over-dried and argon purged apparatus consisting of a 5-L three-necked flask attached to a Friedrichs condenser, a 500-mL addition funnel, and an overhead stirrer is assembled. The flask is charged with 50 g (7.2 mol) of lithium wire (Alfa) pressed from sticks (1-mm diameter), Me_3SiCl (Aldrich) (1080 mL, 8.54 mol) and THF (2800 mL). The resulting mixture is stirred and brought to reflux. The contents of the addition funnel, previously charged with $CHCl_3$ (94.74 mL, 1.176 mol) and THF (400 mL), are added dropwise to the main reaction vessel over a period of 5 h with continuous refluxing.

- **Caution.** *Chloroform is a suspected carcinogen. It should be handled in a well-ventilated fume hood, and protective gloves should be worn.*

After complete addition, refluxing is maintained for 2 days, followed by 2 more days at room temperature with continuous stirring. At the end of this time most of the solvent is removed by distillation (~ 3 L). The precipitate of LiCl is allowed to settle, and the liquid portion is decanted into a 2-L flask. This solution is quenched with water (500 mL), after which the organic layer is transferred to a new flask and dried overnight with 50 g of $MgSO_4$. Vacuum distillation of the resulting liquid affords 197 g (72% yield based on $CHCl_3$) of $(Me_3Si)_3CH$ in pure form (bp 104 °C at 20 torr). ^{1}H NMR: (C_6D_6) δ 1.32 (s, 9H, *p-t*-Bu), 1.60 (s, 18H, *o-t*-Bu), 4.22 (d, 2H, PH, $^1J_{PH} = 209$ Hz), 7.52 (d, 2H, C_6H_2, $^4J_{PH} = 2.3$ Hz); mp 144–146 °C.

F. [TRIS(TRIMETHYLSILYL)METHYL]PHOSPHONOUS DICHLORIDE

$$(Me_3Si)_3CH + MeLi \longrightarrow [(Me_3Si)_3C]Li + CH_4$$

$$[(Me_3Si)_3C]Li + PCl_3 \longrightarrow [(Me_3Si)_3C]PCl_2 + LiCl$$

Procedure

This is a modification of the procedure published by Issleib et al.[6] To a 500-mL round-bottomed flask is added 100 mL of THF, 17 mL of Et_2O, $(Me_3Si)_3CH$ (18.25 g, 78.6 mmol) and (dropwise) 62 mL of 1.4 M (86.8 mmol) MeLi (Aldrich) in Et_2O solution.

The reaction mixture is stirred overnight at room temperature by which time it assumes a clear orange-red color. Refluxing the solution for an additional 4 h destroys any unreacted MeLi and results in the development of a deep red color. Upon cooling to room temperature the $(Me_3Si)_3CLi$ will remain stable for periods up to 24 h.

The solution of $(Me_3Si)_3CLi$ is cooled to 0 °C and is added dropwise *via* a cannula* at 0 °C to a stirred solution of PCl_3 (6.7 mL, 78.6 mmol) in THF (40 mL). The mixture is warmed to room temperature, allowed to stir for 2 h, and the solvents are removed *in vacuo*. The residue is extracted with 100 mL of pentane and the extract is filtered through a fine frit. After removal of pentane, the product is recrystallized from dry (magnesium), deoxygenated methanol. Yield: 17.5 g (67%) of waxy white crystals of $[(Me_3Si)_3C]PCl_2$

---

*Checkers used a dropping funnel.

is obtained. ^{31}P NMR, δ 233; ^1H NMR (C_6D_6) δ 0.44 (d, 27H, Me_3Si, $^5J_{PH} = 1$ Hz); mp, 158 °C (dec).

References

1. A. H. Cowley and N. C. Norman, *Prog. Inorg. Chem.*, **34**, 1 (1986).
2. A. H. Cowley, N. C. Norman, and M. Pakuslki, *Inorg. Synth.*, **27**, 240 (1990).
3. D. E. Pearson, M. G. Frazer, V. S. Frazer, and L. C. Washburn, *Synthesis*, 621 (1976).
4. M. Yoshifuji, I. Shima, N. Inamoto, K. Hirotsu, and T. Higuchi, *J. Am. Chem. Soc.*, **103**, 4587 (1981); **104**, 6167 (1982).
5. R. Merker and M. Scott, *J. Organometal. Chem.*, **4**, 98 (1965).
6. K. Issleib, M. Schmidt, and C. Wirkner, *Z. Chem.*, **20**, 153 (1986).

47. COMPOUNDS WITH UNSUPPORTED P=P BONDS

Submitted by ALAN H. COWLEY,* NICHOLAS C. NORMAN,†
and MAREK PAKULSKI*
Checked by G. BECKER,‡ M. LAYH,‡ E. KIRCHNER,‡ and M. SCHMIDT‡

Recently, compounds with unsupported double bonds between phosphorus and/or arsenic atoms have become available.[1] This development has been achieved by the use of sterically demanding groups such as $2,4,6\text{-}t\text{-}Bu_3C_6H_2$, $(Me_3Si)_3C$, and $(Me_3Si)_2CH$. Essentially two strategies are available for the synthesis of diphosphenes (RP=PR) and diarsenes (RAs=AsR), namely, reductive cleavage of dihalides (REX$_2$, E = P and As), and the base-promoted dehydrohalogenation of equimolar mixtures of REX$_2$ and REH$_2$. The latter method is also appropriate for phosphaarsene (RP=AsR) syntheses; however, better yields are realized if REX$_2$ is treated with the silyllithium reagent, $RP(Li)(SiMe_3)$.

The preparations of the starting materials $[(Me_3Si)_3C]PCl_2$, $(2,4,6\text{-}t\text{-}Bu_3C_6H_2)PCl_2$, $(2,4,6\text{-}t\text{-}Bu_3C_6H_2)PH_2$, and $(2,4,6\text{-}t\text{-}Bu_3C_6H_2)P(SiMe_3)H$ are described in Section 46.[2] The compound $[(Me_3Si)_2CH]PCl_2$ is prepared as described in ref. 3 and 4.

■ **Caution.** *Phosphines and their chloro-substituted derivatives are toxic and air sensitive. Silyllithium phosphides, (naphthalene)sodium, and alkyllithium*

*Department of Chemistry, The University of Texas at Austin, Austin, TX 78712.
†Department of Inorganic Chemistry, University of Newcastle upon Tyne, Newcastle upon Tyne NE1 7RU, United Kingdom.
‡Institut für Anorganische Chemie der Universität Stuttgart Pfaffenwaldring 55, D-7000 Stuttgart 80, Federal Republic of Germany.

reagents are very sensitive to oxygen and moisture. All operations involving these materials should be carried out under an inert atmosphere and in a well-ventilated fume hood.

A. BIS[TRIS(TRIMETHYLSILYL)METHYL]-DIPHOSPHENE AND BIS(2,4,6-TRI-*TERT*-BUTYLPHENYL)-DIPHOSPHENE

1. (Naphthalene)sodium Method

$$2RPCl_2 + 4Na[C_{10}H_8] \longrightarrow RP{=}PR + 4NaCl + 4C_{10}H_8$$

$$R = (Me_3Si)_3C \text{ or } 2,4,6\text{-}t\text{-}Bu_3C_6H_2$$

Procedure

This method is suitable only for small scale (5 mmol or less) preparations of these diphosphenes because of the difficulties involved in removing larger quantities of naphthalene from the product.

A solution of $[(Me_3Si)_3C]PCl_2$ (1.67 g, 5.0 mmol) in dry, degassed THF (100 mL) is added to a nitrogen-filled 150-mL Schlenk flask equipped with a rubber septum. In a separate 50-mL Schlenk flask (also fitted with a rubber septum), a solution of (naphthalene)sodium[5] in tetrahydrofuran (THF) (20 mL, 0.5 M) is cooled to $-78\,°C$ (Dry Ice–acetone). The (naphthalene)sodium solution is transferred dropwise by means of a cannula to the flask containing the $[(Me_3Si)_3C]PCl_2$. The reaction mixture is maintained at $-78\,°C$ during the addition, after which it is allowed to warm gradually to room temperature. Monitoring of the reaction by ^{31}P NMR spectroscopy indicates the quantitative conversion of $[(Me_3Si)_3C]PCl_2$ to $[(Me_3Si)_3C]P{=}P[C(SiMe_3)_3]$. Removal of the solvent *in vacuo* leaves a brown residue, from which the naphthalene is removed by dynamic vacuum sublimation (55 °C, 10^{-4} torr) for 12 h. The residue is then dissolved in 100 mL of hexane and the NaCl is removed by filtration through a 3-cm length of Celite® under N_2. Further purification of the solution is effected by column chromatography under N_2 on a 30-cm length of silica gel, using hexane as eluant.* The leading (orange-yellow) band is collected and evaporated to dryness to afford a 63% (0.83 g) yield of $[(Me_3Si)_3C]P{=}P[C(SiMe_3)_3]$, mp 152 °C (dec). If desired, further purification can be carried out by recrystallization from a saturated solution in toluene at 0 °C.

---

*Checkers indicate that column chromatography is not always required.

Anal. Calcd. for $C_{20}H_{54}P_2Si_6$: C, 45.76; H, 10.37. Found: C, 45.80; H, 10.95. ^{31}P NMR: $\delta + 599.6$.

The corresponding aryl-substituted diphosphene, $(2,4,6\text{-}t\text{-}Bu_3C_6H_2)$ P=P-$(2,4,6\text{-}t\text{-}Bu_3C_6H_2)$, mp 175–176 °C, is made by the same method in 72% yield using $(2,4,6\text{-}t\text{-}Bu_3C_6H_2)PCl_2$ (1.74g, 5.0 mmol).* If necessary, additional purification can be carried out by recrystallization from hexane at -10 °C.

Anal. Calcd. for $C_{36}H_{58}P_2$: C, 78.28; H, 10.58. Found: C, 78.19; H, 10.60. ^{31}P NMR $\delta + 494$.

2. Magnesium Method

$$2RPCl_2 + 2Mg \longrightarrow RP=PR + 2MgCl_2$$

$$R = 2,4,6\text{-}t\text{-}Bu_3C_6H_2$$

Procedure

This is essentially the preparation reported by Yoshifuji et al.,[6] the only significant modification being the use of ultrasonics to minimize the reaction time. A mixture of $(2,4,6\text{-}t\text{-}Bu_3C_6H_2)PCl_2$ (3.47 g, 10 mmol) and Mg turnings (0.27 g, 11 mmol) in 25 mL of THF is placed in a 100-mL Schlenk flask under a nitrogen atmosphere and closed with a rubber septum. The flask is placed in an ultrasonic bath for 30 to 50 min.† Evacuation of the THF leaves an orange-yellow residue that is extracted with 3 × 50-mL portions of hexane. Removal of the $MgCl_2$ and residual Mg is achieved by means of a cannula, the tip of which is covered with filter paper held in place by copper wire. Orange crystals of $(2,4,6\text{-}t\text{-}Bu_3C_6H_2)P=P(2,4,6\text{-}t\text{-}Bu_3C_6H_2)$, mp 175–176 °C, form upon reduction of the solution volume by ~ 40 mL followed by storage at -10 °C. A second crop of the diphosphene is obtained after column chromatographic work-up of the mother liquor as described previously. The combined yield of product is 1.50 g (54%).‡

---

*Checkers recrystallized from hexane at -20 °C. Yield: 68%. Chromatography was not employed.

†The irradiation time is highly dependent on the power of the ultrasonic device. A color change of the reaction mixture from orange-brown to deep purple indicates reduction of the diphosphene to the corresponding anion radial. If this occurs, the flask should be removed from the ultrasonic bath. Addition of few drops of MeOH restores the orange color.

‡Checkers remove $MgCl_2$ by filtration through sintered glass and indicate that column chromatography is not required.

References

1. A. H. Cowley and N. C. Norman, *Prog. Inorg. Chem.*, **34**, 1 (1986).
2. A. H. Cowley, N. C. Norman, and M. Pakulski, *Inorg. Synth.*, **27**, 235 (1990).
3. M. J. S. Gynane, A. Hudson, M. F. Lappert, and P. P. Power, *J. Chem. Soc. Chem. Commun.*, **1976**, 623; M. J. S. Gynane, A. Hudson, M. F. Lappert, P. P. Power, and M. Goldwhite, *J. Chem. Soc. Dalton Trans.*, 1980, 2428.
4. The preparation of the $(Me_3Si)_2CH$ derivatives is facilitated by the one-pot synthesis of the precursor, $(Me_3Si)_2CHCl$: A. H. Cowley and R. A. Kemp, *Synth. React. Inorg. Met. Org. Chem.*, **11**, 591 (1981).
5. Houben-Weyl: *Methoden der Organischen Cheme*, Vol. 13/1, Thieme Verlag Stuttgart 1970, p. 381.
6. M. Yoshifuji, I. Shima, N. Inamoto, K. Hirotsu, and T. Higuchi, *J. Am. Chem. Soc.*, **103**, 4587 (1981); **104**, 6167 (1982).

48. TRIS(TRIMETHYLSILYL)PHOSPHINE AND LITHIUM BIS(TRIMETHYLSILYL)PHOSPHIDE·BIS-(TETRAHYDROFURAN)

Submitted by GERD BECKER,* HELMUT SCHMIDT,* GUDRUN UHL,* and WERNER UHL*
Checked by MANFRED REGITZ,† WOLFGANG RÖSCH,† and UWE-JOSEF VOGELBACHER†

Tris(trimethylsilyl)phosphine[1] and its more reactive derivative lithium bis-(trimethylsilyl)phosphide·2tetrahydrofuran[2] are very useful reagents for the preparation of compounds with a single or a multiple element phosphorus bond. They react readily with various element halides,[3-5] with carboxylic acid chlorides,[6] and with carboxylic esters,[7] as well as with other organic electrophiles[7] *via* a substitution of lithium and/or a cleavage of the weak polar Si-P bonds.

A. TRIS(TRIMETHYLSILYL)PHOSPHINE

$$3M + \tfrac{1}{4}P_4 \xrightarrow{\langle DME \rangle} M_3P \xrightarrow[-3MCl]{+3(H_3C)_3SiCl} [(H_3C)_3Si]_3P$$

$$M = Na, K; DME = 1,2\text{-dimethoxyethane}$$

*Institut für Anorganische Chemie der Universität Stuttgart, Pfaffenwaldring 55, D-7000 Stuttgart 80, Federal Republic of Germany.
†Fachbereich Chemie der Universität Kaiserslautern, Erwin-Schrödinger-Straße, D-6750 Kaiserslautern, Federal Republic of Germany.

■ **Caution.** *Sodium and particularly potassium react violently with water and may ignite in air. The incrustations of potassium may be explosive; only potassium with little or no incrustation should be cut, and care should be taken to ensure anhydrous conditions. Sodium–potassium alloy as well as malodorous tris(trimethylsilyl)phosphine are pyrophoric and extremely sensitive to moisture. Therefore the entire procedure must be carried out in an atmosphere of dry argon in a well-ventilated fume hood. In case of emergency a sand bath should be available to catch the liquid alloy. 1, 2-Dimethoxyethane and tetrahydrofuran (THF) may form explosive peroxides. Only fresh, peroxide-free material should be used.*

These solvents must be deoxygenated and freed from traces of water before use by refluxing over sodium wire in the presence of benzophenone in a slight stream of argon. When the color has changed to *deep* blue, the solvent is distilled off. Pentane is purified in a similar way with lithium tetrahydrido-aluminate.

Procedure

The complete equipment used in this procedure must first be evacuated and then filled with argon. All joints are greased with H-grease (Apiezon Products). When stoppers are removed and compounds are transferred from one flask to the other, a slight stream of argon must be maintained.

A 46-g quantity (2.0 mol) of sodium and 59 g (1.5 mol) of potassium are purified by removing their incrustations under petroleum ether. The metals are cut into pieces and placed in a 500-mL two-necked round-bottomed flask provided with a 500-mL pressure equalizing dropping funnel and connected to the argon–vacuum supply *via* a side arm with a stopcock (see Fig. 1). After evaporation of the residual organic solvent *in vacuo*, the sodium–potassium alloy is formed by warming the metals cautiously with a hair

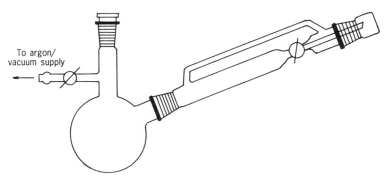

To argon/
vacuum supply

Fig. 1. Apparatus for preparing sodium–potassium alloy.

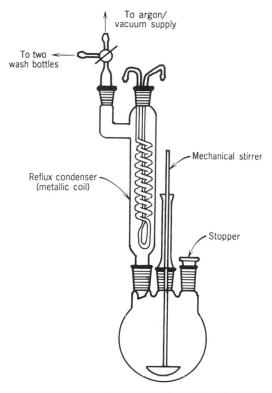

Fig. 2. Apparatus for preparing sodium–potassium phosphide and reacting it with chlorotrimethylsilane.

dryer, while the flask is shaken occasionally. The alloy is then poured into the dropping funnel.

■ **Caution.** *White phosphorus is extremely toxic and causes severe burns; gloves should be worn when handling the solid or solution.*

A 2-L three-necked flask is equipped with a reflux condenser* and a mechanical stirrer, which is sealed with paraffin oil. The condenser is fitted with a three-way stopcock connecting to an argon–vacuum supply and to two wash bottles filled with paraffin oil (see Fig. 2). Small pieces of white phosphorus (31 g, 1 mol) are placed in the flask and freed from traces of water *in vacuo.*† After adding 1 L of DME the dropping funnel with the liquid alloy

---

*The use of a reflux condenser with a metallic coil is strongly recommended.
†The checkers report white phosphorus can also be freed from traces of water by washing with 50 mL of DME.

is mounted on the free neck of the flask, and the stopcock is switched from the argon supply to the wash bottles. The solvent is heated to 50 °C and the alloy is added slowly, in the beginning drop by drop, over a period of 1 h to the vigorously stirred emulsion. To complete the formation of black sodium–potassium phosphide the suspension must be heated at reflux with a heating mantle for 24 h.

A slight excess of freshly distilled chlorotrimethylsilane (353 g, 3.25 mol) is added carefully to the phosphide suspension while stirring and refluxing. These conditions are maintained for another 24 h. During this period, the reaction mixture becomes increasingly viscous,* and its color changes from black to gray. The dropping funnel is exchanged for an argon inlet; the reflux condenser and stirrer are removed in a slight stream of argon. After filtration (see Fig. 3) through sintered glass (porosity 50 μm), the residue is washed several times with a total of 2 L of DME. Finally, the solvent is distilled off at atmospheric pressure, and the remaining tris(trimethylsilyl)phosphine is purified by vacuum distillation (30-cm-Vigreux column). Bp 30–35 °C/ 10^{-3} mbar. Yield: 188 g (75%).

Anal. Calcd. for $C_9H_{27}Si_3P$: C, 43.1; H, 10.9; Si, 33.6; P, 12.4. Found: C, 43.3; H, 11.0; Si, 33.5; P, 12.0.

■ **Caution.** *On cooling, the pure distillation product may crystallize rapidly and block the apparatus. The residues from the filtration and distillation may contain either pyrophoric sodium–potassium alloy or tris(trimethylsilyl)-phosphine and trimethylsilyl-substituted cyclic polyphosphines.[8] For safety reasons all residues must be treated very cautiously with a mixture of ethanol and petroleum ether (1:10) under an inert atmosphere in a well-ventilated fume hood.*

Properties

Tris(trimethylsilyl)phosphine is a colorless compound with a melting point near room temperature. Characteristic NMR data for a DME solution are $\delta_{(^1H)}$ 0.30; $^3J_{P-H} = 4.4$ cps; $\delta_{(^{31}P)}$-251. In reactions with alcohols and water bis(trimethylsilyl) and (trimethylsilyl)phosphine can be synthesized;[9] acetyl chloride in a molar ratio of 1:3 gives triacetylphosphine.[10] For the important reaction with 2,2-dimethylpropionyl chloride see Section 49.

---

*In case of insufficient stirring, incrustation and subsequent decomposition of tris(trimethylsilyl)-phosphine may occur. Therefore, the whole process should be continually supervised.

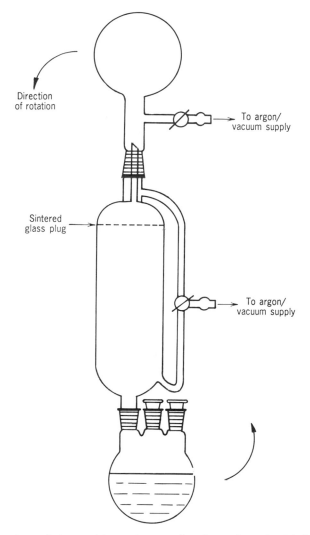

Fig. 3. Filtration of the reaction mixture after formation of tris(trimethylsilyl)-phosphine.

B. LITHIUM BIS(TRIMETHYLSILYL)PHOSPHIDE·BIS(TETRAHYDROFURAN)[2]

$$[(H_3C)_3Si]_3P \xrightarrow[-(H_3C)_4Si]{+\frac{1}{4}(H_3CLi)_4;\langle THF \rangle} [(H_3C)_3Si]_2PLi \cdot 2\,THF$$

THF = tetrahydrofuran

■ **Caution.** *Methyllithium and lithium bis(trimethylsilyl)phosphide·bis-(tetrahydrofuran) are very sensitive to moisture, and even in solution they are flammable in air. Therefore the entire procedure must be carried out in an atmosphere of dry argon in a well-ventilated fume hood. Solvents should be dried and deoxygenated as described previously.*

Procedure

In a single-necked 1-L flask connected to the argon–vacuum supply *via* a side arm with a stopcock and equipped with a pressure equalizing dropping funnel and a magnetic stirring bar, 62.6 g (0.25 mol) of tris(trimethylsilyl)-phosphine is dissolved in 300 mL of THF and cooled to 0 °C. An ~ 1 *M* solution of methyllithium* in diethyl ether (0.245 mol[†]), the molarity of which has been determined by titration,[‡] is added over a period of 1 h. After allowing the reaction mixture to warm to 20 °C and stirring if for another 8 h, all volatile compounds are distilled off under reduced pressure at room temperature. The brown residue is covered with 400 mL of pentane. While stirring the suspension, THF (~ 20 mL) is slowly added until complete dissolution has been achieved. At − 30 °C colorless to pale yellow crystals of lithium bis(trimethylsilyl)phosphide·bis(tetrahydrofuran) precipitate. The liquid phase is carefully removed from the cold solution with a pipette. The precipitate is ground under argon with a flattened glass rod, and the remaining solvent is removed *in vacuo* (minimum pressure 0.3 mbar). The mother liquor is worked-up as above to get another crop of crystals. The exact THF content must be determined by ¹H NMR spectroscopy. Yield: 64 g (80%).

---

*Solutions of methyllithium are prepared before use from methyl chloride and pieces of lithium in diethyl ether.[11]

 With solutions of methyllithium in diethyl ether, either commercially available or prepared from lithium and methyl chloride, the checkers obtained lithium bis(trimethylsilyl)phosphide·bis(tetrahydrofuran) in only 40 or 50% yield. Therefore they advise preparing the compound with *n*-butyllithium as described in the literature.[2] The submitters report reproducible yields of 80%.

[†]Since excess tris(trimethylsilyl)phosphine can easily be removed by recrystallization of the product, a slight deficiency of methyllithium is recommended.

[‡]A 10-mL quantity of the solution is cautiously hydrolized in 25-mL of 1 *M* HCl. Excess acid is determined by back titration with a 1 *M* solution of sodium hydroxide.

Properties

Lithium bis(trimethylsilyl)phosphide·bis(tetrahydrofuran) is a colorless crystalline compound that has been used by several research groups for introducing the bis(trimethylsilyl)phosphinyl group.[3-6] Characteristic NMR data of the THF solution are $\delta_{(^1H)}$ 0.53, $^3J_{P-H} = 4.0$ cps; $\delta_{(^{13}P)} - 298$.

References

1. G. Becker and W. Hölderich, *Chem. Ber.*, **108**, 2484 (1975).
2. G. Fritz and W. Hölderich, *Z. Anorg. Allg. Chem.*, **422**, 104 (1976); E. Hey, P. B. Hitchcock, M. F. Lappert, and A. K. Rai, *J. Organomet. Chem.*, **325**, 1 (1987).
3. G. Fritz and W. Hölderich, *Z. Anorg. Allg. Chem.*, **431**, 61 (1977).
4. G. Fritz and W. Hölderich, *Z. Anorg. Allg. Chem.*, **431**, 76 (1977).
5. L. Weber and D. Bungardt, *J. Organomet. Chem.*, **311**, 269 (1986).
6. G. Becker, W. Becker, and O. Mundt, *Phosphorus Sulfur*, **14**, 267 (1983); T. Allspach, M. Regitz, G. Becker, and W. Becker, *Synthesis*, **1986**, 31.
7. G. Becker, W. Becker, R. Knebl, H. Schmidt, U. Weeber, and M. Westerhausen, *Nova Acta Leopold.*, **59**, 55 (1985); G. Becker, W. Becker, R. Knebl, H. Schmidt, U. Hildenbrand, and M. Westerhausen, *Phosphorus Sulfur*, **30**, 349 (1987).
8. G. Fritz and W. Hölderich, *Naturwissenschaften*, **62**, 573 (1975).
9. H. Bürger and U. Goetze, *J. Organomet. Chem.*, **12**, 451 (1968).
10. G. Becker, *Z. Anorg. Allg. Chem.*, **480**, 21 (1981).
11. Houben-Weyl, *Methoden der Organischen Chemie*, 4th ed., Vol. 13/1, Thieme Verlag, Stuttgart, 1970, p. 135.

49. (2, 2-DIMETHYLPROPYLIDYNE)PHOSPHINE

Submitted by GERD BECKER,* HELMUT SCHMIDT,* GUDRUN UHL,* and WERNER UHL*
Checked by MANFRED REGITZ,† WOLFGANG RÖSCH,† and UWE-JOSEF VOGELBACHER†

(2, 2-Dimethylpropylidyne)phosphine is the first compound with a phosphorus–carbon triple bond stable at room temperature.[1] Since the electronegativities of phosphorus (2.2) and carbon (2.5) differ substantially from the value of nitrogen (3.0), the compound is more likely to react as the analog of an alkine than of a nitrile. Thus far, the reactivity of the phosphine

*Institut für Anorganische Chemie der Universität Stuttgart, Pfaffenwaldring 55, D-7000 Stuttgart 80, Federal Republic of Germany.
†Fachbereich Chemie der Universität Kaiserslautern, Erwin-Schrödinger-Straße, D-6750 Kaiserslautern, Federal Republic of Germany.

towards transition metal complexes with the metals in low oxidation states[2] and towards organic 1, 3-dipolar[3] or Diels–Alder reagents,[3,4] has been studied. With halides of main group and subgroup elements[5] numerous insertion reactions have been observed; from the addition of lithium bis(trimethylsilyl)phosphide·2tetrahydrofuran (THF) in 1, 2-dimethoxethane (DME), lithium 3, 5-di-*tert*-butyl-1, 2, 4-triphospholid·3DME can be obtained.[5,6]

A. [2, 2-DIMETHYL-1-(TRIMETHYLSILILOXY)PROPYLIDENE]-(TRIMETHYLSILYL)PHOSPHINE[7,*]

$$[(H_3C)_3Si]_3P \xrightarrow[-(H_3C)_3SiCl]{+(H_3C)_3CCOCl} [(H_3C)_3Si]_2PCOC(CH_3)_3$$

$$\longrightarrow \quad \begin{array}{c} (H_3C)_3Si \\ \diagdown \\ P{=}C \\ \diagup \diagdown \\ \end{array} \begin{array}{c} O{-}Si(CH_3)_3 \\ \diagup \\ \\ C(CH_3)_3 \end{array}$$

■ **Caution.** *[2, 2-Dimethyl-1-(trimethylsiloxy)propylidene](trimethyl-silyl)phosphine, a malodorous liquid, is extremely sensitive to moisture and ignites in air. Therefore, all procedures must be carried out under an atmosphere of dry argon in a well-ventilated fume hood.*

Procedure

In a 500-mL round-bottomed flask connected to the argon–vacuum supply *via* a side arm with a stopcock, 50.0 g (0.20 mol) of tris (trimethylsilyl)phosphine (see Section 48) is dissolved in 200 mL of LiAlH₄-dried pentane. At room temperature, 26.5 g (0.22 mol; 10% excess) of 2, 2-dimethylpropionyl chloride distilled over calcium hydride before use,[†] is slowly added to the stirred solution. Although the reaction is usually complete after 24 h, its progress should be followed by NMR spectroscopy[‡] in order to ensure

---

[*]Starting materials for the corresponding adamant-1-yl derivative are lithium bis(trimethylsilyl)-phosphide·2tetrahydrofuran and adamantoyl chloride.[8]

[†]This purification is necessary to remove traces of hydrogen chloride that otherwise would catalyze a rapid formation of tris(2, 2-dimethylpropionyl)phosphine.[7]

[‡]Since the ¹H chemical shift values depend strongly on the solvent used, the compounds tri(trimethylsily)phosphine (I) and [2, 2-dimethyl-1-(trimethylsiloxy)propylidene] (trimethyl-silyl)phosphine (II) can best be identified by their $^3J_{P-H}$ coupling constants (I:4.4 cps; II:3.8 cps), or by their $\delta_{(^{31}P)}$ values (I: −251; II: +120).

complete consumption of tris(trimethylsilyl)phosphine and to avoid further substitution.* After removing the volatile compounds under reduced pressure, the residual yellow alkylidenephosphine is purified by vacuum distillation. Bp 45–48 °C/10^{-3} mbar. Yield: 47.2 g (90%).

Anal. Calcd. for $C_{11}H_{27}OSi_2P$: Si, 21.4; P, 11.8. Found: Si, 21.5; P, 11.7.

Properties

[2, 2-Dimethyl-1-(trimethylsiloxy)propylidene] (trimethylsilyl)phosphine is a yellow viscous liquid. Characteristic NMR data for a solution in benzene are $(H_3C)_3C$ $\delta_{(^1H)}$ 1.25, $^4J_{P-H} = 1.5$; $(H_3C)_3Si$—O $\delta_{(^1H)}$ 0.27; $(H_3C)_3Si$—P $\delta_{(^1H)}$ 0.31, $^3J_{P-H} = 3.8$ cps; $\delta_{(^{31}P)} + 120$. Substitution of another trimethylsilyl group by the 2, 2-dimethylpropionyl moiety leads to (2, 2-dimethylpropionyl)-[2, 2-dimethyl-1-(trimethylsiloxy)propylidene]phosphine, which is the starting material for the keto–enol isomeric bis(2, 2-dimethylpropionyl)-phosphine,[9] whereas reaction with alcohols gives (2, 2-dimethylpropionyl)-phosphine.[10]

B.   (2, 2-DIMETHYLPROPYLIDYNE)PHOSPHINE[1,11]

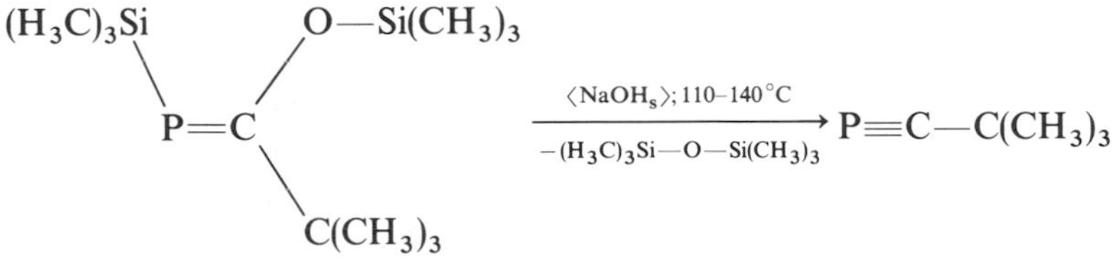

■ **Caution.** *[2, 2-Dimethyl-1-(trimethylsiloxy)propylidene](trimethyl-silyl)phosphine is extremely sensitive to moisture and ignites in air. Therefore, all procedures must be carried out under an atmosphere of dry argon in a well-ventilated fume hood.*

Procedure

The apparatus (Fig. 1) consists of a two-necked 250-mL flask, one neck of which is equipped with a 100-mL pressure equalizing dropping funnel and a stopcock connected to the argon–vacuum supply, while the other neck is connected to a 200-mL cooling trap *via* a bent glass tube.† After adding

---

*The checkers report reaction times of 6 to 7 h using cyclohexane as the solvent and heating under reflux.

†Instead of a bent glass tube the checkers used a reflux condenser with running water.

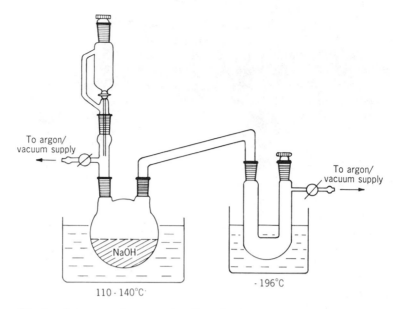

Fig. 1. Apparatus for preparing (2, 2-dimethylpropylidyne)phosphine.

50.0 g (0.19 mol) of [2, 2-dimethyl-1-(trimethylsiloxy)propylidene](trimethyl-silyl)phosphine to the dropping funnel and ~ 10 g of granular sodium hydroxide to the 250-mL flask, the pressure within the apparatus is reduced to ~ 1 mbar, and the catalyst is heated to 110–140 °C with an oil bath. The alkylidenephosphine is then added very slowly over a period of 2 h in order to achieve a complete elimination of hexamethyldisiloxane. All volatile compounds are collected at − 196 °C and separated later by fractional distillation under normal pressure.* Bp 52–57 °C. Yield after distillation: 14.0 g (74%).

Anal. Calcd. for C_5H_9P: C, 60.0; H, 9.1; P, 30.9. Found: C, 60.2; H, 9.0; P, 30.3.

Properties

(2, 2-Dimethylpropylidyne)phosphine is a fairly volatile compound that is scarcely oxidized by air. Characteristic IR[1,12] and NMR data[1] are $\tilde{\nu}_{C\equiv P}$ 1533 cm^{-1}; $\delta_{(^1H)}$ 1.15, $^4J_{P-H} = 0.9$ cps; $\delta_{(^{31}P)}$ − 69. Its chemical properties have recently been summarized.[2,13]

---

*Because of its high vapor pressure at room temperature, the distillate should be cooled with an ice bath.

References

1. G. Becker, G. Gresser, and W. Uhl, *Z. Naturforsch. Teil B*, **36**, 16 (1981).
2. J. F. Nixon, *Chem. Rev.*, **88**, 1327 (1988).
3. M. Regitz, W. Rösch, T. Allspach, U. Annen, K. Blatter, J. Fink, M. Hermesdorf, H. Heydt, U. Vogelbacher, and O. Wagner, *Phosphorus Sulfur*, **30**, 479 (1987).
4. U. Annen and M. Regitz, *Tetrahedron Lett.*, **28**, 5141 (1987).
5. G. Becker, W. Becker, R. Knebl, H. Schmidt, U. Weeber, and M. Westerhausen, *Nova Acta Leopold.*, **59**, 55 (1985); G. Becker, W. Becker, R. Knebl, H. Schmidt, U. Hildenbrand, and M. Westerhausen, *Phosphorus Sulfur*, **30**, 349 (1987).
6. R. Bartsch, P. B. Hitchcock, and J. F. Nixon, *J. Chem. Soc. Chem. Commun.*, **1987**, 1146.
7. G. Becker, *Z. Anorg. Allg. Chem.*, **430**, 66 (1977).
8. T. Allspach, M. Regitz, G. Becker, and W. Becker, *Synthesis*, **1986**, 31.
9. G. Becker, M. Rössler, and G. Uhl. *Z. Anorg. Allg. Chem.*, **495**, 73 (1982).
10. G. Becker, M. Rössler, and W. Uhl, *Z. Anorg. Allg. Chem.*, **473**, 7 (1981).
11. W. Rösch, U. Hees, and M. Regitz, *Chem. Ber.*, **120**, 1645 (1987).
12. I. I. Patsanovskii, A. Kh. Pliamovatyi, I. Kh. Shakirov, Iu. Z. Stepanova, G. Becker, R. Knebl, U. Weeber, E. A. Ishmaeva, R. R. Shagidullin, and A. N. Pudovik, *Dokl. Akad. Nauk SSSR*, **290**, 400 (1986); *Chem. Abstr.*, **107**, 59106v, 744 (1987).
13. M. Regitz and P. Binger, *Angew. Chem. Int. Ed. Engl.*, **27**, 1484 (1988).

50. TRIPHOSPHENIUM SALTS

Submitted by ALFRED SCHMIDPETER,* and SIEGFRIED LOCHSCHMIDT*
Checked by ALAN H. COWLEY,[†] and MAREK PAKULSKI[†]

$$2R_3P + PCl_3 + 2e^- \longrightarrow R_3P{=}P{-}Pr_3{}^+ + 3Cl^-$$

No defined P(I) chloride is obtained from the reduction of PCl_3[1-3] at ambient temperature.[4‡] In the presence of phosphanes, however, a clear-cut two-electron reduction is observed, and triphosphenium ions are formed.[5-7] They may be viewed as R_3P complexes of P^+. Although the triphosphenium chlorides are unstable, stable representatives are obtained with the help of a chloride acceptor such as $AlCl_3$ (Section A) or by exchange of the chloride ion by a nonbasic anion such as $[BPh_4]^-$ (Section C). The phosphane (Section A and C) or $SnCl_2$ serves as the reducing agent; the oxidation of the latter gives

*Institut für Anorganische Chemie, Universität München, Meiserstraße 1, D-8000 München 2, Federal Republic of Germany.

†Department of Chemistry, The University of Texas at Austin, Austin, TX 78712.

‡Reduction of PCl_3 by magnesium or lithium hydride in polar solvents at $-40\,°C$ results in $(PCl)_6$ which is stable however only in diluted solutions at low temperature.[4]

$SnCl_4$ which acts as a chloride acceptor (Section B).

$$R_3P + Cl^- \longrightarrow R_3PCl^+ + 2e^-$$
$$SnCl_2 + 4Cl^- \longrightarrow SnCl_6^{2-} + 2e^-$$

Five-membered cyclic and amino substituted triphosphenium ions (Section B and C, respectively) are much less sensitive towards heat and chloride than the hexaphenyl triphosphenium ion Section (A).

■ **Caution.** *AlCl$_3$ and PCl$_3$ cause severe skin and eye burns; the latter is also a respiratory irritant. Chloroform is a suspected carcinogen. Work in a well-ventilated fume hood and wear protective gloves and goggles.*

A. 1,1,1,3,3,3-HEXAPHENYLTRIPHOSPHENIUM TETRACHLOROALUMINATE[6]

$$3Ph_3P + PCl_3 + 2AlCl_3 \longrightarrow (Ph_3P{=}P{-}PPh_3)[AlCl_4] + (PClPh_3)[AlCl_4]$$

Procedure

Dichloromethane is dried by passing it through a column (80-cm long, 2.5-cm i.d.) of 4-Å molecular sieves at a rate of $1\,L\,h^{-1}$. Phosphorus trichloride is distilled, Ph_3P and $AlCl_3$ are used as commercially available. The equipment is oven dried, evacuated and flushed with Ar or N_2.

A 1000-mL round-bottomed flask with an inert gas inlet is equipped with a magnetic stirrer ($\sim$ 2-cm long) and closed with a paraffin oil bubbler. To a solution of 53.0 g (204.0 mmol) Ph_3P in 500 mL CH_2Cl_2 at 0 °C is added 22.0 g (163.0 mmol) $AlCl_3$ without agitation. Immediately thereafter, 6.00 mL (68.8 mmol) PCl_3 is added with stirring within 5 min. (To avoid reactions in the system $Ph_3P/CH_2Cl_2/AlCl_3$, stirring should be started only when the PCl_3 is added.) Successful reaction may be checked by $^{31}P\{^1H\}$ NMR: Ph_3PCl^+ (singlet, δ 65.3) and $(Ph_3P)_2P^+$ (A_2B, δ_A 30, δ_B -174, $J_{AB} = 502\,Hz$, $J_{AA} = -26\,Hz$) in a 1:1 ratio. The reaction is complete after all of the $AlCl_3$ has dissolved, approximately within 10 min. The solution is evaporated under reduced pressure to $\frac{2}{3}$ of its volume and kept under Ar or N_2 at 3 °C for 2 weeks. During this time its volume is further reduced, by successive evaporations of not more than 25 mL each time, to $\sim$ 200 mL. The compound $[(Ph_3P)_2P][AlCl_4]\cdot\frac{1}{2}CH_2Cl_2$ separates as large pale yellow crystals, which float in the solution. (Their size depends on the rate of evaporation. The larger the crystals, the smaller is the loss during the subsequent washing; see below.) Also, some $(PClPh_3)[AlCl_4]$ precipitates as very fine crystals. The two products are easily and cleanly separated by filtration through a coarse glass frit (pore size $\sim 100\,\mu m$). The recovered $[(Ph_3P)_2P][AlCl_4]$ is rapidly washed with 5 mL of precooled $CHCl_3$ and immediately dried in a vacuum. Yield: 36.1 g (68.5%).

Anal. Calcd. for $C_{36}H_{30}P_3[AlCl_4]\cdot\frac{1}{2}CH_2Cl_2$ (766.8): C, 57.17; H, 4.07. Found: C, 56.92; H, 4.13.

Properties

The crystals of $[(Ph_3P)_2P][AlCl_4]\cdot\frac{1}{2}CH_2Cl_2$, mp 78 °C, can be stored under Ar or N_2 without decomposition at 0 °C for months and at -20 °C for a year or longer, but they decompose at room temperature to an orange powder within a week or so.

The crystals can be redissolved in CH_2Cl_2. In this solution $[(Ph_3P)_2P][AlCl_4]$ decomposes with a half-life of 10–12 h at 25 °C under Ar or N_2, giving an orange precipitate roughly according to the equation

$$4[(Ph_3P)_2P][AlCl_4] \longrightarrow 6Ph_3P + 2AlCl_3 + (PClPh_3)[AlCl_4]$$
$$+ (R_3PP_4Cl)[AlCl_4]$$

The decomposition can be prevented by adding $\frac{1}{4}$-mol equivalent of $AlCl_3$. Solutions prepared this way remain unchanged at room temperature for 1 month. Crystals of $[(Ph_3P)_2P][AlCl_4]$ can be handled in air for up to 30 min without obvious change. On longer exposure to air the crystals darken and partially decompose; when dissolved in dichloromethane they unavoidably give some orange precipitate.

More basic phosphanes will replace Ph_3P in $[(Ph_3P)_2P][AlCl_4]$ to produce other triphosphenium tetrachloroaluminates.[6,8] The triphosphenium ions can be protonated and alkylated to triphosphane-1,3-diium ions.[6,9]

B. 1,1,3,3-TETRAPHENYL-1,2,3-TRIPHOSPHOLENYL HEXACHLOROSTANNATE[5]

Procedure

A dry 50-mL round-bottomed flask with an inert gas inlet is equipped with a magnetic stirrer (1-cm long), closed with a paraffin oil bubbler, and flushed with Ar or N_2. Anhydrous $SnCl_2$ (0.57 g, 2.8 mmol) and $Ph_2PC_2H_4PPh_2$ (1.69 g, 4.2 mmol) in 15 mL CH_2Cl_2 (dried as in Section A) react to yield a voluminous white precipitate. This precipitate redissolves when 0.24 mL (2.8 mmol) PCl_3 is added with stirring. After 20 min the solution is evaporated under reduced pressure (oil pump) to $\frac{2}{3}$ its volume and is then kept at $-25\,°C$ for 7 days. The white precipitate of $Ph_2PC_2H_4PPh_2 \cdot SnCl_4$,[9] δ_{31P} 17.3, $J_{(117,119SnP)} = 827$, 865 Hz, is removed, and the filtrate is concentrated to ~ 5 mL. The product separates as colorless crystals within a few days. The crystals are collected on a glass frit, washed quickly twice with 2 mL of cold CH_2Cl_2, and dried in a vacuum. Yield: 0.61 g (45%).

Anal. Calcd. for $[C_{52}H_{48}P_6][SnCl_6] \cdot 2CH_2Cl_2$ (1360.1): C, 47.69; H, 3.85; Cl, 26.07. Found: C, 47.90; H, 3.87; Cl, 24.60.

Properties

The crystals, mp 230 °C, are stable at room temperature in air for at least 1 week, and they may be heated to 120 °C for a short time without decomposition. The $^{31}P\{^1H\}$ NMR: A_2B, $\delta_A + 64$, $\delta_B - 232$, $J_{AB} = 454$.

The corresponding tetrachloroaluminate can be obtained from $[(Ph_3P)_2P][AlCl_4]$ (see Section A) and bis(diphenylphosphino)ethane.[8]

C. 1,1,1,3,3,3-HEXAKIS(DIMETHYLAMINO)TRIPHOS-PHENIUM TETRAPHENYLBORATE[7]

$$3P(NMe_2)_3 + PCl_3 + 2Na[BPh_4]$$

$$\longrightarrow [(Me_2N)_3P{=}P{-}P(NMe_2)_3][BPh_4]$$

$$+ [(Me_2N)_3PCl][BPh_4] + 2NaCl$$

Procedure

■ **Caution.** *1,2-Dimethoxyethane (DME) may form explosive peroxides. Use fresh, peroxide-free material.*

1,2-dimethoxymethane, with some benzophenone (2.0 g) added, is heated at reflux with excess sodium (N_2 atmosphere, copper condenser) until the blue color persists and is then distilled. Dichloromethane is dried as in Section A. Phosphorus trichloride is freshly distilled under normal pressure.

Hexamethylphosphorous triamide is used as commercially available, for larger quantities it is economically prepared from PCl_3 and Me_2NH. All equipment is oven-dried, evacuated, and flushed with N_2 or Ar.

A solution of 6.00 g (17.4 mmol) $Na[BPh_4]$ and 4.74 mL (26.1 mmol) $P(NMe_2)_3$ in 120 mL of DME is prepared in a 250-mL round-bottomed flask equipped with a nitrogen inlet. Under a slight pressure of N_2 (or Ar), 0.78 mL (8.4 mmol) PCl_3 is added dropwise with stirring at room temperature within 1 min (magnetic stirring is sufficient; however, as a thick precipitate forms, a large magnetic bar is necessary). After 2 h of stirring under inert gas, the colorless precipitate is removed by filtering (glass frit pore size $\sim 30\,\mu m$, filtration takes 2–3 h) and then returned to the reaction flask. It is stirred with 100 mL of CH_2Cl_2 for 15 min and filtered through the same frit. The dichloromethane is then concentrated to $\frac{1}{10}$ of its volume under reduced pressure at room temperature. The precipitate that forms is separated by filtering through a glass frit, washed with a mixture of 3 mL of diethyl ether and 3 mL of dichloromethane, and dried *in vacuo*. Yield: 3.92 g (69%).

Anal. Calcd. for $[C_{12}H_{36}N_6P_3]C_{24}H_{20}B$ (676.6): C, 63.91; H, 8.34; N, 12.42. Found: C, 63.61; H, 8.46; N, 12.02.

Properties

The compound $\{[Me_2N)_3P]_2P\}[BPh_4]$ forms colorless crystals, mp 161 °C, soluble in dichloromethane and benzonitrile, less soluble in tetrahydrofuran (THF) and chlorobenzene. $^{31}P\{^1H\}$ NMR: A_2B, $\delta_A +85$, $\delta_B -194$, $J_{AB} = 513$ Hz, $J_{AA} = -30$ Hz. It is air stable for a short time and can be heated in solution to 120 °C for 5 days without change. It is not sensitive to even a high Cl^- concentration. More basic anions X^-, such as CN^-, $SnPh_3^-$, PPh_2^-, $POPh_2^-$, will replace one or both $P(NMe_2)_3$ groups to give neutral $(Me_2N)_3P = PX$ or anionic PX_2^-, respectively.[11] The compound $\{[(Me_2N)_3P]_2P\}[BPh_4]$ transfers P^+ to very electron-rich olefins giving 2-phosphaallylic cations.[12]

The preparation of the corresponding chloroaluminate $\{[(Me_2N)_3P]_2P\}$-$[AlCl_4]$, by analogy to Section A, is complicated by side reactions.[7] It is best prepared from $[(Ph_3P)_2P][AlCl_4]$ (see Section A) by $Ph_3P/P(NMe_2)_3$ exchange.[6]

References

1. S. F. Spangenberg and H. H. Sisler, *Inorg. Chem.*, **8**, 1006 (1969).
2. F. Ramirez and E. A. Tsolis, *J. Am. Chem. Soc.*, **92**, 7553 (1970).
3. M. Veith and Huch, *Tetrahedron Lett.*, **24**, 4219 (1983).

4. M. Baudler, D. Grenz, U. Arndt, H. Budzikiewicz, and M. Feher, *Chem. Ber.*, **121**, 1717 (1988).

5. A. Schmidpeter, S. Lochschmidt, and W. S. Sheldrick, *Angew. Chem.*, **94**, 72 (1982); *Angew. Chem. Int. Ed. Engl.*, **21**, 63 (1982).

6. A. Schmidpeter, S. Lochschmidt, and W. S. Sheldrick, *Angew. Chem.*, **97**, 214 (1985); *Angew. Chem. Int. Ed. Engl.*, **24**, 226 (1985).

7. A. Schmidpeter and S. Lochschmidt, *Angew. Chem.*, **98**, 271 (1986); *Angew. Chem. Int. Ed. Engl.*, **25**, 253 (1986).

8. S. Lochschmidt and A. Schmidpeter, *Z. Naturforsch. Teil B*, **40**, 765 (1985).

9. A. Schmidpeter, S. Lochschmidt, K. Karaghiosoff, and W. S. Sheldrick, *J. Chem. Soc. Chem. Commun.*, **1985**, 1447.

10. A. J. Carty, T. Hinsperger, L. Mihichuk, and H. D. Sharma, *Inorg. Chem.*, **9**, 2573 (1970).

11. A. Schmidpeter and S. Lochschmidt, unpublished.

12. A. Schmidpeter, S. Lochschmidt, and A. Willhalm, *Angew. Chem.*, **95**, 561 (1983); *Angew. Chem. Int. Ed. Engl.*, **22**, 545 (1983); *Angew. Chem. Suppl.*, **1983**, 710.

51.   *cis*-1,3-DI-*TERT*-BUTYL-2,4-DICHLORO-1,3,2,4-DIAZADIPHOSPHETIDINE

$$2PCl_3 + 6t\text{-}C_4H_9NH_2 \xrightarrow[-78\,°C]{(C_2H_5)_2O} \quad \begin{array}{c} t\text{-}C_4H_9 \\ | \\ Cl \diagdown \quad N \quad \diagup Cl \\ P \quad \quad P \\ \diagdown \quad N \quad \diagup \\ | \\ t\text{-}C_4H_9 \end{array} \quad + 4(t\text{-}C_4H_9NH_3)Cl$$

Submitted by WAN AHMAD KAMIL* and JEAN'NE M. SHREEVE*
Checked by ROBERT R. HOLMES[†] and KUMARA SWAMY[†]

Although the first preparation of a four-membered P(III)–N ring system was reported in 1894,[1] it was not until 1969 that *cis*-1, 3-di-*tert*-butyl-2, 4-dichloro-1, 3, 2, 4-diazadiphosphetidine,[2,‡] I, was fully characterized. Its derivatives have been studied extensively as precursors to other interesting compounds as well as for structural reasons. Because of the lability of the phosphorus–chlorine bond, a large number of derivatives can be obtained conveniently.

*Department of Chemistry, University of Idaho, Moscow, ID 83843.
[†]Department of Chemistry, University of Massachusetts, Amherst, MA 01003.
[‡]Cyclodiphosph(III)azanes is commonly used nomenclature, but the more systematic ring name 1, 3-di-*tert*-butyl-2, 4-dichloro-1, 3, 2, 4-diazadiphosphetidine is used for indexing purposes.

Although other routes to **I** have been reported,[2,3] the use of such inexpensive starting materials as PCl_3 and $t\text{-}C_4H_9NH_2$ in a relatively simple procedure makes this synthesis particularly attractive.[4] The X-ray crystal structure shows that the cis isomer is formed exclusively.[5]

■ **Caution.** *Phosphorus trichloride and tert-butylamine cause severe skin and eye burns and are respiratory irritants. The reaction should be carried out in a well-ventilated fume hood, and gloves and safety goggles should be worn.*

Procedure

Into a 250-mL three-necked flask, equipped with a mechanical stirrer, reflux condenser and dropping funnel, is added 15 mL (169 mmol) of phosphorus trichloride in 50 mL of anhydrous diethyl ether (distilled over $Li[AlH_4]$) under an anhydrous nitrogen stream. The flask is cooled to $-78\,°C$ (Dry Ice–ethanol), and 53 mL (507 mmol) of *tert*-butylamine dissolved in 120 mL of anhydrous diethyl ether is added dropwise from the dropping funnel over a period of 1 h. The mixture is stirred continuously during this time. Using standard Schlenk techniques,[6] the mixture is filtered through a medium porosity frit to remove the solid *tert*-butylammonium chloride, $t\text{-}C_4H_9NH_3Cl$. The filtrate is concentrated to a volume of 50 mL under vacuum. It is necessary to redissolve the concentrate in diethyl ether (~ 75 mL) and filter again in order to remove additional $t\text{-}C_4H_9NH_3Cl$ before completing the removal of solvent. The remainder of the solvent is separated by passing anhydrous nitrogen over the surface of the liquid. The crystals form after 1 day. Yield: 10.1–10.5 g ($\sim 45\%$) of product with purity $\geqslant 95\%$. A small amount of $t\text{-}C_4H_9NH_3Cl$ may be present.

Anal. Calcd. for $C_8H_{18}Cl_2N_2P_2$: C, 34.90; H, 6.55. Found: C, 34.10; H, 7.01.

Properties

*cis-*1, 3-Di-*tert*-butyl-2, 4-dichloro-1, 3, 2, 4-diazadiphosphetidine, **I**, is a colorless crystalline solid (mp 40–41 °C) that is soluble in diethyl ether, pentane, and benzene.[2] It is mildly sensitive to hydrolysis and oxidation. The NMR spectra taken in $CDCl_3$ show for $^{31}P\{^1H\}$ δ 207.7 (s) and ^1H δ 1.39 (t); $J_{P-H} = 0.98$ Hz. The electron impact mass spectrum has a molecular ion m/e at 274, 1.3%. Other fragments found are 239 ($M^+ - Cl$) 2.4%, 104 ($M^+ - 2Cl$) 1%, 102 ($PNC_4H_9^+$) 10.4%, 71 ($NC_4H_9^+$) 4%, 57 (C_4H_9) 100%.

References

1. A. Michaelis and G. Schroeter, *Chem. Ber.*, **27**, 490 (1894).

2. O. J. Scherer and P. Klusmann, *Angew. Chem. Int. Ed. Engl.*, **8**, 752 (1969).

3. R. Keat, *Top. Curr. Chem.*, **102**, 89 (1982) and references therein.

4. J. F. Nixon, R. Jefferson, T. M. Painter, R. Keat, and L. Stobbs, *J. Chem. Soc. Dalton Trans.*, **1973**, 1414.

5. K. W. Muir and J. F. Nixon, *J. Chem. Soc. Chem. Commun.*, **1971**, 1405.

6. D. F. Shriver, *The Manipulation of Air Sensitive Compounds*, McGraw-Hill, New York, 1969.

Chapter Seven

TRANSITION METAL COMPLEXES OF BIOLOGICAL INTEREST

52. REVERSIBLE COBALT(II) AND IRON(II) DIOXYGEN CARRIERS OF TOTALLY SYNTHETIC LACUNAR CYCLIDENE COMPLEXES:

(2, 3, 10, 11, 13, 19-Hexamethyl-3, 10, 14, 18, 21, 25-hexaazabicyclo[10.7.7]-hexacosa-1, 11, 13, 18, 20, 25-hexaene-$\kappa^4 N^{14,18,21,25}$)cobalt(II) Hexafluorophosphate and (3, 11-Dibenzyl-14, 20-dimethyl-2, 12-diphenyl-3, 11, 15, 19, 22, 26-hexaazatricyclo[11.7.7.1^{5,9}]octacosa-1, 5, 7, 9(28), 12, 14, 19, 21, 26-nonaene-$\kappa^4 N^{15,19,22,26}$)iron(II) Hexafluorophosphate

Submitted by COLIN J. CAIRNS and DARYLE H. BUSCH*
Checked by CAROL A. BESSEL[†] and KENNETH J. TAKEUCHI[†]

The most common dioxygen carriers of nature, the heme proteins hemoglobin and myoglobin, bind O_2 in a characteristic structure. The O_2 molecule binds to the metal atom through one of its atoms, and the M—O—O unit is angular in shape. Studies on small molecules that bind O_2 in this way have concentrated on two classes of ligands, porphyrins and Schiff bases.[1-3] In the first case, bulky superstructures are usually appended to the porphyrin ligand; this superstructure inhibits autoxidation of the metal atom in the

*Department of Chemistry, The Ohio State University, Columbus, OH 43210.
[†]Department of Chemistry, University at Buffalo, Buffalo, NY 14214.

same way as the protein in the natural product. Only the cobalt complexes of the Schiff bases react reversibly with O_2 without such protective structural features, under ordinary conditions. The most common Schiff base ligands are bis(salicylal)ethylenediimine and bis(acetylacetonyl)ethylenediimine, and related substances.

The lacunar cyclidene ligands are the first family of totally synthetic ligands to form both iron(II) and cobalt(II) O_2 carriers.[4-7] The molecular structure of the lacunar cyclidene complexes is most simply represented by the flat projection (**Ia**), while the stereochemistry is shown in **Ib**. *Cyclidene* refers to the parent macrocycle encircling the metal ion and the *lacuna* is the permanent void created by bridging the cleft arising from the saddle shape of the cyclidene macrocycle.

Ia   Ib

The syntheses of two complexes are described here: the cobalt(II) lacunar cyclidene complex in which $R^1 = -(CH_2)_6-$ and $R^2 = R^3 = -CH_3$, and the iron(II) complex in which $R^1 = m$-xylylene, $R^2 = -CH_2C_6H_5$, and $R^3 = -C_6H_5$. The former is the most studied example of a lacunar cyclidene oxygen carrier, and the latter represents the first nonporphyrin iron complex capable of forming a stable dioxygen adduct at room temperature.

The starting material for both of these target molecules is the neutral nickel(II) complex [3, 11-diacetyl-4, 10-dimethyl-1, 5, 9, 13-tetraazacyclohexa-deca-1, 3, 9, 11-tetraenato(2 —)-$\kappa^4N^{1,5,9,13}$]nickel(II), ([Ni(Ac$_2$Me$_2$[16]-tetra-enatoN$_4$)]). This complex is prepared by the procedure described earlier.[8] The lacunar cyclidene ligands are constructed on the nickel(II) ion, and the formation of the second macrocyclic ring is a template reaction. The first two reactions in the Scheme are required unless R^3 is to be methyl. Thus, the synthesis of the cobalt complex does not require these steps, whereas that for the iron complex does. The third reaction creates the novel functional group that makes these syntheses possible. Reaction with a powerful methylating agent* converts the acyl function into a methoxy substituted vinyl group. Because the vinyl group is in conjugation with the imine groups, the electron-withdrawing effect of the metal ion affects the vinyl carbon atom to which the methoxy group is attached. This enhances the electrophilic character of the vinyl carbon, leading to facile addition–elimination reactions that closely resemble those of carboxylate esters. Thus, the fourth reaction involves nucleophilic substitution of amino groups for the methoxy groups.

At this point, the ligand is a cyclidene. Several crystal structures have shown that the conformations of the 16-membered cyclidene rings, in their complexes, are saddle shaped. This facilitates the fifth reaction, the bridging process. Since the presence of the metal ion is required to produce this saddle conformation, the process is indeed a template reaction. The lacunar ligand is then removed from the nickel(II) ion (sixth reaction) and used to form a cobalt or iron complex (last reaction). The procedures for forming the iron and cobalt complexes are different; the scheme exemplifies the reactions used to synthesize the iron(II) complex.

SCHEME

*In place of methyl trifluoromethanesulfonate, the checkers used triethyloxonium tetrafluoroborate, obtained from Lancaster Synthesis Inc., as the alkylating agent.

■ **Caution.** *Triethyloxonium tetrafluoroborate is a corrosive, moisture sensitive solid. Contact with this reagent should be avoided.*

SCHEME (*Continued*)

A.   A LACUNAR CYCLIDENE COMPLEX OF COBALT(II)

1.   [2, 12-Dimethyl-3, 11-bis(1-methoxyethylidene)-1, 5, 9, 13-tetraaza-cyclohexadeca-1, 4, 9, 12-tetraene-$\kappa^4 N^{1,5,9,13}$nickel(II)] Hexafluorophosphate, [Ni((MeOEthi)$_2$Me$_2$[16]tetraeneN$_4$)](PF$_6$)$_2$

Procedure

The starting diacetylated nickel(II) complex **I** (24 g, 0.102 mol)* is dissolved in 400 mL of dry dichloromethane (distilled from CaH$_2$) in a 1-L one-necked round-bottomed flask equipped with a stirring bar. To this red solution is added 30 g (0.052 mol) of methyl trifluoromethanesulfonate.

- **Caution.** *Methyl trifluoromethanesulfonate is an extremely dangerous chemical: Inhaling the vapors and/or absorption of a single drop through the skin can cause a potentially fatal pulmonary edema if not promptly treated. Skin contact should be avoided by the use of proper clothing. Avoid breathing the vapors by carrying out the reaction in a well-ventilated laboratory hood. Small amounts of unreacted alkylating agent; for example, residue in the glass vial in which the chemical is sold, can be destroyed by addition to (immersion, in the case of the vial) alcoholic KOH. We have found that a solution of ~ 20 g KOH in 500 mL of methanol or ethanol hydrolyzes the residual reagent. It is advisable to let the hydrolyzing solution and its contents (e.g., the vial) stand for at least 1 day. Since Aldrich sells the reagent in units of 10 and 50 g we usually carry out the methylation reaction on a scale such that a complete vial is used. If a smaller amount is used, a gas-tight syringe should be used to transfer the appropriate amount of reagent to the solution of nickel(II) reagent.*

A calcium chloride drying tube is inserted into the ground glass joint of the flask, and the solution is stirred at room temperature for 16 h (overnight is convenient), during which time the color of the reaction mixture changes from deep red to a very dark olive green.

Excess alkylating agent is destroyed by the addition of ~ 25 mL of methanol, followed by stirring for an additional 45 min. The solvent is removed by rotary evaporation, using an aspirator and a protecting trap. (*Note:* Placing ~ 25 mL of methanolic KOH in the receiving flask guards against the presence of inadvertently undestroyed methyl trifluoromethanesulfonate, which might transfer during rotary evaporation.) The residual thick green oil is redissolved in the minimum amount of methanol. Ammonium hexafluorophosphate[†] (30 g, 0.183 mol, dissolved in 100 mL of methanol) is slowly added to this mixture, with vigorous stirring. The light green crystalline product should form within a few minutes. The reaction mixture is refrigerated (~ 5 °C) for ~ 6 h, and the product is collected by suction filtration. This product is adequate for use without purification, but it can be recrystallized from 1:2 acetonitrile–methanol to give bright green needles. A small amount of deacetylated material is removed in this way. Yield: 24 g (55%).

---

*Checkers ran the reactions with one fifth the amounts of reactants specified here.
[†] Checkers report that the compound can be isolated as the tetrafluoroborate.

Anal. Calcd. for $NiC_{20}H_{32}N_4O_2P_2F_{12}$: C, 33.87; H, 4.56; N, 7.90. Found: C, 34.07; H, 4.63; N, 7.77.

Properties

Recrystallized samples of this diamagnetic square planar nickel(II) complex, **II**, are reasonably stable, but they gradually decompose over a period of several months, presumably resulting from a slow hydrolysis reaction. The compound is soluble in acetonitrile, less so in methanol or ethanol, and insoluble in less polar solvents. Its IR spectrum, measured with a KBr pellet, shows two characteristic bands of almost equal intensity at 1615 and 1565 cm^{-1}, due to C=C and C=N stretching vibrations. This pattern is useful in judging the purity of the product. The most likely impurity is the deacetylated nickel(II) complex, which gives a rather different IR spectral pattern in this region, with a strong band at 1675 cm^{-1} and a much weaker one at 1640 cm^{-1}. An X-ray crystal structure has been reported on the analogous 15-membered cyclidene derivative.[9] The ^{13}C NMR spectroscopy is also useful in identifying the product and in judging its purity; in fact, it is the most effective method of characterizing all of the nickel(II) complexes discussed here. The following resonances are observed in CD_3NO_2: δ 181.1, 174.3, 164.4, 118.2, 58.3, 57.7, 52.2, 30.5, 29.8, 22.5, and 15.7 relative to TMS. Almost identical values are observed in CD_3CN as solvent, although the resonance at δ 118.2 is obscured by the broad solvent peak.

2. **[2, 12-Dimethyl-3, 11-bis[1-(methylamino)ethylidene]-1, 5, 9, 13-tetraazacyclohexadeca-1, 4, 9, 12-tetraene-$\kappa^4 N^{1,5,9,13}$nickel(II)] Hexafluorophosphate, [Ni((MeNHEthi)$_2$Me$_2$[16]tetraeneN$_4$)](PF$_6$)$_2$**

II   III

Procedure

A sample (10 g, 0.014 mol)* of the methoxycyclidene complex **II**, prepared as described in Section **A1**, is dissolved in 400 mL of acetonitrile in a 1-L Erlenmeyer flask. An excess of methylamine hydrochloride[†] (4 g, 0.042 mol) is added, followed by 6 mL (0.043 mol) of triethylamine. The color of the reaction mixture changes immediately from light green to a dark red-brown. The mixture is stirred at room temperature for 45 min, during which time the color lightens somewhat to an orange-red. Filtration, to remove excess methylamine hydrochloride, is followed by concentrating the filtrate by rotary evaporation at reduced pressure. Upon reduction of the volume below ~ 50 mL most of the triethylamine hydrochloride by-product precipitates. This is removed by suction filtration, and the filtrate is diluted with an equal volume of methanol. The mixture is concentrated again; the aminocyclidene product usually begins to crystallize from the mixture during this process. If crystals do not appear, crystallization can be induced by repeating the procedure of diluting with methanol and reducing the volume. Refrigeration of the reaction mixture, followed by suction filtration, gives the yellow crystalline product, **III**, which is washed with methanol and dried *in vacuo* over P_4O_{10}. This product can be used without further purification; however, recrystallization from an acetonitrile–methanol mixed solvent removes any residual triethylamine hydrochloride and gives analytically pure material. Yield: 7.5 g (76%).

Anal. Calcd., for $NiC_{20}H_{34}N_6P_2F_{12}$: C, 33.96; H, 4.86; N, 11.89. Found: C, 33.63; H, 4.88; N, 12.22.

Properties

Product **III** is a square planar diamagnetic complex, which is readily soluble in acetonitrile or acetone, but less so in methanol or ethanol. Its IR spectrum (KBr pellet) shows a strong characteristic absorption at 3390 cm^{-1} associated with the N—H stretching vibration, and a rich pattern in the double bond region (1640–1570 cm^{-1}). The presence of triethylamine hydrochloride impurity can be assessed by the presence of bands in the 2400–2800-cm^{-1} region of the spectrum. The ^{13}C NMR (CD$_3$CN)[‡]: δ 170.0, 168.3, 159.9, 112.4, 56.2, 51.5, 31.8, 30.6, 30.2, 20.9, and 15.5 relative to TMS. The peaks at δ 170.0, 112.4, 31.8, and 15.5 are broad as a result of rotation about the C—N bond.

---

*Checkers used 4.7 g instead of 10 g.

[†]Monomethylamine gas was used by checkers instead of methylamine hydrochloride.

[‡]Checkers report ^{1}H NMR (CD$_2$CN): δ 7.9 (s, 2H), 5.5 (q, 4H), 3.1 (m, 8H), 2.5 (s, 6H), 2.3 (s, 6H), 1.2 (t, 68).

3. (2, 3, 10, 13, 19-Hexamethyl-3, 10, 14, 18, 21, 25-hexaazabicyclo[10.7.7]hexacosa-1, 11, 13, 18, 20, 25-hexaene-$\kappa^4 N^{14,18,21,25}$)nickel(II) Hexafluorophosphate, [Ni((CH$_2$)$_6$(MeNEthi)$_2$[16]tetraeneN$_4$)](PF$_6$)$_2$

III → IV

1. NaOMe/MeOH
2. TsO(CH$_2$)$_6$OTs/MeCN

Procedure

The following standard procedure can be adapted to synthesize poly-methylene-bridged nickel(II) complexes where R^1 ranges from —(CH$_2$)$_4$— to —(CH$_2$)$_{12}$—. Five grams (7.1 mmol)* of the aminocyclidene product **III** from Section A.2 is dissolved in 300 mL of acetonitrile (500-mL Erlenmeyer flask) with stirring under an atmosphere of dry nitrogen, forming a yellow solution. A sodium methoxide solution prepared by reaction of 0.33 g (14.3 mmol) of sodium metal in 10 mL of methanol is added. The yellow solution turns a deep red immediately upon addition of the sodium methoxide, which removes the amine hydrogen atoms. Another solution, containing 3.20 g (7.5 mmol) of 1, 6-hexanediyl bis(p-toluenesulfonate)[10] in 300 mL of acetonitrile (500-mL Erlenmeyer flask) under nitrogen, is prepared, and these two solutions are added at a rate of 1 mL min^{-1} to a 1-L three-necked round-bottomed flask containing 200 mL of stirred acetonitrile maintained under N$_2$ at reflux temperature. A peristaltic pump is suitable for this transfer. Pressure equalizing dropping funnels can be used but require almost constant attention. After addition is complete (~ 5 h) the deep red reaction mixture is maintained at a gentle boil for an additional 5 h, after which time a copious precipitate of sodium p-toluenesulfonate is observed.

The mixture is cooled to room temperature and filtered. The solvent is removed by rotary evaporation under reduced pressure to give a viscous dark red-brown oily residue. This residue is redissolved in the minimum amount of acetonitrile, and the solution, filtered if necessary, is applied to a

*Checkers reduced amount to 3.6 g.

8×25-cm neutral alumina column (activity grade 1) loaded using acetonitrile. The mixture is eluted slowly with acetonitrile, and the deep yellow band is collected. The eluate is concentrated to $\sim 25\,mL$ and diluted with an equal volume of methanol. Repetition of this concentration–dilution procedure leads to crystallization of the bridged product, **IV**, which is collected by suction filtration, washed with methanol, and dried *in vacuo* over P_4O_{10}. Yield: 3.5 g (62%).

Anal. Calcd. for $NiC_{26}H_{44}N_6P_2F_{12}$: C, 39.56; H, 5.62; N, 10.65. Found: C, 39.46; H, 5.53; N, 10.60.

Properties

Like virtually all of the lacunar aminocyclidene complexes, this nickel(II) product is soluble in polar solvents such as acetonitrile. Its IR spectrum is useful in verifying that complete reaction has occurred, as there should be no N—H stretching band in the 3200–3500-cm^{-1} region. The spectral properties of these complexes have been studied thoroughly.[10–12] The ^{13}C NMR is most useful for identification of the pure product: δ 173.3, 167.5, 160.1, 110.9, 57.5, 56.1, 51.3, 40.2, 30.7, 30.3, 25.0, 24.0, 20.7, and 19.7 in CD_3CN relative to TMS.

4.   [2, 3, 10, 11, 13, 19-Hexamethyl-3, 10, 14, 18, 21, 25-hexaazabicyclo[10.7.7]hexacosa-1, 11, 13, 18, 20,25-hexaene] Hexafluorophosphate, $[H_3((CH_2)_6(MeNEthi)_2Me_2[16]$-tetraeneN$_4$)](PF$_6$)$_3$

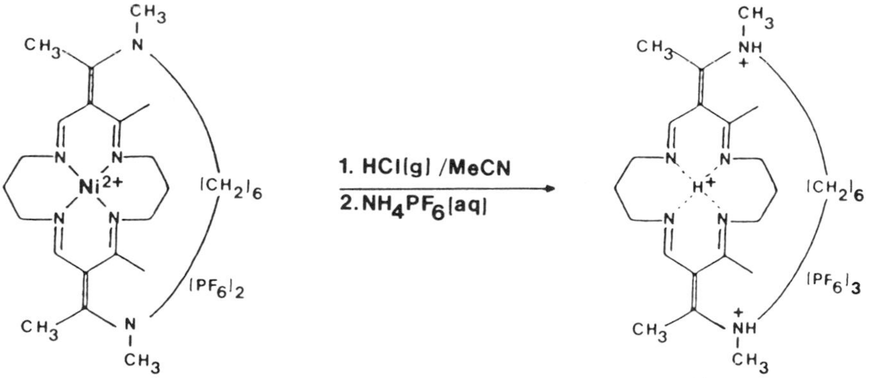

Procedure

Two and $\frac{5}{10}$ g* (3.2 mmol) of the nickel(II) complex, **IV**, prepared as described

*Checkers reduced amount to 0.95 g.

in Section A.3, is dissolved in 75 mL of acetonitrile in a 250-mL flask, and anhydrous hydrogen chloride is bubbled through the solution. After ~ 10 min of bubbling the solution has changed in color from yellow-orange to a turquoise blue. The solvent is removed by rotary evaporation, and the gummy residue is redissolved with stirring in 50 mL of water containing 5 mL of ethanol. The mixture is cooled to $\sim 5\,°C$ in an ice bath, and a solution containing 5 g (0.03 mol) of ammonium hexafluorophosphate in 20 mL of water is added in small portions over 1 h. The cream colored solid that precipitates during this addition is collected by suction filtration and washed with a small amount of cold water followed by diethyl ether, to remove residual acid. Drying *in vacuo* over P_4O_{10} gives 1.6 g (58%) of an off-white ligand salt, **V**, which, although not analytically pure, is suitable for preparation of the cobalt(II) complex.

Properties

When using the procedure reported here, ligand **V** is isolated as a tris(hexafluorophosphate) salt. In some procedures with other complexes the ligand can be isolated as a mixed chloride tris(hexafluorophosphate) salt. In general, this does not interfere with the synthesis of new metal complexes. These ligand salts are relatively fragile materials that have commonly been used without thorough characterization.*

5. **(2, 3, 10, 11, 13, 19-Hexamethyl-3, 10, 14, 18, 21, 25-hexaazabicyclo[10.7.7]hexacosa-1, 11, 13, 18, 20, 25-hexaene-$\kappa^4 N^{14,18,21,25}$)cobalt(II) Hexafluorophosphate, $[Co((CH_2)_6(MeNEthi)_2[16]tetraeneN_4)](PF_6)_2$**

1. Co(OAc)$_2$/MeOH
2. NaOAc/MeOH

*Checkers report ^{1}H NMR (CD$_3$CN): δ 7.6 (s, 2H), 3.3 (s, 18H), 2.5 (s, 6H), 2.0 (s, 6H), 1.6 (m, 12H).

Procedure

This reaction is carried out under oxygen-free nitrogen in an inert atmosphere glove box using predried and degassed solvents. The ligand salt **V** (1.7 g, 1.95 mmol),* prepared as described in Section A.4, is slurried in 50 mL of methanol in a 125-mL flask, and the mixture is heated to boiling. A solution containing 0.49 g (1.95 mmol) of cobalt(II) acetate tetrahydrate and 0.27 g (1.95 mmol) of sodium acetate trihydrate in 30 mL of hot methanol is added to the suspension. The color of the mixture changes quickly to a deep orange as the ligand salt dissolves, and the product begins to crystallize within a few minutes. The reaction mixture is boiled gently for 10 min and cooled to room temperature. Suction filtration gives an orange microcrystalline solid, **VI**, which is washed with a little methanol and dried. Yield: 1.30 g (83%).

Anal. Calcd. for $CoC_{26}H_{44}N_6P_2F_{12}$: C, 39.55; H, 5.62; N, 10.64; Co, 7.46. Found: C, 39.31; H, 5.75; N, 10.47; Co, 7.41.

Properties

The cobalt(II) cyclidene product **VI** is reasonably stable toward autoxidation in the solid state, and a solid state IR spectrum can be obtained under aerobic conditions.[†] The cobalt(II) complex is best identified[7] by its ESR spectrum in an acetonitrile–1.5 M N-methylimidazole solution, frozen to a glass at $-196\,°C$; $g_\perp \approx 2.3$ and $g_\parallel \approx 2.00$. Exposure of the thawed solution to O_2 then causes replacement of this spectrum with that of the 1:1 O_2 adduct; $g_1 \approx g_2 \approx 2.01$ and $g_3 \approx 2.08$. At $20\,°C$ in 2.5 M aqueous N-methylimidazole the complex binds dioxygen reversibly with an equilibrium constant $K_{O_2} = 1.6\,torr^{-1}$. An X-ray structure of the cobalt(II) complex confirms the presence of the commodious lacuna, which protects the coordinated dioxygen from autoxidation in polar solvents.[13] An X-ray structure determination of the dioxygen adduct, prepared by recrystallization of the complex from aqueous N-methylimidazole exposed to the atmosphere, confirms the existence and nature of the O_2 adduct.[14]

---

*Amount was reduced to 1.54 g by checker.
[†]Checker reports that the cobalt(II) product gives a quasireversible voltammogram in CH_3CN with $E_{1/2}$ of $+0.83\,V$ versus silver wire reference electrode.

B.   A LACUNAR CYCLIDENE COMPLEX OF IRON(II)

1.   [4, 10-Dimethyl-1, 5, 9, 13-tetraazacyclohexadeca-1, 3, 9, 11-tetraenato(2-)-$\kappa^4 N^{1,5,9,13}$]nickel(II), [Ni(Me$_2$[16]tetraenatoN$_4$)]

Procedure

The diacetylated nickel(II) complex **I** used in Section A.1 is also the starting material for the series of steps leading to the lacunar iron(II) cyclidene complex. The first reaction removes the original R^3 group. This procedure can be carried out in two steps with isolation of an intermediate deacetylated, protonated nickel(II) complex. However, a higher overall yield is realized in the one-step procedure that is described here. The diacetylated nickel(II) Complex **I** (19.5 g, 0.050 mol)* is dissolved in 250 mL of dry methanol (dried using magnesium turnings) in a 500-mL round-bottomed flask under nitrogen. *p*-Toluenesulfonic acid (16.3 g, 0.085 mol) is added to this solution, and the reaction mixture is heated at reflux for 45 min. After cooling the mixture in a Dry Ice–acetone bath, a solution of 8 g (0.2 mol) of sodium hydroxide in 100 mL of dry methanol is added under nitrogen from a pressure equalizing funnel. The mixture is then removed from the cooling bath and stirred for 30 min. A greenish-red solution and a precipitate result.

The methanol is removed by rotary evaporation at reduced pressure under a nitrogen atmosphere,[†] and 50 to 100 mL of dry benzene[‡] (sodium dried) is added.

■   **Caution.** *Benzene is a human carcinogen. It should be used only in a well-ventilated fume hood, and protective gloves should be worn.*

* Checker reduced amount to 5 g.

[†] All subsequent manipulations in this procedure are performed in an inert atmosphere.

[‡] Checkers used toluene in place of benzene in this and all subsequent reactions.

This mixture is again taken to dryness to remove residual methanol, and a further 100-mL aliquot of dry benzene is added. The mixture is filtered quickly under nitrogen through a pad of Celite ® filter aid, using dry benzene to aid in the product transfer and to wash the sodium *p*-toluenesulfonate by-product. The filtrate is concentrated under nitrogen to ~ 50 mL and chromatographed on a column of ~ 5-cm diameter, using no more than ~ 8 cm of activity grade 1 neutral alumina and dry benzene as the eluant. The fastest moving band is collected, and the benzene is removed by rotary evaporation under nitrogen to give 10–12 g (65–75%) of a deep red-black product, **VII**.

Properties

The deacetylated Product **VII** is moisture sensitive and should not be exposed to the atmosphere. Analytical data are not normally obtained, and the intermediate is used immediately for the following step. The chemical and physical properties of the compound have been extensively studied.[15]

2. **[3, 11 Bis(benzoyl)-2, 12-dimethyl-1, 5, 9, 13-tetraazacyclohexadeca-1, 3, 9, 11-tetraenato(2 −)-κ⁴*N*¹,⁵,⁹,¹³]nickel(II), [Ni(Me₂Bzyl₂[16]-tetraenatoN₄)]**

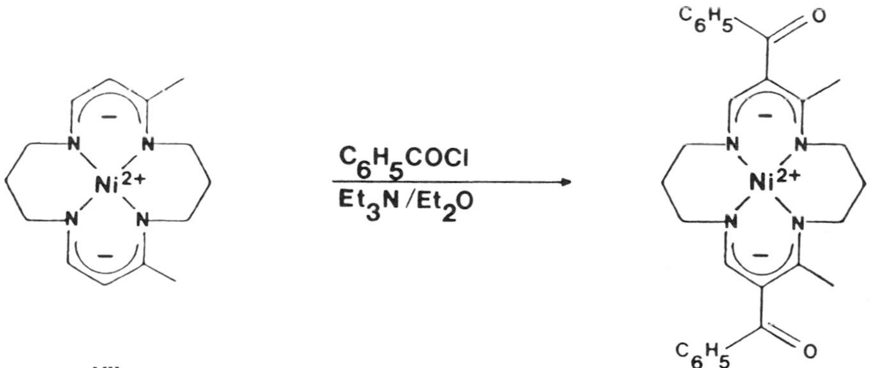

VII

VIII

Procedure

Ten grams* (0.033 mol) of the deacetylated nickel(II) Complex **VII**, freshly prepared as described in Section B.1, is dissolved in 400 mL of dry degassed

---

*Checkers reduced amount to 1.2 g.

diethyl ether (distilled from CaH_2) in a 1-L three-necked round-bottomed flask equipped with a stirring bar. An 11.5-mL volume (0.082 mol) of triethylamine is added to the mixture. A solution containing 8.2 mL (0.070 mol) of benzoyl chloride in 250 mL of dry diethyl ether is added to the reaction mixture, dropwise via a pressure equalizing funnel, over a period of 30 min. As the acid chloride is added, an orange precipitate forms. The mixture is stirred for an additional 1 h, and the precipitate is collected by suction filtration. This crude product is washed three times with 150-mL portions of water to remove the triethylamine hydrochloride, followed by three washings with 100-mL volumes of diethyl ether containing 10 mL of ethanol. The solid is dried for 12 h in a vacuum oven at ~ 50 °C. To purify the material it is dissolved in the minimum amount of chloroform* and chromatographed on a 8×30 cm, activity grade 1, neutral alumina column using chloroform as the eluant.

▪ **Caution.** *Chloroform is a suspected carcinogen and may cause adverse reproductive effects. It should be used only in a well-ventilated fume hood, and protective gloves should be worn.*

A dark brown band remains on the column while the broad orange band is collected. The eluate is concentrated by rotary evaporation, and ethanol is added to induce precipitation of the product. The Product **VIII** is collected by suction filtration and dried *in vacuo* over P_4O_{14}. Yield: 11.9 g (70%).

Anal. Calcd. for $NiC_{28}H_{30}N_4O_2$: C, 65.52; H, 5.89; N, 10.91. Found: C, 65.32; H, 5.89; N, 10.89.

Properties

The neutral nickel(II) complex **VIII** has similar physical properties to those of the analog where $R^3 = -CH_3$. Infrared and ^{13}C NMR data are useful for identification. The IR spectrum (KBr disk) has characteristic bands at 1615 ($v(C=O)$), 1545 ($v(C=N)$), and 1599, 1581, and 1498 cm^{-1} (v (C=C)). The ^{13}C NMR (CD$_3$CN): δ 192.9, 167.8, 160.9, 143.1, 129.7, 128.4, 128.1, 115.2, 54.7, 50.3, 30.5, 29.2, and 19.9 relative to TMS. The compound should be dried thoroughly because it has a tendency to retain benzene as a solvate.

---

*Checker substituted dichloromethane for chloroform in this and all subsequent reactions.

3. **[3, 11-Bis(α-methoxybenzylidene)-2, 12-dimethyl-1, 5, 9, 13-tetraazacyclohexadeca-1, 4, 9, 12-tetraene-κ^4$N^{1,5,9,13}$]nickel(II) Hexafluorophosphate, [Ni((MeOBzi)$_2$Me$_2$[16]tetraeneN$_4$)](PF$_6$)$_2$**

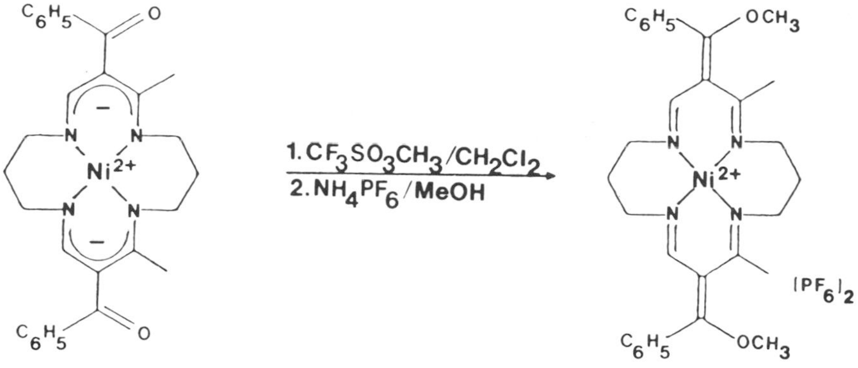

Procedure

A 1-L one-necked round-bottomed flask containing a stirring bar is charged with 10.5 g (0.020 mol)* of the product **VIII** from Section B.2, which has been dried in a vacuum oven overnight. A 250-mL volume of dry dichloromethane is added, giving a red solution. To this solution is added 10 g (0.06 mol) of methyl trifluoromethanesulfonate.†

- **Caution.** *This is a dangerous reagent; see instructions in Section A.1.*

A calcium chloride drying tube is used to protect the contents of the flask from atmospheric moisture, and the reaction mixture is stirred for 18 h. Subsequently, ~ 50 mL of methanol is added to the brown mixture, which is stirred for an additional 45 min. The mixture is concentrated to ~ 50 mL by rotary evaporation using an aspirator and a protecting trap, and a solution of 8.2 g (0.05 mol) of ammonium hexafluorophosphate in 40 mL of methanol is added by gravity filtration. The mixture is stirred for 45 min, and the yellow solid that precipitates is collected by suction filtration, washed with a little cold methanol, and dried *in vacuo* over P$_4$O$_{10}$. Recrystallization may be effected from acetonitrile–methanol if desired, although the product **IX** is sufficiently pure for the following reaction. Yield: 13.4 g (79%).

---

*Checkers used 2.9 g.
†Checkers again used triethyloxonium tetrafluoroborate (see note in Section A.1).

Properties

The physical and chemical properties of this material **IX** parallel those of the analog where $R^3 = -CH_3$. Analytical data are not obtained as the complex is rather less stable toward acid and moisture than the aforementioned analog. ^{13}C NMR (CD_3CN)*: δ 175.9, 171.2, 162.4, 130.9, 130.0, 127.7, 127.5, 116.6, 57.7, 53.6, 48.9, 27.2, 26.2, and 19.7 relative to TMS.

4. **[3, 11-Bis(α-(benzylamino)benzylidene)-2, 12-dimethyl-1, 5, 9, 13-tetraazacyclohexadeca-1, 4, 9, 12-tetraene-κ^4$N^{1,5,9,13}$]nickel(II) Hexafluorophosphate, [Ni((BzlNBzi)$_2$Me$_2$[16]tetraeneN$_4$)](PF$_6$)$_2$**

Procedure

A 6.0-g (7.2 mmol)† sample of the methoxycyclidene **IX** prepared as described in Section B.3 is dissolved in 100 mL of acetonitrile in a 250-mL flask, and 1.6 g (15 mmol) of benzylamine is added dropwise over 5 min to give a deep orange solution. The mixture is stirred at room temperature for 30 min, and solvent is removed (rotary evaporator) to give a yellow residue. The product can be recrystallized by dissolution in boiling methanol‡ to give deep yellow crystals **X** in 95% yield (6.73 g).

---

*Checkers report ^{1}H NMR (CD_3CN): δ 7.6 (m, 10H), 4.1 (q, 4H), 3.1 (s, 6H), 1.4 (t, 6H).
†Checkers reduced amount to 3.76 g.
‡Checkers isolated this material as a yellow solid by dissolving in a minimum amount of hot methanol and dropping into stirring water.

Properties

The product **X** exhibits similar physical and solubility properties to those of the analogous complex with $R^3 = -CH_3$. The ^{13}C NMR spectroscopy is used to demonstrate complete reaction: δ 170.1, 169.7, 162.8, 137.2, 133.2, 131.3, 130.3, 130.1, 129.6, 127.6, 116.0, 56.1, 52.0, 51.7, 30.3, 30.1, and 21.2 relative to TMS. Only seven of the eight peaks expected in the aromatic region, 125–140 ppm, are resolved.

5. **[3, 11-Dibenzyl-14, 20-dimethyl-2, 12-diphenyl-3, 11, 15, 19, 22, 26-hexaazatricyclo[11.7.7.1^{5,9}]octacosa- 1, 5, 7, 9(28), 12, 14, 19, 21, 26-nonaene-κ$^4N^{15,19,22,26}$]nickel(II) Hexafluorophosphate, [Ni((*m*-xylylene)(BzlNBzi)$_2$Me$_2$[16]tetraeneN$_4$)](PF$_6$)$_2$**

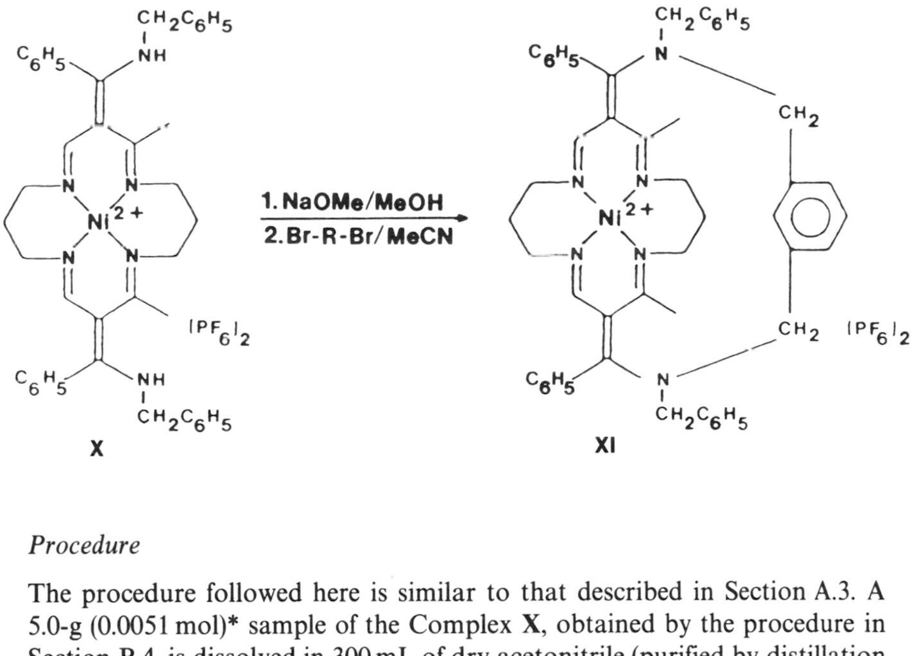

Procedure

The procedure followed here is similar to that described in Section A.3. A 5.0-g (0.0051 mol)* sample of the Complex **X**, obtained by the procedure in Section B.4, is dissolved in 300 mL of dry acetonitrile (purified by distillation from CaH$_2$) in a 500-mL Erlenmeyer flask under nitrogen, and a solution prepared by the reaction of 0.24 g (0.01 mol) of sodium in 10 mL of methanol

*Checkers reduced amount to 1.32 g.

is added. The resulting solution, and one containing 1.35 g (0.0051 mol) of α, α'-dibromo-*m*-xylene in 300 mL of acetonitrile, are added dropwise at a rate of ∼ 1 mL min⁻¹ to 200 mL of refluxing acetonitrile repidly stirred under nitrogen in a three-necked round-bottomed flask. The addition takes ∼ 5 h, and the reaction mixture is maintained at reflux temperature for a further 26 h, after which time the mixture is deep orange and a precipitate of sodium bromide is evident. The mixture is cooled to room temperature, filtered through Celite, and the solvent is removed by rotary evaporation. The residual gum is redissolved in the minimum amount of acetonitrile and applied to an alumina column (8 × 25 cm, activity grade 1, neutral alumina). Acetonitrile is used as the eluant and the fast-moving yellow band is collected. The eluate is removed by rotary evaporation, and the residue **XI** is recrystallized from acetonitrile–ethanol. Yield: 2.6 g (47%) of large deep yellow crystals.

Anal. Calcd. for $NiC_{50}H_{52}N_6P_2F_{12}$: C, 55.32; H, 4.83; N, 7.74; Ni, 5.41. Found: C, 55.47; H, 4.89; N, 7.58; Ni, 5.37.

Properties

This lacunar nickel(II) cyclidene Complex **XI** possesses similar physical properties to those already described. Although IR spectroscopy can be used to verify that the reaction has gone to completion, ¹³C NMR spectroscopy is the most useful characterization technique. An X-ray crystal structure determination on this complex illustrates the demanding steric requirements placed upon the lacuna by the bulky peripheral substituents.[11] ¹³C NMR (CD_3CN)*: δ 172.7, 168.3, 162.4, 137.8, 135.4, 134.0, 133.4, 132.4, 131.2, 130.9, 130.5, 130.2, 130.0, 128.4, 126.4, 115.5, 57.8, 56.5, 30.1, 29.4, and 21.2 relative to TMS.

6. **[3, 11-Dibenzyl-14, 20-dimethyl-2, 12-diphenyl-3, 11, 15, 19, 22, 26-hexaazatricyclo[11.7.7.1⁵,⁹]octacosa-1, 5, 7, 9(28), 12, 14, 19, 21, 26-nonaene] Hexafluorophosphate, $[H_3((m\text{-xylylene})(BzlNBzi)_2Me_2[16]tetraeneN_4)](PF_6)_3$**

Procedure

The ligand salt of the nickel(II) Complex **XI**, the preparation of which is described in the preceding section, is made in a manner entirely analogous

---

*Checkers report ¹H NMR (CD_3CN): δ 7.4 (m, 24H), 2.4 (s, 8H), 2.0 (s, 6H).

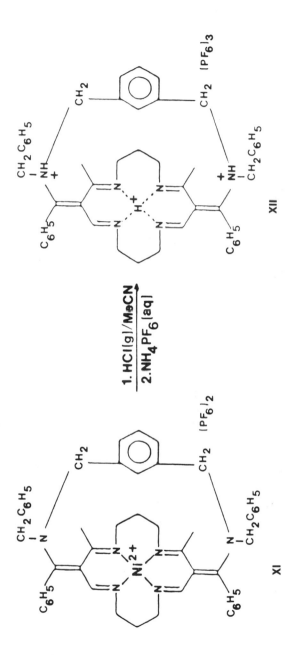

to that used in Section A.4. Two grams (1.842 mmol)* of the nickel(II) Complex **XI** prepared as described in Section B.5 is dissolved in 100 mL of acetonitrile, and anhydrous hydrogen chloride is bubbled through the solution for 10 min. At the end of this time, the solution has a blue-green color. Solvent is removed by rotary evaporation, and the residue is dissolved in the minimum amount of water. This solution is stirred and cooled in an ice bath, and an aqueous solution containing 1.20 g (7.37 mmol) of ammonium hexafluorophosphate is added by gravity filtration. The resulting off-white precipitate[†] is collected by suction filtration and washed, first with water and then with 95% diethyl ether–5% ethanol. The Product **XII** is dried *in vacuo* over P_4O_{10} and used for metal insertion without further purification. Yield: 1.95 g (90%).

7. **[3, 11-Dibenzyl-14, 20-dimethyl-2, 12-diphenyl-3, 11, 15, 19, 22, 26-hexaazatricyclo(11.7.7.1^{5,9}]octacosa-1, 5, 7, 9(28), 12, 14, 19, 21, 26-nonaene-κ^4$N^{15,19,22,26}$]iron(II) Hexafluorophosphate, [Fe((*m*-xylylene)(BzlNBzi)$_2$Me$_2$[16]tetraeneN$_4$)](PF$_6$)$_2$**

Procedure

This reaction is carried out under oxygen-free conditions in an inert atmosphere glove box, using dry degassed solvents. One gram (0.85 mmol)[‡] of the ligand salt **XII**, prepared as described in Section B.6, is slurried in 20 mL of acetonitrile along with 0.29 g (1.0 mmol) of bis(pyridine)iron(II) chloride in a 125-mL Erlenmeyer flask.[16,§] Addition of 0.38 g (3.76 mmol) of triethylamine produces a clear red solution, which is boiled gently for 10 min, cooled, and filtered through Celite. The solvent is removed, and the residue is dissolved in the minimum amount of hot methanol. The solution is stirred as it cools. Red crystals of the desired iron(II) Product **XIII** form by this procedure; they are removed by filtering and dried *in vacuo*. An additional crop of the product can be obtained by slow addition of a saturated ethanolic solution of ammonium hexafluorophosphate. Yield: 0.59 g (64%).

Anal. Calcd. for $FeC_{52}H_{55}N_7P_2F_{12}$.2.5 CH_3OH: C, 54.41; H, 5.43; N, 8.16; Fe, 4.65. Found: C, 54.28; H, 5.17; N, 8.25; Fe, 4.54.

---

*Checkers reduced amount to 0.6 g.
[†] Checkers report that the precipitate appeared yellow in color.
[‡] Checkers reduced amount to 0.3 g.
[§] Checkers substituted white $FeCl_2$ for bis(pyridine)iron(II) chloride.

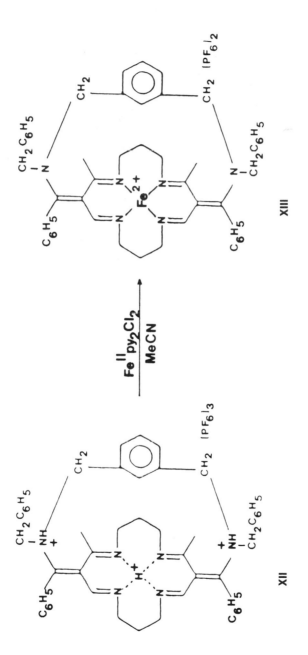

$Fe^{II}py_2Cl_2$
MeCN

XII

XIII

Properties

The iron(II) complex **XIII** crystallizes as the bis(hexafluorophosphate) salt with acetonitrile as the fifth, axial, ligand. This composition appears to be dictated largely by the relative solubilities of the salts of the possible anions in the mixed solvent. Most of the iron(II) lacunar complexes crystallize as the chloride hexafluorophosphate mixed salts, in which the chloride occupies the axial site. The complex is high spin and five coordinate in the solid state and in coordinating solvents.[4] In *N*-methylimidazole at 20 °C the title complex binds dioxygen reversibly with $K_{O_2} = 0.0012\,torr^{-1}$. In 3:1:1 acetone–*N*-methylimidazole–water the half-life toward irreversible autoxidation is ∼ 24 h at room temperature.[5] This complex constitutes the first example of a nonporphyrin iron(II) complex that can act as a dioxygen carrier at room temperature.*

References

1. E. C. Niederhoffer, J. H. Timmons, and A. E. Martell, *Chem. Rev.*, **84**, 137 (1984).

2. D. H. Busch, *Critical Care Medicine*, **10**, 246, (1982).

3. R. D. Jones, D. A. Summerville, and F. Basolo, *Chem. Rev.*, **79**, 139, (1979).

4. N. Herron, L. L. Zimmer, T. T. Grzybowski, D. J. Olszanski, S. C. Jackels, R. W. Callahan, J. H. Cameron, G. G. Christoph, and D. H. Busch, *J. Am. Chem. Soc.*, **105**, 6585 (1983).

5. N. Herron, J. H. Cameron, G. L. Neer, and D. H. Busch, *J. Am. Chem. Soc.*, **105**, 298 (1983).

6. N. Herron and D. H. Busch, *J. Am. Chem. Soc.*, **103**, 1236 (1981).

7. J. C. Stevens and D. H. Busch, *J. Am. Chem. Soc.*, **102**, 3285 (1980).

8. D. P. Riley and D. H. Busch, *Inorg. Synth.*, **18**, 36 (1978).

9. P. W. R. Corfield, J. D. Mokren, C. J. Hipp, and D. H. Busch, *J. Am. Chem. Soc.*, **95**, 4465 (1973).

10. D. H. Busch, D. J. Olszanski, J. C. Stevens, W. P. Schammel, M. Kojima, N. Herron, L. L. Zimmer, K. A. Holter, and J. Mocak, *J. Am. Chem. Soc.*, **103**, 1472 (1981).

11. B. Korybut-Daszkiewicz, M. Kojima, J. H. Cameron, N. Herron, M. Y. Chavan, A. J. Jircitano, B. K. Coltrain, G. L. Neer, N. W. Alcock, and D. H. Busch, *Inorg. Chem.*, **23**, 903 (1984).

12. D. H. Busch, S. C. Jackels, R. W. Callahan, J. J. Grzybowski, L. L. Zimmer, M. Kojima, D. J. Olszanski, W. P. Schammel, J. C. Stevens, K. A. Holter, and J. Mocak, *Inorg. Chem.*, **20**, 2834 (1981).

13. J. C. Stevens, P. J. Jackson, W. P. Schammel, G. G. Christoph, and D. H. Busch, *J. Am. Chem. Soc.*, **102**, 3283 (1980).

14. N. Matsumoto, N. W. Alcock, and D. H. Busch, unpublished results.

15. J. A. Streeky, D. G. Pillsbury, and D. H. Busch, *Inorg. Chem.*, **19**, 3148 (1980).

16. G. J. Long, D. L. Whitney, and J. E. Kennedy, *Inorg. Chem.*, **10**, 1406 (1971).

*Checkers report that cyclic voltammogram of this iron(II) complex in CH_3CN gives quasireversible behavior with $E_{1/2} = +0.35\,V$ versus a silver wire reference electrode.

53. [Pt(1,2-CYCLOHEXANEDIAMINE)(ASCORBATO-C^2,O^5)]

$$K_2[PtCl_4] + 4KI + dach \longrightarrow [Pt(dach)I_2] + 2KI + 4KCl$$

$$[Pt(dach)I_2] + 2AgNO_3 \xrightarrow{H_2O} [Pt(dach)(H_2O)_2](NO_3)_2 + 2AgI$$

$$[Pt(dach)(H_2O)_2](NO_3)_2 + Na(ascorbate)$$

$$\longrightarrow [Pt(dach)(ascorbato\text{-}C^2,O^5)]$$

dach = 1,2-cyclohexanediamine

Submitted by L. STEVEN HOLLIS,* SHERYL L. DORAN,* ALAN R. AMUNDSEN,* and ERIC W. STERN*

Checked by KAZI J. AHMED† and STEPHEN J. LIPPARD†

The importance of metal–carbon binding in the reactions of vitamin C with transition metals was first demonstrated from structural studies on [Pt(diamine)(ascorbato)] complexes.[1,2] In these studies, complexes of the general formula [Pt(dach)(ascorbato-C^2,O^5)], were isolated and shown, through X-ray crystallography and NMR spectroscopy, to contain an ascorbate ligand bound to platinum through the C^2 carbon and the deprotonated O^5 hydroxyl group. The structure of the diamine platinum–ascorbate chelate is depicted in **I**.

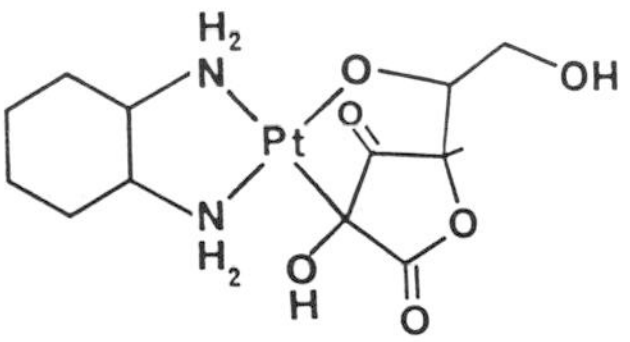

I

Each of the three isomeric forms of the dach ligand, *trans*-(R, R)-dach, *trans*-(S, S)-dach, and *cis*-dach, gives a platinum–ascorbate chelate that has unique physical properties. A detailed procedure is given here only for the synthesis of [Pt(*trans*-(R, R)-dach)(ascorbato-C^2,O^5)]·3H_2O. With minor variations, the same method is used in the syntheses of the platinum–ascorbate complexes of *trans*-(S, S)-dach and *cis*-dach.

*Research and Development, Engelhard Corporation, Menlo Park, Edison, NJ 08818.
†Department of Chemistry, Massachusetts Institute of Technology, Cambridge, MA 02139.

Procedure

A. [Pt(*trans*-(*R*, *R*)-dach)I$_2$]

[Pt(*trans*-(*R*, *R*)-dach)I$_2$] is formed by slow addition of *trans*-(*R*, *R*)-dach (1.14 g, 10 mmol) (Alfa) to a freshly prepared solution of K$_2$[PtCl$_4$] (4.15 g, 10 mmol in 100 mL of water in a 250-mL beaker) and KI (6.64 g, 40 mmol in 10 mL of water) at room temperature. The iodide complex precipitates from solution during the reaction. After ∼45 min, the precipitate is collected by vacuum filtration on a glass frit (10–15 μm), washed with water and ethanol, and dried under vacuum. Typical yields are in excess of 95%.

B. [Pt(*trans*-(*R*, *R*)-dach) (ASCORBATO-C^2, O^5)]·3H$_2$O

A 0.2 *M* solution of [Pt(*trans*-(*R*, *R*)-dach) (H$_2$O)$_2$] (NO$_3$)$_2$ is prepared by the reaction of [Pt(*trans*-(*R*, *R*)-dach)I$_2$] (5.63 g, 10 mmol) with AgNO$_3$ (3.40 g, 20 mmol) in 50 mL water in a 100-mL beaker. The mixture is stirred at 45 °C for 1.5 h, and the resulting AgI is removed by filtration through a nylon filter (0.45 μm). The filtrate is placed under nitrogen, and ascorbic acid (3.52 g, 20 mmol) and NaOH (0.80 g, 20 mmol in 5 mL H$_2$O) are added with stirring. The solution is stirred for 24 h at room temperature, during which time the product precipitates from solution. The mixture is placed in the refrigerator overnight at 4 °C, and the resulting crystals are collected by filtration. A second crop is obtained after concentrating the filtrate through rotary evaporation at 50 °C. The resulting products are recrystallized from water (80–90 °C) at a concentration of 20–25 mg mL^{-1}. Hot filtration, through a nylon filter (0.45 μm), is recommended to remove traces of platinum metal. A yield of 3.35 g (64%) is obtained after recrystallization. HPLC data: retention time = 1.6 min, at a flow rate of 6 mL min^{-1} in H$_2$O on a Novapak C18 column (radial-compression Z module), $\lambda = 254$ nm.

Anal. Calcd. for C$_{12}$H$_{26}$N$_2$O$_9$Pt: C, 26.82; H, 4.88; N, 5.21; Pt, 36.30. Found: C, 26.21; H, 4.96; N, 5.00; Pt, 36.16.

C. MODIFICATIONS FOR *cis*-DACH AND *trans*-(*S*, *S*)-DACH PLATINUM ASCORBATE COMPLEXES

The individual isomers of dach [*cis*-, *trans*-(*R*, *R*)-, and *trans*-(*S*, *S*)-] may be purchased (Alfa) or can be resolved from an isomeric mixture using the following methods: *cis*- and *trans*-1, 2-dach can be separated using NiCl$_2$·6H$_2$O, according to published methods,[3] while the separate *trans*-*R*, *R*- and *trans*-*S*, *S*-dach isomers can be isolated by allowing the racemic *trans*-dach mixture to react with either D- (for *S*, *S*) or L- (for *R*, *R*) tartaric acid.[4]

In the resolution procedure, racemic *trans*-dach (60 g) is dissolved in 200 mL of water along with 20 g of the appropriate (D or L) tartaric acid and 40 mL of acetic acid. The resulting dach-tartrate precipitate is collected and recrystallized two to three times from hot water. The resolved dach isomer is then liberated by treating the salt with excess 20% NaOH (aq) and extracting the free base into $CHCl_3$.

■ **Caution.** *Chloroform is a suspected carcinogen. Work in well ventilated fume hood and wear gloves.*

The $CHCl_3$ is evaporated under reduced pressure, and the product is distilled under vacuum (bp 79–81 °C/15 mm). The pure, S,S- or R,R-dach isomers are obtained as colorless liquids that crystallize as white solids at room temperature. $[\alpha]_D^{25} = +35°$ (S,S), $-35°$ (R,R).

The *trans*-(S,S)-dach complex, [Pt(*trans*-(S,S)-dach)(ascorbato-C^2, O^5)], can be prepared by using the method described in Section B. This complex is more soluble in water than the *trans*-(R,R)-dach complex, and it is necessary to concentrate the solution of the S,S-dach complex to induce crystallization. Yields of up to 45% are typical for this isomer. HPLC data: retention time = 1.8 min (under the conditions reported previously).

The reaction of [Pt(*cis*-dach)(H$_2$O)$_2$]$^{2+}$ with sodium ascorbate gives a white precipitate that contains two isomeric forms of the [Pt(*cis*-dach)-(ascorbato-C^2,O^5)] chelate (due to a lack of rotational symmetry in the *cis*-dach ligand).[1] The two isomers have different HPLC retention times and solubility properties, and may be separated by using the following method.

The mixture of [Pt(*cis*-dach)(ascorbato-C^2,O^5)] isomers (1.5 g) is dissolved with stirring in 13 mL of ethylene glycol, and the solution is filtered. The filtrate is poured into 50 mL of water and the resulting solution is placed in the refrigerator (4 °C). A white precipitate (0.8 g, isomer **1**) is collected by filtration after ～30 min. HPLC data: retention time = 1.7 min (with above conditions). The second isomer is obtained by treating the resulting filtrate with ～500 mL of acetone and leaving this mixture in the freezer (−5 °C) for several h. The resulting precipitate (0.4 g, isomer **2**) is collected by filtration. HPLC data: retention time = 1.5 min. Both isomers are characterized by using a combination of NMR (^{195}Pt and ^{13}C) and single crystal X-ray diffraction techniques (isomer **2**).[1,2]

Properties

The compound [Pt(*trans*-(R,R)-dach)(ascorbato-C^2,O^5)] and its analogs containing other amines are white, crystalline solids that are stable in air. They are only slightly soluble in water, with the exception of the complex of the (S,S)-dach analog (solubility: 18 mg mL^{-1}). The complexes are soluble in dimethyl sulfoxide (DMSO), and they are stable in this solvent for at least 24 h.

The ^{13}C NMR spectra, obtained in d_6-DMSO, show the following chemical shifts: [Pt(*trans*-(R, R)-dach) (ascorbato)], 198.7 (C1), 69.1 (C2), 175.6 (C3), 80.7 (C4), 85.8 (C5), 64.9 (C6); [Pt(*trans*-(S, S)-dach) (ascorbato)], 198.0 (C1), 69.5 (C2), 175.4 (C3), 80.5 (C4), 85.5 (C5), 64.8 (C6); [Pt(*cis*-dach)-(ascorbato)]-isomer **1**, 196.5 (C1), 69.5 (C2), 175.7 (C3), 80.7 (C4), 85.2 (C5), 64.8 (C6); [Pt(*cis*-dach) (ascorbato)]-isomer **2**, 199.5 (C1), 69.1 (C2), 174.6 (C3), 80.5 (C4), 85.5 (C5), 64.8 (C6). Chemical shifts are measured in ppm relative to d_6-DMSO at 39.5 ppm.

References

1. L. S. Hollis, A. R. Amundsen, and E. W. Stern, *J. Am. Chem. Soc.*, **107**, 274 (1985).

2. (a) L. S. Hollis, E. W. Stern, A. R. Amundsen, A. V. Miller, and S. L. Doran, *J. Am. Chem. Soc.*, **109**, 3596, (1987). (b) L. S. Hollis, S. L. Doran, A. R. Amundsen, and E. W. Stern, in *Platinum and Other Metal Coordination Compounds in Cancer Chemotherapy*, M. Nicolini (ed.), Nijhoff, Boston, 1988, p. 538.

3. R. Saito and Y. Kidani, *Chem. Lett.*, **1976**, 123.

4. (a) Y. Kidani, K. Inagaki, R. Saito, and S. Tsukagoshi, *J. Clin. Hematol. Oncology*, **7**, 197 (1977). (b) W. K. Anderson and D. A. Quagliato, U. S. Patent 4,550,187 (1985).

Chapter Eight

MISCELLANEOUS TRANSITION METAL COMPLEXES

54. (DITHIOCARBONATO)[1, 1, 1-TRIS(DIPHENYLPHOS-PHINOMETHYL)ETHANE]RHODIUM(III) TETRA-PHENYLBORATE SOLVATE

$$[(1, 5\text{-}C_8H_{12})RhCl]_2 + 2H_3CC(CH_2PPh_2)_3 + 2Et_3PCS_2$$

$$Triphos^*$$

$$\longrightarrow 2RhCl(S_2CPEt_3)(triphos) + 2(1, 5\text{-}C_8H_{12})$$

$$RhCl(S_2CPEt_3)(triphos) + O_2 \longrightarrow RhCl(S_2CO)(triphos) + OPEt_3$$

$$RhCl(S_2CO)(triphos) + Na[BPh_4] \longrightarrow [Rh(S_2CO)(triphos)][BPh_4] + NaCl$$

Submitted by C. BIANCHINI[†] and A. MELI[†]
Checked by Y. OHGOMORI[‡]

(Dithiocarbonato)[1, 1, 1-tris(diphenylphosphinomethyl)ethane]rhodium(III) tetraphenylborate has been prepared in 39% yield by a four-step reaction involving, as a starting point, the synthesis of the η^2-CS_2 complex

---

*Triphos = 1,1,1-tris(diphenylphosphinomethyl)ethane. IUPAC name:[2-[(diphenylphosphino)-methyl]-2-methyl-1,3-propanediyl]bis(diphenylphosphine).

[†]Istituto per lo Studio della Stereochimica ed Energetica dei Composti di Coordinazione, CNR, Via J. Nardi 39, 50132, Firenze, Italy.

[‡]Central Research Laboratory, Mitsubishi Petrochemical Co. Ltd., 8-3 Chu-ou, Ami, Ibaraki 300-03, Japan.

RhCl(η^2-CS$_2$) (triphos).[1] In the present preparation, the number of steps has been reduced to three while the yield has been increased to 73%.

(Dithiocarbonato)[1, 1, 1-tris(diphenylphosphinomethyl)ethane]rhodium-(III) tetraphenylborate is a coordinatively and electronically unsaturated complex containing an electrophilic metal center and three potential nucleophilic sites, namely, the sulfur and oxygen atoms of the chelate dithiocarbonate ligand. Because of its dual nature the compound is capable of cleaving hydrogen in heterolytic fashion[2] and can react by metathesis with polarizable molecules such as the heteroallenes COS, CS$_2$, and SCNR.[3] Also, (dithiocarbonato)-[1, 1, 1-tris-(diphenylphosphinomethyl)ethane]rhodium(III) tetraphenylborate undergoes photochemically,[4] thermally, or chemically[5] the chelotropic elimination either of CO or of COS. As a result, highly reactive[6] homo- and heterobimetallic bis(μ-S) and bis(μ-S$_2$) complexes can be obtained.[4]

Procedure

■ **Caution.** *Benzene is a human carcinogen. It should be handled only in a well-ventilated fume hood, and gloves should be worn.*

Under a nitrogen atmosphere, a 500-mL Schlenk flask is charged successively with 250 mL of deoxygenated benzene (freshly distilled under nitrogen from sodium benzophenone), triphos[7] (Strem) (1.24 g, 2 mmol), and dichlorobis(1, 5-cyclooctadiene)dirhodium(I)[8] (0.49 g, 1 mmol). The mixture is stirred at room temperature for 10 min. Solid Et$_3$PCS$_2$ (0.43 g, 2.2 mmol)[9,10,*] is added to the resultant yellow solution, which immediately turns to deep blue. Within 20 min RhCl(S$_2$CPEt$_3$)(triphos)·C$_6$H$_6$ begins to precipitate as air-sensitive blue crystals. The compound may separate as an oil which, however, rapidly crystallizes by addition of very few preformed crystals. When the color of the solution becomes light green ($\sim$ 3 h), the crystals are collected by filtration on a Schlenk frit, washed with a deoxygenated 2:1 mixture of benzene and petroleum ether and deoxygenated petroleum ether, and dried in a stream of nitrogen. Yield: 1.84 g (89%). IR (Nujol) 1030(m) and 1045(m) cm^{-1}.

Anal. Calcd. for C$_{48}$H$_{54}$ClP$_4$RhS$_2$·C$_6$H$_6$: C, 62.64; H, 5.84; Rh, 9.93; S, 6.19. Found: C, 62.38; H, 5.78; Rh, 9.91; S, 6.23.

A slow stream of oxygen is passed through the deep blue solution of RhCl(S$_2$CPEt$_3$)(triphos)·C$_6$H$_6$ (1.03 g, 1 mmol) in 100 mL of CH$_2$Cl$_2$ for 1 h, during which time the color turns to orange. The solution is concentrated under

---

*Et$_3$PCS$_2$ immediately precipitates as red crystals in the time of pipetting neat PEt$_3$ (0.73 mL, 5 mmol), under a nitrogen atmosphere, into a solution of CS$_2$ (0.3 mL, 5 mmol) in 30 mL of diethyl ether. Yield: 98%.

vacuum to ~30 mL and treated with 100 mL of benzene to precipitate the product. The yellow microcrystalline precipitate of RhCl(S₂CO)(triphos) is collected by filtration on a Büchner filter, washed with benzene and petroleum ether, and dried in vacuum. Yield: 0.74 g (87%). IR (Nujol) 1720(m), and 1600(vs) cm⁻¹. ^{31}P{^{1}H} NMR spectrum (32.2 MHz, CH₂Cl₂, 20 °C) AM₂Q spin system with δ_A 23.30 (d of t, $J_{PP} = 28$ Hz, $J_{PRh} = 107.1$ Hz) and δ_M −5.04 (d of d, $J_{PRh} = 96.6$ Hz).

Anal. Calcd. for C₄₂H₃₉ClOP₃RhS₂: C, 58.99; H, 4.59; Rh, 12.03; S, 7.49. Found: C, 58.81; H, 4.51; Rh, 11.87; S, 7.48.

A solution of Na[BPh₄] (0.68 g, 2 mmol) in 20 mL of ethanol is added to a stirred solution of RhCl(S₂CO)(triphos) (0.85 g, 1 mmol) in 100 mL of CH₂Cl₂. There is an immediate color change from yellow to red-brown. The resulting mixture is stirred for a further 15 min, and then 80 mL of ethanol is added. On gentle concentration under reduced pressure to ~100 mL, analytically pure [Rh(S₂CO)(triphos)] [BPh₄]·CH₂Cl₂ precipitates as red crystals. These are collected by filtration on a sintered glass frit, and washed with ethanol, water, ethanol, and petroleum ether. Yield: 1.16 g (95%). The compound can be recrystallized from 1:3 THF/butanol or acetone/butanol mixtures to give the corresponding monosolvate (THF or acetone) derivatives.

Anal. Calcd. for C₆₆H₅₉BOP₃RhS₂·CH₂Cl₂: C, 65.75; H, 5.02; Rh, 8.40; S, 5.23. Found: C, 65.67; H, 4.99; Rh, 8.29; S, 5.08.

Properties

(Dithiocarbonato)[1, 1, 1-tris(diphenylphosphinomethyl)ethane]rhodium(III) tetraphenylborate is air stable both in the solid state and in solution at ambient temperature. It is soluble in chlorinated solvents, acetone, THF, and DMF. The IR (Nujol) shows two bands in the C=O stretching region at 1680(vs) and 1600(vs) cm⁻¹, which are indicative of a chelated dithiocarbonate ligand. The ^{31}P{H} NMR (32.2 MHz, 20 °C) spectrum in CH₂Cl₂, DMF, and acetone consists of doublets centered at 35.22 ppm ($J_{PRh} = 98.7$ Hz), 12.19 ppm ($J_{PRh} = 102.7$ Hz) and 37.07 ppm ($J_{PRh} = 98.1$ Hz), respectively. Λ_M(nitroethane) 44 Ω⁻¹ cm²-mol⁻¹.

References

1. C. Bianchini, C. Mealli, A. Meli, and M. Sabat, *J. Chem. Soc. Chem. Commun.*, 1024 (1985).
2. C. Bianchini and A. Meli, *Inorg. Chem.*, **26**, 4268 (1987).
3. C. Bianchini, A. Meli, and F. Vizza, *Angew. Chem. Int. Ed. Engl.*, **26**, 767 (1987).

4. C. Bianchini and A. Meli, *Inorg. Chem.*, **26**, 1345 (1987).

5. C. Bianchini, A. Meli, F. Laschi, A. Vacca, and P. Zanello, *J. Am. Chem. Soc.*, **110**, 3913 (1988).

6. C. Bianchini, C. Mealli, A. Meli, and M. Sabat, *Inorg. Chem.*, **25**, 4617 (1986).

7. W. Hewertson and H. R. Watson, *J. Chem. Soc.*, 1490 (1962).

8. G. Giordano and R. H. Crabtree, *Inorg. Synth.*, **19**, 218 (1979).

9. A. W. Hofmann, *Liebigs Ann. Chem. Suppl.*, **1**, 1 (1861).

10. C. U. Pittman, Jr., and M. Narita, *Bull. Chem. Soc. Jpn.*, **49**, 1996 (1976).

55. THE VASKA-TYPE RHODIUM COMPLEXES, *trans*-RhX(CO)L$_2$

Submitted by YUJI OHGOMORI* and YOSHIHISA WATANABE*
Checked by MARTIN A. CUSHING[†] and STEVEN D. ITTEL[†]

The complex *trans*-RhX(CO)L$_2$ (ref. 1) is an important type of rhodium complex, which is used as an air stable catalyst precursor in organic syntheses.[2] A number of syntheses for these rhodium complexes have been summarized.[3] Among recent methods of synthesis, the most useful, using RhCl(CO)L$_2$, Ag$_2$CO$_3$, NH$_4$F, and MX (alkali metal salt), was reported by Vaska et al.[4]

We describe here a more convenient method for preparing the rhodium complexes. A series of the rhodium(I) complexes can be prepared by direct combination of the components in an aprotic solvent at room temperature. The scope of this reaction is very large. The complexes are obtained in good yields by a simple procedure and can be isolated easily. Among the products, the hydrogen phthalato and 1-methoxy-1,3-butanedionato-*O* complexes possess an unusual anionic ligand. The former complex is the first example of RhX(CO)L$_2$ in which the X moiety has a carboxyl group, and the latter complex is the first example of an enolato rhodium complex.

■ **Caution.** *The following syntheses involve the use or evolution of carbon monoxide, metal carbonyls, and tertiary phosphines. These are toxic materials, and the syntheses must therefore be carried out in a well-ventilated fume hood.*

Reagent Preparation

Dodecacarbonyltetrarhodium was used as the starting material for syntheses of *trans*-RhX(CO)L$_2$. This cluster compound can be purchased (Strem), or

*Tsukuba Research Center, Mitsubishi Petrochemical Co., Ltd., 8-3 Chu-ou, Ami, Ibaraki 300-03, Japan.
[†]Central Research and Development Department, E. I. du Pont de Nemours and Co., Wilmington, DE 19898.

it can be prepared from rhodium trichloride trihydrate by the method given by Martinengo, Chini, and Giordano.[5]

A 1000-mL two-necked flask equipped with a frit diffuser tube (CO inlet), a two-way stopcock (CO outlet) and a magnetic stirring bar is used as the reactor. Rhodium trichloride trihydrate (2.63 g, 10.0 mmol) and potassium chloride (2.24 g, 30.0 mmol) are dissolved in 500 mL of water in the reactor. Copper powder (2 g) is added, and the solution is bubbled with CO for 12 h with stirring at room temperature. To the resultant solution is added 2.94 g of trisodium citrate dihydrate, and stirring is continued for an additional 12 h under a CO atmosphere. The precipitate is collected under N_2 and extracted with CO saturated hexane five times. Dodecacarbonyltetrarhodium (1.46 g, 1.96 mmol) is obtained by evaporation of the hexane on a rotary evaporator (78% yield). IR (in hexane): 2072, 2070, 2043, and 1886 cm^{-1}.[6]

A. CARBONYL(HYDROGEN PHTHALATO)BIS(TRICYCLO-HEXYLPHOSPHINE)RHODIUM(I)

$$Rh_4(CO)_{12} + 4 \underset{}{\overset{CO_2H}{\diagdown}} -CO_2H + 8P(cyclo\text{-}C_6H_{11})_3$$

$$\longrightarrow 4Rh\left(\underset{}{\overset{CO_2H}{\diagdown}} -CO_2 \right)(CO)[P(cyclo\text{-}C_6H_{11})_3]_2 + 2H_2 + 8CO$$

Procedure

A 50-mL two-necked flask equipped with a dry nitrogen inlet, a stopper, and a magnetic strring bar is used as the reactor. To this reactor, 0.56 g (0.75 mmol) of dodecacarbonyltetrarhodium and 30 mL of freshly distilled 1, 3-dimethylimidazolidin-2-one (DMI) are added under a nitrogen atmosphere. To the resulting solution, 1.68 g (6.0 mmol) of tricyclohexylphosphine and 0.50 g (3.0 mmol) of phthalic acid are added, and the mixture is stirred for 5 h at room temperature. A yellow precipitate is slowly formed. The precipitate is collected, washed with 30 mL of cold diethyl ether, and dried *in vacuo*. The product is dissolved in 60 mL of hot (~ 70 °C) toluene–hexane (1/2 vol ratio), followed by filtration on a Büchner funnel under N_2. Yellow micro crystals are obtained by allowing the filtrate to stand under N_2 overnight at 0 °C.* The crystals are collected and dried under vacuum (2 torr) at room temperature for 6 h. The yield is 2.05 g (80%).

---

*Checkers used 40 mL of hot 1:1 toluene–hexane for the recrystallization. After filtering, the solution was placed in a − 30 °C freezer with an N_2 atmosphere. No product formed. The solution was then concentrated until cloudy (~ 15 mL) and returned to the freezer. Yellow crystals formed overnight. Yield: 1.60 g (62%).

TABLE I   Syntheses of *trans*-RhX(CO)L$_2$

X	L^a	Reaction Solvent^b	Recrystallization Solvent (v/v)	Yield (%)^c
PhCO$_2$	Pcy$_3$	DMI	Toluene–hexane (1:2)	79
3-F—C$_6$H$_4$CO$_2$	PPh$_3$	TMU^d	Toluene–hexane (1:5)	62
CH$_3$CO$_2$	Pi-Pr$_3$	Diethyl ether^d	Pentane	50
PhO	Pi-Pr$_3$	Diethyl ether^d	Hexane	53
p-TsO	Pcy$_3$	DMI	Toluene	75
I	Pcy$_3$	DMI	Toluene	75
C$_5$H$_4$N—4-CO$_2$	Pi-Pr$_3$	Diethyl ether		79

^{a}cy = cyclohexyl.
^{b}DMI = 1, 3-dimethylimidazolidin-2-one; TMU = 1, 1, 3, 3-tetramethylurea.
^{c}Based on rhodium.
^{d}Solvent was evaporated under reduced pressure, followed by recrystallization.

Anal. Calcd. for Rh(HO$_2$C—C$_6$H$_4$CO$_2$)(CO)[P(C$_6$H$_{11}$)$_3$]$_2$: C, 63.07; H, 8.35; P, 7.23; Rh, 12.01. Found: C, 63.10; H, 8.84; P, 7.35; Rh, 11.65.
Similar procedures are applicable to the preparation of a number of complexes, as listed in Table I.

B.   CARBONYL(1-METHOXY-1, 3-BUTANEDIONATO-*O*)BIS-(TRIISOPROPYLPHOSPHINE)RHODIUM(I)

$$Rh_4(CO)_{12} + 4CH_3COCH_2CO_2CH_3 + 8P(i\text{-}C_3H_7)_3$$

$$\xrightarrow{NEt_3} 4Rh[OC(CH_3)\!\!=\!\!CHCO_2CH_3](CO)[P(i\text{-}C_3H_7)_3]_2$$

$$+ 2H_2 + 8CO$$

Procedure

The same reactor described in Section A is used. A 15-mL volume of a diethyl ether solution of Rh$_4$(CO)$_{12}$ (0.56 g, 0.75 mmol), triisopropylphosphine (0.96 g, 6.0 mmol), methyl acetoacetate (0.52 g, 4.5 mmol), and triethylamine (1.5 mL) are stirred for 3 h at room temperature under a nitrogen atmosphere. The initially deep red solution turns to yellow, and yellow microcrystals are formed.* The crystals are collected, washed with 15 mL of diethyl ether, and

*Checkers report that the solution turned brown-yellow; they obtained a yellow precipitate (700 mg, 41%).

dried (2 torr, room temperature, 6 h). Yield: 1.00 g (59%). This product gives a satisfactory analysis without further purification.

Anal. Calcd. for $Rh[OC(CH_3)CHCO_2CH_3](CO)[P(C_3H_7)_3]_2$: C, 50.88; H, 8.72; P, 10.94; Rh, 18.17. Found: C, 50.97; H, 8.98; P, 11.5; Rh, 18.4.

Properties

Carbonyl(hydrogenphthalato)bis(tricyclohexylphosphine)rhodium(I) is stable in air; it can be handled in air not only in the solid state but also in solution. Its principal IR absorption in CH_2Cl_2 solution is $v_{(CO)} = 1945$ cm^{-1}. The ^1H NMR spectrum in $CDCl_3$ exhibits a single sharp resonance at δ 18.5 (standard: TMS). The ^{31}P NMR spectrum in $CDCl_3$ solution shows a doublet at 38.30 ppm with a $J_{(Rh-P)}$ value of 123.1 Hz (standard: H_3PO_4). The major fragments in the mass spectrum at 70 eV are MH$^+$ ($m/e = 857$), MH$^+$-CO$_2$H ($m/e = 812$), MH$^+$-[CO$_2$H + P(C$_6$H$_{11}$)$_3$] ($m/e = 532$), and P(C$_6$H$_{11}$)$_3^+$ ($m/e = 280$).

Carbonyl(1-methoxy-1,3-butanedionato-O)bis(triisopropylphosphine)-rhodium(I) is stable in air in the solid state. An IR band at 1949 cm^{-1} is observed for $v_{(CO)}$. The ^1H NMR spectrum in $CDCl_3$ shows: δ 4.75 [1H, s, CH], 3.55 [3H, s, OCH$_3$], 2.22 [6H, m, CH(CH$_3$)$_2$], 2.12 [3H, s, CH$_3$], and 1.34 [36H, dd, $J_{(CH)} = 7$ Hz, $J_{(PH)} = 14$ Hz, CH(CH_3)$_2$]. The ^{13}C NMR spectrum shows: δ 187.57 [—O—C(CH$_3$)=CH—], 170.84 [=CH—CO—CO$_2$CH$_3$], 90.90 [—C=CH—CO—], 49.35 [—CO$_2$$CH_3$], 24.76 [—O—C($CH_3$)=CH—], 24.48 [—$CH$(CH$_3$)$_2$], and 20.07 [—CH($CH_3$)$_2$] (standard: TMS). The ^{31}P NMR spectrum in $CDCl_3$ (standard: H_3PO_4) shows a doublet at δ 47.99 with a $J_{(Rh-P)}$ value of 127.0 Hz. This complex is reduced to afford [Rh(CO)$_4$]$^-$ under an atmospheric pressure of CO–H$_2$ (1:1) in hexamethylphosphoric triamide (HMPA) solution at room temperature.[1]

References

1. Y. Ohgomori, S. Yoshida, and Y. Watanabe, *J. Chem. Soc. Dalton Trans.*, **1987**, 2969.

2. I. Wender and P. Pino, *Organic Synthesis via Metal Carbonyls*, Wiley, Vol. 2, 1977, p. 136.

3. R. P. Hughes, *Comprehensive Organomet. Chem.*, Vol. 5, 1982, p. 277.

4. L. Vaska and J. Peone, Jr., *Inorg. Synth.*, **15**, 65 (1974).

5. S. Martinengo, P. Chini, and G. Giordano, *J. Organomet. Chem.*, **27**, 389 (1971).

6. W. Beck and K. Lottes, *Chem. Ber.*, **94**, 2578 (1961).

56. POTASSIUM HEXAIODORHENATE(IV)

Submitted by M. C. CHAKRAVORTI* and T. GANGOPADHYAY*
Checked by R. J. ANGELICI[†] and M. G. CHOI[†]

$$KReO_4 + KI + 8HI \longrightarrow K_2[ReI_6] + \tfrac{3}{2}I_2 + 4H_2O$$

Hexahalorhenates(IV), $M_2^I[ReX_6]$, where $X = Cl$, Br, and I, are among the commonest complex compounds of rhenium(IV). They can be prepared readily, are stable in air, and serve as starting materials for the preparation of many Re(IV) and Re(V) complex compounds. Two preparations of $K_2[ReCl_6]$ and one of $K_2[ReBr_6]$ have appeared in *Inorganic Syntheses.*[1,2] As a source material of Re(IV), potassium hexaiodorhenate(IV) offers some advantages over the chloro complex because of its solubility in many organic solvents and the labile nature of the iodide groups. The compound $K_2[ReI_6]$ has been prepared by reducing potassium perrhenate with hydriodic acid in the presence of potassium iodide or hypophosphite, or by evaporating $K_2[ReCl_6]$ with successive portions of hydriodic acid.[3-4] A simple procedure for the preparation of $K_2[ReI_6]$ in high yield is described here.

Procedure

■ **Caution.** *Hydriodic acid is toxic and corrosive; all operations with it should be carried out in a well-ventilated fume hood.*

A 22-mL sample of freshly distilled hydriodic acid (57%) is placed in a 100-mL beaker (tall form), and 3.5 g (21.1 mmol) of finely powdered potassium iodide is added to it. The beaker is covered with a watch glass and gently heated over a hot plate for ~ 10 min, until all of the potassium iodide dissolves. After removing the watch glass, 3.0 g (10.4 mmol) of finely powdered potassium perrhenate is added with stirring. Immediately, the color of the solution turns deep brown, iodine vapor is evolved, and a deep brown solid separates. The beaker is partly covered with the watch glass, and the mixture is gently boiled for ~ 15 min with occasional stirring. Within this period, evolution of iodine vapor ceases, and the volume is reduced to ~ 12 mL.[‡] The mixture is allowed to stand overnight at room temperature and then filtered on a sintered glass funnel (porosity 4) under suction. The suction is continued for 2 h, and then the funnel along with the product is kept *in vacuo* over fused calcium chloride. After 2 days the product is transferred to a

*Department of Chemistry, Indian Institute of Technology, Kharagpur 721302, India.
[†]Department of Chemistry, Iowa State University, Ames, IA 50011.
[‡]The checkers found that 30 min of boiling is required to reduce the volume to 12 mL.

container and dried further under vacuum in the same desiccator for 2 days.*
Yield: ~ 90%.

Anal. Calcd. for $K_2[ReI_6]$: K, 7.56; Re, 18.13; I, 74.23. Found: K, 7.54; Re, 17.65; I, 73.72.

Properties

Potassium hexaiodorhenate(IV) is a deep brown, nearly black, crystalline substance. It dissolves in methanol, ethanol, and acetone to give dark purple solutions. It can be recrystallized from hot hydriodic acid (57%). It is stable when kept in a desiccator, but in moist air it undergoes slow hydrolysis. The hydrolysis is rapid in neutral or alkaline medium. The magnetic moment at room temperature is 3.43 BM. The high Θ value suggests antiferromagnetic interaction.[5] The absorption spectrum in methanol gives bands at 425, 465, and 510 nm and shoulders at 540 and 645 nm. The spectrum in hydriodic acid (57%) shows bands[6] at 434(6980), 444(6920), 481(4700), 520(6850), 572(5090), and 659(2280) nm and shoulders at 423(6280), 457(5850), 533(6140), 592(4310), and 690(1860) nm (the values in parenthesis are the molar extinction coefficients).

References

1. L. C. Hurd and V. A. Reinders, *Inorg. Synth.*, **1**, 178 (1939).
2. G. W. Watt and R. J. Thompson, *Inorg. Synth.*, **7**, 187 (1963).
3. R. D. Peacock, in *Comprehensive Inorganic Chemistry*, J. C. Bailar, Jr. (ed), Vol. 3, Pergamon Press, 1973, p. 974.
4. G. K. Schweitzer and D. L. Wilhelm, *J. Inorg. Nucl. Chem.*, **3**, 1 (1956).
5. B. W. Figgis, J. Lewis, and F. E. Mabbs, *J. Chem. Soc.*, **1961**, 3138.
6. C. K. Jorgensen and K. Schwochau, *Z. Naturforsch. Teil A*, **20**, 65 (1965).

57. GROUP 6 PENTACARBONYL ACETATES AS THEIR μ-NITRIDO-BIS(TRIPHENYLPHOSPHORUS)(1+) SALTS

Submitted by DONALD J. DARENSBOURG,† SUSAN A. MORSE, CESAR OVALLES, and HOLLY C. PICKNER
Checked by WAYNE L. GLADFELTER‡ and DANIEL R. MANTELL

The chemistry of low valent Group 6 metal acetate complexes has been extensively investigated in the past several years.[1-4] It has been proposed[1,4]

---

*The checkers observed the sublimation of I_2 during the drying process.

†Department of Chemistry, Texas A & M University, College Station, TX 77843.
‡Department of Chemistry, University of Minnesota, Minneapolis, MN 55455.

that the pentacarbonyl acetato derivatives of the Group 6 metals are possible intermediates in the reactions of the metal hexacarbonyls with acetic acid, which eventually afford metal dimers and trimers. More importantly, acetato complexes have been key species in studies involving the activation of carbon dioxide.[5-7] In particular, the insertion of CO_2 into the $M—CH_3$ bond to produce the corresponding acetato complexes has been studied in detail in our laboratories.[5-7] The CO_2 insertion reaction into $M—C$ bonds is of significance since it represents one of the primary steps[8] in the catalytic conversion of CO_2 (the cheapest and most abundant of the carbon oxides) into interesting and useful chemicals. In addition, tungsten and chromium pentacarbonyl acetate anions have been used for the stoichiometric production of methyl,[9] ethyl,[9] and butyl[10] acetates and as precursors for the catalytic production of alkyl formates (methyl[9] or butyl formates[10]).

The two methods of synthesis previously reported in the literature involved the reaction of $[M(CO)_5Cl]^-$ with thallium(I)[3] or silver(I)[1] acetates. These methods not only require the synthesis of the metal halide from $M(CO)_6$, but they also have the disadvantages of the extreme toxicity of thallium(I) salts and the high price of silver salts.

The method presented herein possesses the following advantages in comparison with those available in the literature[1,3]: (1) One-step synthesis, (2) cheaper starting materials (alkali salts vs Ag salts) and high yields of products (80–90%), (3) (PPN) acetate where PPN = μ-nitrido-bis(triphenyl-phosphorus)$(1+)$* is a stable compound and no special procedures are needed, and (4) no by-products are generated in any of the steps described.

A.   μ-NITRIDO-BIS(TRIPHENYLPHOSPHORUS)(1+) ACETATE

$$(PPN)Cl + KO_2CCH_3 \longrightarrow KCl + (PPN)O_2CCH_3$$

Procedure

The procedure followed was first reported by Martinsen and Söngstang and was performed with minor modifications.[11] In a general preparation, 11.64 g (0.02028 mol) of (PPN)Cl (Aldrich) is dissolved by heating in 120 mL of distilled and deionized water in a 250-mL Erlenmeyer flask. A 40-g quantity (0.4 mol) of potassium acetate in 50 to 60 mL of water is added to the hot (PPN)Cl solution. The reaction mixture is stirred for ∼ 2 h and then placed in an ice bath for another hour. After filtration, the white solid is washed with cold water (3 × 1 mL) and then with cold diethyl ether (3 × 1 mL). The solid is

---

*The common name of this cation is bis(triphenylphosphine)iminium.

dried under vacuum overnight. Typically, 10.45 g (0.01751 mol) of product is isolated [86.2% based on (PPN)Cl]. Integration of the aromatic to methyl protons of the ^{1}H NMR spectrum was found to be 30:2.0, and a drop of $AgNO_3$ solution when added to an aqueous solution of (PPN) acetate yielded a positive test for chloride ion.

To decrease the chloride ion contamination, the (PPN)acetate was reprecipated with excess potassium acetate in the following manner. A 10-g (0.01 mol) sample of (PPN) acetate was dissolved in a minimum amount of hot deionized and distilled water (~ 12 mL at 90 °C). Approximately 40 g (0.4 mol) of potassium acetate was dissolved in 14 mL of deionized and distilled water. This solution was added dropwise, while stirring, to the hot (PPN) acetate solution. The reaction mixture was allowed to reach room temperature slowly, then placed in an ice–water bath for about 1 h. After filtering, the solid was washed with cold water (3 × 1 mL) and then cold diethyl ether (3 × 1 mL). After drying under vacuum overnight, the complex was ground into a powder. The solid was further dried by heating at 70 °C under vacuum for a few days, to yield a white powder with a decomposition point greater than 190 °C.

Although an aqueous solution of this solid still gave a positive chloride ion test with $AgNO_3$, the NMR aromatic–methyl integrations were 30:2.7, indicating less chloride ion contamination. A sample of (PPN) acetate was examined for chloride ion by elemental analysis (Galbraith) and it was determined to contain 3% (PPN)Cl by weight.

IR(KBr pellet): 1568 (C—O), 1439 (P—Ph), 1259, 1116, and 1005 (aromatic) cm^{-1}. ^{1}H NMR: 1.95 ppm ($\leftarrow$ TMS).

B. μ-NITRIDO-BIS(TRIPHENYLPHOSPHORUS)(1+) PENTACARBONYL ACETATO GROUP 6 METAL

$$M(CO)_6 + (PPN)O_2CCH_3 \longrightarrow (PPN)[M(CO)_5(O_2CCH_3)] + CO$$

$$M = Cr, Mo, \text{ or } W$$

Procedure

■ **Caution.** *Metal carbonyls are volatile and toxic. The thermolysis of metal carbonyls in solution produces carbon monoxide. Hence, all reactions should be carried out in a well-ventilated fume hood.*

All operations are carried out under nitrogen, with rigorous exclusion of air and moisture. Solvents are dried and deoxygenated by distillation from sodium benzophenone solution and collected under an N_2 atmosphere.

One gram (0.0017 mol) of $(PPN)O_2CCH_3$ and 0.71 g (0.0020 mol) of $W(CO)_6$ (or 0.53 g of $Mo(CO)_6$ or 0.44 g of $Cr(CO)_6$] are placed in a 100-mL

Schlenk flask equipped with a condenser. [All $M(CO)_6$ compounds were purchased from Strem Chemicals.] After evacuating and backfilling with N_2 three times, 35 mL of degassed DME (1, 2-dimethoxyethane) is added, and the solution is degassed again. The reaction mixture is stirred under nitrogen and heated at reflux for 25 min. The resulting orange solution is cooled to room temperature, and one half of the solvent is removed under vacuum. After the slow addition of 30 mL of hexane, the solution is left in a freezer overnight. The resultant solid is collected by filtration under nitrogen and dried under vacuum overnight. For the chromium or molybdenum derivatives, the procedure is exactly the same with the exception that the refluxing time is only 12 min for molybdenum. The yields and melting points for the acetate complexes are shown in Table I.

The IR absorption bands for the three pentacarbonyl acetato compounds can be seen in Table II. Four bands are found in the CO stretching region, and two bands are attributed to the acetate moiety. The proton and ^{13}C NMR resonance peaks are listed in Table III for this family of compounds.

TABLE I   **Percent Yields and Melting Points for the Acetato Complexes, $(PPN)[M(CO)_5(O_2CCH_3)]$**

M	Yield %	Melting Point Range[a] (°C)
Cr	89	139–141
Mo	89	137–142
W[b]	85	137–142

[a] Sealed capillary tube.
[b] Elemental analysis (Galbraith). *Anal.* Calcd. for [PPN]-[$W(CO)_5(O_2CCH_3)$]: C, 56.05; H, 3.61. Found: C, 55.40; H, 3.60.

TABLE II   **Infrared Frequencies for the Acetato Complexes[a]**

M	$\nu_{(CO)}$ (cm^{-1})				$\nu_{(CO_2)}$ (cm^{-1})		
	$A_1{}^1$(w)	B_2(w)	E(s)	$A_1{}^2$(m)	Asym	Sym	$\Delta\nu_{(CO_2)}{}^b$
Cr	2053	1964	1915	1845	1606	1373	233
Mo	2060	1967	1917	1845	1608	1370	238
W	2059	1955	1907	1842	1616	1370	246

[a] Spectra taken in THF solutions. The peak positions are accurate to ± 1 cm^{-1}. IR taken on a FT–IR IBM-32.
[b] In unidentate complexes the separation between the two $\nu_{(CO_2)}$ vibrations is larger than that in the free ion, where $\Delta\nu_{(CO_2)} = 200$ cm^{-1}, whereas the opposite trend is seen in bidentate complexes. See, for example, Nakamoto, K. *Infrared and Raman Spectra of Inorganic and Coordination Compounds*, 3rd ed., Wiley-Interscience, New York, 1977.

TABLE III NMR Frequencies for the Acetato Compounds[a]

M	^{1}H NMR ($\leftarrow$ TMS) $\delta_{(MO_2CCH_3)}$	13C NMR ($\leftarrow$ TMS) $\delta_{(^{13}CO)}$	
Cr	1.80	224.0[b]	216.2[c]
Mo	1.95	217.8[b]	205.9[c]
W	1.90	206.4[b]	200.9[c]

[a]$CDCl_3$ solution, TMS as reference, measured on a Varian EM-390 spectrometer (^{1}H NMR) and a Varian XL-200 (^{13}C NMR) at ambient temperatures.
[b]Trans with respect to the acetate ligand.
[c]Cis with respect to the acetate ligand.

In general, the acetato complexes of the Group 6 metals are stable in the solid state, but they must be stored in a freezer for prolonged periods. Their solutions are very air sensitive, and after several days the $M(CO)_6$ complex is formed even under a N_2 atmosphere. The Group 6 acetato complexes are soluble in organic ethers [tetrahydrofuran (THF), diethyl ether, etc.], acetonitrile, dichloromethane, and in other polar solvents such as methanol and ethanol. These complexes are insoluble in hydrocarbon solvents, for example, hexane or benzene.

The acetate moiety is found to behave as a *cis* CO labilizing ligand,[1] and the reaction of exchange of CO for ^{13}CO occurs at a relatively fast rate. In the same way, THF solutions of the acetato complexes (where M = W) lose CO under photolysis conditions to generate the species $[M(CO)_4(O_2CCH_3)-(THF)]^-$ or $[M(CO)_4(O_2CCH_3)]^-$, where the acetate ligand is chelating.[12] These adducts are believed to be active species for the hydrogenation of aldehydes and ketones under high H_2 pressure.[12] Addition of PR_3 to the solvent adduct, the tetracarbonyl chelate, or the pentacarbonyl acetate complex produces the phosphine substituted *cis*-$[M(CO)_4(O_2CCH_3)(PR_3)]^-$. One of these compounds (where M = W and R = ethyl) has been synthesized, and its crystal structure has been reported.[2]

Group 6 pentacarbonyl acetato complexes have been also found to undergo addition of alkyl halides (MeI) to produce alkyl acetates and the pentacarbonyl halide metal complex. When allyl halides are used instead of alkyl halides, an oxidative addition product is formed,[3] which contains the allyl, halide, and acetate ligands bonded to the metal.

Alcohol solvents have been found[9] to interact strongly with the acetate ligand producing a shift in the IR frequencies, for example, the strong band of the tungsten derivative shifts from 1906 to 1920 cm^{-1}. Interactions like this serve to neutralize the negative charge on the acetate moiety in much the same manner as alkali metal ions.[13] Disruption of this weak interaction was observed upon vacuum removal of the solvent methanol and redissolving in aprotic solvents.

References

1. F. A. Cotton, D. J. Darensbourg, and B. W. S. Kolthammer, *J. Am. Chem. Soc.*, **103**, 398 (1981).

2. F. A. Cotton, D. J. Darensbourg, B. W. S. Kolthammer, and R. A. Kudaroski, *Inorg. Chem.*, **21**, 1656 (1982).

3. G. A. Doyle, *J. Organomet. Chem.*, **84**, 323 (1975).

4. A. Bino, M. Ardon, I. Maor, M. Kaftory, and Z. Dori, *J. Am. Chem. Soc.*, **98**, 7093 (1976).

5. (a) D. J. Darensbourg and R. A. Kudaroski, *Adv. Organomet. Chem.*, **106**, 3672 (1983). (b) D. J. Darensbourg, A. Rokicki, and M. Y. Darensbourg, *J. Am. Chem. Soc.*, **103**, 3223 (1982).

6. D. J. Darensbourg and R. A. Kudaroski, *J. Am. Chem. Soc.*, **106**, 3672 (1984).

7. D. J. Darensbourg, R. K. Hanckel, C. G. Bauch, M. Pala, D. Simmons, and J. N. White, *J. Am. Chem. Soc.*, **107**, 7463 (1985).

8. D. J. Darensbourg and C. Ovalles, *Chem. Tech.*, **15**, 636 (1985).

9. D. J. Darensbourg and C. Ovalles, *J. Am. Chem. Soc.*, **106**, 3750 (1984).

10. D. J. Darensbourg and C. Ovalles, *J. Am. Chem. Soc.*, **109**, 3330 (1987).

11. A. Martinsen and J. Söngstang, *Acta Chem. Scand. Ser. A*, **31**, 645 (1977).

12. P. A. Tooley, C. Ovalles, S. C. Kao, D. J. Darensbourg, and M. Y. Darensbourg, *J. Am. Chem. Soc.*, **108**, 5465 (1986).

13. M. Y. Darensbourg, *Progress in Inorganic Chemistry*, Vol. 33, S. J. Lippard (ed.), Wiley, New York, 1985, p. 221.

58. BIS(ORGANOIMIDO) COMPLEXES OF TUNGSTEN(VI)

Submitted by A. J. NIELSON*
Checked by B. L. HAYMORE[†]

Organoimido complexes are currently of interest in studies of ligands multiply bonded to transition metal centers. It is now possible to prepare a variety of monoimido complexes for a diversity of the early transition metals,[1,2] but bis(organoimido) complexes are less well known, inasmuch as routes for their synthesis have been less well developed. The bis(organoimido) moiety is isoelectronic with the *cis*-dioxo group that is commonly found for higher oxidation states, and it shows potential utility as a stable source of nitrenes in organic syntheses.[1]

*Department of Chemistry, University of Auckland, Private Bag, Auckland, New Zealand.
[†]Monsanto Co., 800 N. Lindbergh Boulevard, St. Louis, MO 63167.

Detailed below are procedures for preparing bis(organoimido) complexes of tungsten(VI) containing two different organoimido ligands. The initial synthesis makes use of the silylamine $Me_3SiNHCMe_3$, which reacts with tetrachloro(phenylimido)tungsten dimer to generate two cis oriented amido ligands from which imido and amine ligands form by an intramolecular proton transfer. The resulting complex is the dimer $[W(NCMe_3)(\mu\text{-}NPh)Cl_2(NH_2CMe_3)_2]_2$ (refs. 3 and 4) which, when allowed to react with 2, 2′-bipyridine or trimethylphosphine, gives the monomeric complexes $[W(NCMe_3)(NPh)Cl_2L_2]$ in which the two chloro ligands are oriented trans or cis.[5] On a smaller scale than that given here, the monoimido dimer complexes $[W(NR)Cl_4]_2$ ($R = Ph$, $CHMe_2$, and Me) and the silylamines Me_3SiNHR' ($R' = Ph$, CMe_3, $CHMe_2$, and Et) have given the complexes $[W(NR)(NR')Cl_2(NH_2R')]_2$,[4] producing a range of bis(organoimido) complexes that are useful materials for further syntheses.

Starting Materials and General Procedure

The compound $[W(NPh)Cl_4]_2$ is prepared by heating $WOCl_4$ with isocyanoatobenzene in benzene.[6] Petroleum ether (bp fraction 40–60 °C), benzene, and toluene are distilled from sodium wire under dry oxygen-free nitrogen. 1, 1-Dimethyl-N-(trimethylsilyl)ethanamine is prepared by reaction of 1, 1-dimethylethanamine with chlorotrimethylsilane (see Section. 65).[7] Trimethylphosphine (Strem) is prepared by reaction of MeMgI with $P(OPh)_3$ in diethyl ether.[8] All manipulations are carried out under moisture- and oxygen-free nitrogen, using normal techniques for air sensitive compounds.* When solutions are transferred between flasks, a stainless steel transfer tube is used. Each flask is fitted with a gas inlet tap and a serum cap through which the transfer tube passes. The nitrogen supply to the receiving vessel is turned off and a vent needle placed through the septum. With the transfer tube placed below the level of liquid, a positive nitrogen pressure is used to force the solution into the receiving flask.

A. TETRACHLOROBIS(1, 1-DIMETHYLETHANAMINE)-BIS[(1, 1-DIMETHYLETHYL)IMIDO]BIS(μ-PHENYLIMIDO)-DITUNGSTEN(VI)

$$[W(NPh)Cl_4]_2 + 4Me_3SiNHCMe_3$$
$$\longrightarrow [W(NCMe_3)(\mu\text{-}NPh)Cl_2(NH_2CMe_3)]_2 + 4Me_3SiCl$$

---

*The checker worked in a good dry box, and obtained yields similar to those reported by the submitter.

Procedure

■ **Caution.** *Benzene is a human carcinogen. It should be used only in a well-ventilated hood, and protective gloves should be worn.*

Tetrachloro(phenylimido)tungsten(VI) (5 g, 12.0 mmol) is placed in a 250-mL two-necked flask fitted with a gas inlet tap and a rubber serum cap. Benzene (60 mL) is added with a syringe, and the mixture is cooled to just above the freezing point of benzene with an ice–water bath. 1, 1-Dimethyl-*N*-(trimethylsilyl)ethanamine (4.8 mL, 24.8 mmol) is added with a syringe to benzene (50 mL) in a 100-mL two-necked round-bottomed flask also fitted with a gas inlet tap and serum cap. The solution is cooled in an ice–water bath and then added slowly to the chilled suspension of tetrachloro(phenyl-imido)tungsten(VI) in benzene using a stainless steel transfer tube. The serum cap is replaced by a greased glass stopper, the ice–water is removed, and the reaction mixture is stirred for 17–18 h at ambient temperature, during which time the yellow product precipitates. The complex is filtered under N_2, washed first with benzene (3 mL) chilled in an ice–water bath and then with petroleum ether (40 mL), and dried under vacuum for 1 h with the flask immersed in a hot-water bath maintained at 60–70 °C. The complex is analytically pure, but it contains varying amounts of benzene solvate, depending on the drying period. Yield: 4.8–5.0 g (80–83%).

Anal. Calcd. for $C_{28}H_{25}Cl_4N_6W_2 \cdot \frac{1}{3}C_6H_6$: C, 35.8; H, 5.2; N, 8.4. Found: C, 35.5; H, 5.7; N, 8.5.

Properties

The complex is not particularly air or moisture sensitive, but it is best handled and stored under nitrogen. The melting point is 85–90 °C. The complex is insoluble in petroleum ether, slightly soluble in benzene, and very soluble in chloroform or dichloromethane. An X-ray crystal structure determination[3] has shown the complex to be dimeric, with terminal (1, 1-dimethylethyl)imido and 1, 1-dimethylethanamine ligands, bridging phenylimido ligands, and cis-oriented dichloro groups. In the IR spectrum, the *cis*-dichlorides give rise to strong absorption at 308 and 280 cm^{-1}. The ^1H NMR spectrum (CDCl$_3$)* shows the phenyl group protons at δ (ppm from TMS) = 7.50–8.66, the *tert*-butyl group methyl resonances as singlets at 1.28 and 1.37, and the NH proton resonances at 3.4 and 5.8. ^{13}C NMR spectrum in CDCl$_3$ (ppm from TMS): 30.0 (Me$_3$), 54.9 (H$_2$NC), 71.3 (N—C), 120.6 (C-ortho, C$_6$H$_5$), 124.8 (C-para, C$_6$H$_5$), 127.2 (C-meta, C$_6$H$_5$), and 162.9 (C-ipso, C$_6$H$_5$).[4]

---

*The checker determined ^1H NMR spectra in CD$_2$Cl$_2$. His results were similar to the reported spectra in CDCl$_3$.

Using the preparative procedure above, yields of the complexes $[W(NR)(\mu\text{-}NR')Cl_2(NH_2R')]_2$ (R = Ph, $CHMe_2$, Me; R' = Ph, CMe_3, Et) range from 60 to 80% when ~ 1 g of monoorganoimido complex is used.[4] In cases where the reaction product is soluble in benzene, the solution volume is reduced until precipitation occurs. The IR spectra show *cis*-W—Cl absorptions in the vicinity of 310 and 260 cm^{-1}, and the ^{13}C NMR spectra show alkylimido α-carbon resonances at 68–71 ppm with alkylamine α carbon atoms at 54–66 ppm. No distinction between bridging and terminal alkylimido groups is indicated by the ^{13}C NMR spectra, but the less sterically hindering group is expected to form the bridge. For phenyl groups, a terminal imido ligand is indicated by an aromatic ring ipso-carbon resonance in the vicinity of 162 ppm, a bridging imido ligand at 150 ppm, and a terminal amine ligand at 148 ppm.

B.   (2, 2'-BIPYRIDINE)DICHLORO[(1, 1-DIMETHYLETHYL)-IMIDO] (PHENYLIMIDO)TUNGSTEN(VI)

$$[W(NCMe_3)(\mu\text{-}NPh)Cl_2(NH_2CMe_3)]_2 + 2C_{10}H_8N_2$$
$$\longrightarrow 2[W(NCMe_3)(NPh)Cl_2(C_{10}H_8N_2)] + 2Me_3CNH_2$$

Procedure

■ **Caution.** *Benzene is a human carcinogen. It should be used only in a well-ventilated hood, and protective gloves should be worn.*

Tetrachlorobis(1, 1-dimethylethanamine)bis[(1, 1-dimethylethyl)imido]bis-(phenylimido)ditungsten(VI) (2 g, 2 mmol) is placed in a 250-mL two-necked round-bottomed flask fitted with a gas inlet tap and a rubber serum cap, and benzene (60 mL) is added via a syringe. Benzene (40 mL) is added with a syringe to 2, 2'-bipyridine (0.7 g, 4.5 mmol) contained in a 100-mL round-bottomed flask also fitted with a gas inlet tap and serum cap, and the solution is transferred to the flask containing the tungsten complex. Several boiling chips are added, and a reflux condenser, which is fitted at the top with a gas inlet tap, is connected. The system is attached to a nitrogen line containing a gas bubbler, and the entire system is thoroughly purged with nitrogen before heating to reflux. After several minutes, the gas inlet tap fitted to the condenser is closed off, and a vacuum is applied to the solution via the gas tap fitted to the flask. Nitrogen is readmitted to the flask, the gas tap attached to the condenser is opened, and refluxing under nitrogen is continued. This evacuation process is repeated three more times at 15-min intervals during the reflux, in order to remove liberated 1, 1-dimethylethan-amine (bp 46 °C). Refluxing is continued for 2 h, and then the deep red solution is cooled and filtered and the solvent is removed under vacuum. The red microcrystalline product is washed several times with hot petroleum ether to remove excess 2, 2'-bipyridine. Pure *trans*-dichloro complex is obtained by dissolving the solid in hot toluene (100 mL), filtering, and allowing the solution to stand at $-20\,°C$. After removing the product by filtration, the volume of the filtrate is reduced under vacuum and the crystallisation procedure is repeated. The yield of dark red crystalline complex after successive crops is up to 1.6 g (70%).

Anal. Calcd. for $C_{20}H_{22}Cl_2N_4W$:C, 41.9; H, 3.9; N, 9.8. Found: C, 41.6; H, 3.9; N, 9.4.

Properties

The complex is not particularly air or moisture sensitive, but it is best handled and stored under nitrogen. The melting point is 120 °C. The complex is insoluble in petroleum ether and soluble in benzene and chloroform. An X-ray crystal structure determination[9] shows cis- oriented organoimido ligands, *trans* chloro groups, and the 2, 2'-bipyridine nitrogen atoms coordinated trans to the imido functions. In the IR spectrum, the *trans*-W—Cl absorption occurs at $210\,cm^{-1}$. The ^{1}H NMR spectrum (CDCl₃) of pure *trans*-dichloro complex shows the *tert*-butyl methyl group resonance as a singlet at 1.42 ppm (from TMS), the phenyl group and 2, 2'-bipyridine γ protons as a multiplet between 6.72 and 7.60, the 2, 2'-bipyridine β protons as a multiplet between 7.80 and 8.20 ppm, and the two 2, 2'-bipyridine α protons as doublets centered at 9.2 and 9.48 ppm. ^{13}C NMR spectrum in CDCl₃ (ppm from TMS): 31.8 (Me₃); 67.5 (N—C); 121.2 (C-ortho, C_6H_5); 122.3, 122.5, 122.9 (C-β, bipy); 126.5 (C-para, C_6H_5); 128.1 (C-meta, C_6H_5); 139.7, 139.9 (C-γ, bipy); 150.6, 151.6 (C-ipso; bipy); 152.5, 154.1 (C-α, bipy); 158.0 (C-ipso, C_6H_5).[5]

C. DICHLORO[(1,1-DIMETHYLETHYL)IMIDO](PHENYL-IMIDO)BIS(TRIMETHYLPHOSPHINE) TUNGSTEN(VI)

$$[W(NCMe_3)(\mu\text{-}NPh)Cl_2(NH_2CMe_3)]_2 + 4PMe_3$$

$$\longrightarrow 2[W(NCMe_3)(NPh)Cl_2(PMe_3)_2] + 2Me_3CNH_2$$

Procedure

■ **Caution.** *Benzene is a human carcinogen and should be handled in a well-ventilated hood with suitable gloves. Trimethylphosphine is spontaneously* *flammable in air. It has a vile odor and is very toxic by inhalation. It must be kept under an inert atmosphere and manipulated in a well-ventilated hood.*

Petroleum ether (100 mL) is added with a syringe to tetrachlorobis(1,1-dimethylethanamine)bis[(1,1-dimethylethyl)imidobis(phenylimido)ditungsten-(VI) (2 g, 2 mmol) contained in a 250-mL two-necked round-bottomed flask fitted with a gas inlet tap and a rubber serum cap. A solution of trimethyl phosphine (1.2 mL, 10.9 mmol) in petroleum ether (40 mL) in a 100-mL two-necked round-bottomed flask fitted with gas inlet tap and rubber septum is transferred to the suspension of the tungsten complex in petroleum ether. The serum cap is replaced by a greased glass stopper, and the mixture is stirred for at least 24 h. The dark red complex is removed by filtering, washed twice with petroleum ether (30 mL), and dried under vacuum. Yield: 2.1 g (91%).

Anal. Calcd. for $C_{16}H_{32}Cl_2N_2P_2W$:C, 33.8; H, 5.7; N, 4.9. Found: C, 33.1; H, 5.9; N, 5.5.

Properties

The complex is air and moisture sensitive, but it can be stored for long periods under N_2 without decomposition. It melts at 128–130 °C and is soluble in benzene and chloroform. In the IR spectrum $v_{(W-Cl)}$ occurs at 240 cm^{-1}. The ^1H NMR spectrum in $CDCl_3$ shows the *tert*-butylimido methyl protons at 1.40 (ppm from TMS), the trimethylphosphine methyl protons as a 1:2:1 triplet centered at 1.78 ($^2J_{P-H} = 5$ Hz), and the phenyl group protons as a multiplet between 6.78 and 7.50. ^{13}C NMR spectrum (ppm from TMS): 14.9, 15.9, 17.0 (1:2:1 triplet, $^1J_{C-P} = 16.1$ Hz, PMe$_3$); 32.6 (Me$_3$); 69.4(NC); 121.7, 124.5, 128.2 (C-ortho, meta, para, C_6H_5); 156.9 (C-ipso, C_6H_5). The ^{31}P NMR spectrum shows a singlet at $\delta = 9.89$ (external H_3PO_4), indicating a trans configuration of phosphines.[5]

References

1. W. A. Nugent and B. L. Haymore, *Coord. Chem. Rev.*, **31**, 123 (1980).
2. D. C. Bradley, M. B. Hursthouse, K. M. A. Malik, A. J. Nielson, and R. L. Short, *J. Chem. Soc. Dalton Trans.*, **1983**, 2651; T. C. Jones, A. J. Nielson, and C. E. F. Rickard, *J. Chem. Soc. Chem. Commun.*, **1984**, 206; P. A. Bates, A. J. Nielson, and J. M. Waters, *Polyhedron*, **4**, 1391 (1985).
3. D. C. Bradley, R. J. Errington, M. B. Hursthouse, A. J. Nielson, and R. L. Short, *Polyhedron*, **2**, 843 (1983).
4. B. R. Ashcroft, A. J. Nielson, D. C. Bradley, R. J. Errington, M. B. Hursthouse, and R. L. Short, *J. Chem. Soc. Dalton Trans.*, **1987**, 2059.

5. D. C. Bradley, R. J. Errington, M. B. Hursthouse, R. L. Short, B. R. Ashcroft, G. R. Clark, A. J. Niclson, and C. E. F. Rickard, *J. Chem. Soc. Dalton Trans.*, **1987**, 2067

6. A. J. Nielson, R. E. McCarley, S. L. Laughlin, and C. D. Carlson, *Inorg. Synth.*, **24**, 194 (1986).

7. A. J. Nielson and B. L. Haymore, *Inorg. Synth.*, **27**, 327 (1990).

8. W. Wolfsberger and H. Schmidbauer, *Synth. React. Inorg. Met. Org. Chem.*, **4**, 149 (1974).

9. G. R. Clark, A. J. Nielson, and C. E. F. Rickard, *Polyhedron*, **7**, 117 (1988).

59.   RELATIVELY AIR-STABLE M(II) SACCHARINATES, M = V, or Cr

Submitted by F. ALBERT COTTON,* EDUARDO LIBBY,[†] CARLOS A. MURILLO,[†] and GRETTEL VALLE[†] Checked by: M. BAKIR,[‡] D. R. DERRINGER,[‡] and R. A. WALTON[‡]

In studies of the chemistry of vanadium(II) compounds, it has been found that one of the major limitations for the advancement of the field is the lack of suitable starting materials, particularly those that could be used in nonaqueous solvents.[1,2] It is desirable for a good starting material to be reactive, soluble, easily accessible and stable enough to be handled without complex manipulations.

The synthesis of a vanadium(II) saccharinate complex, namely, tetraaquabis(1, 2-benzisothiazol-3(2H)-one 1, 1-dioxidato)vanadium(II) dihydrate that satisfies all of these characteristics[3] and can be stored even in the presence of air for a few weeks, without apparent decomposition, is presented here. However, if it is to be kept for long periods of time, it is better to store it under an inert gas.

Also included in this section are procedures for the synthesis of the chromium analog and a pyridine substituted vanadium(II) saccharinate. All of these compounds are potentially useful in the synthesis of other metal(II) compounds. They are prepared in good yields and with high purity.

General Procedures and Techniques

The majority of the solutions are oxygen sensitive. Therefore, all reactions are done under a nitrogen or argon atmosphere, using standard Schlenk-type techniques. Water and 95% ethanol are refluxed under nitrogen prior to

*Contribution from Department of Chemistry and Laboratory for Molecular Structure and Bonding, Texas A & M University, College Station, TX 77843.
†Department of Chemistry, University of Costa Rica, Ciudad Universitaria, Costa Rica.
‡Department of Chemistry, Purdue University, West Lafayette, IN 47907.

distillation. The rest of the solvents are dried and deoxygenated thoroughly prior to use.

Starting Materials

Vanadium(II)sulfate hexahydrate, $VSO_4 \cdot 6H_2O$, is prepared according to the reported procedure,[2] as is also the Cr(II) stock solution,[4] except that for the latter, the chromium metal is allowed to stand in contact with the 0.6 M HCl for 5 h at 70 °C while stirring.

All other reagents are used as received.

A. TETRAAQUABIS(1, 2-BENZISOTHIAZOL-3(2H)-ONE 1, 1-DIOXIDATO)VANADIUM(II) DIHYDRATE

$$VSO_4 \cdot 6H_2O + 2NaC_7H_4NO_3S$$

$$\longrightarrow [V(C_7H_4NO_3S)_2(H_2O)_4] \cdot 2H_2O + Na_2SO_4$$

Procedure

A solution of 8.0 g (31 mmol) of $VSO_4 \cdot 6H_2O$ is prepared in 25 mL of water at 0 °C. Separately, a solution of 16.6 g (68 mmol) of sodium saccharinate dihydrate is made in 27 mL of water. The latter solution is cooled to 10–15 °C and is added to the vanadium sulfate. Upon mixing, the color changes to red. To the reaction mixture is added 20 mL of chilled 95% ethanol. After 24 h at 0 °C a crop of crystals is obtained. The red solid is thoroughly washed with five portions of 20 mL of cooled water and is dried under vacuum for 2 h. Yield: 4.3 g (26%).

The mother liquor which does not include the water from the washings, is stored at 0 °C for 6 days. Another crop of crystals is obtained. They are cleaned as before. Yield: 2.5 g (15%).

Higher yields of $\sim 75\%$ can be secured if the reaction mixture is allowed to crystallize at -18 °C for 2 weeks. However, it is more difficult to free the product of the large crystals of sodium sulfate that are also formed, and more washings with water are needed to obtain the pure product.

Anal. Calcd. for $V(C_7H_4NO_3S)_2 \cdot 6H_2O$: V, 9.73; saccharin, 69.6, sulfate, 0.00,; Found: V, 9.65; saccharin, 69.0; sulfate, 0.00.

Properties

Tetraaquabis(1, 2-benzisothiazol-3(2H)-one 1, 1-dioxidato)vanadium(II) di-hydrate is a compound with a remarkable stability toward air, particularly when it is crystalline. After some weeks exposed to the air, a green coating begins to form on the compound. This may be eliminated by washing with a mixture of water: ethanol: 18 M H_2SO_4 in a 5:5:1 ratio, until the green color disappears. The product is then washed with water, ethanol, and diethyl ether, and finally it is dried.

In spite of its d^3 electronic configuration, the substitution reactions of the complex are relatively fast. It is soluble in ethanol but is virtually insoluble in water. However, it sometimes forms supersaturated solutions in water, and may take very long periods of time to crystallize.

Single crystal X-ray analysis shows that the compound crystallizes in the monoclinic space group $P2_1/c$ with $a = 7.936(2)$Å, $b = 16.149(5)$Å, $c = 7.731(2)$Å, $\beta = 99.84(2)$, $V = 976.2(8)$Å^3, and $Z = 2$. There is a centrosymmetric arrangement of two saccharinate nitrogen atoms and four water molecules around the vanadium center.

Solid state (KBr pellet) IR spectra[5] show the following main bands: 3600–3000 (s, b), 1620 (s), 1580 (s), 1460 (m), 1345 (m), 1285 (s), 1260 (w), 1150 (s), 1120 (m), 1050 (m), 950 (m), 880 (w), and 750 (m) cm^{-1}.

B. BIS(1, 2-BENZISOTHIAZOL-3(2H)-ONE 1, 1-DIOXIDATO)-TETRAKIS(PYRIDINE)VANADIUM(II)-DIPYRIDINE

$$[V(C_7H_4NO_3S)_2(H_2O)_4] \cdot 2H_2O + 6C_5H_5N$$
$$\longrightarrow [V(C_7H_4NO_3S)_2(C_5H_5N)_4] \cdot 2(C_5H_5N) + 6H_2O$$

Procedure

■ **Caution.** *To be done in a hood*

In a distillation apparatus, 140 mL of pyridine is added to 7.73 g (14.8 mmol) of $V(C_7H_4NO_3S)_2(H_2O)_4] \cdot 2H_2O$. The deep red solution is warmed, and ~ 60 mL of the solvent is removed by distillation at slightly above ambient pressure (regulated by a mercury bubbler). The flask and its contents are cooled to room temperature, and a layer of 135 mL of hexane is carefully placed on the reaction solution. The flask is cooled to − 20 °C, and after 1 day at this temperature a large crop of crystals forms. The mixture is shaken to mix the liquid phases and is returned to the freezer. The next day, the resulting black crystals are filtered and washed with four 20-mL portions of hexane and are dried under vacuum for 5 min. Yield: 13.0 g (99%).

Anal. Calcd. for $V(C_7H_4NO_3S)_2 \cdot 6C_5H_5N$:V, 5.72; saccharin, 40.9. Found: V, 5.75; saccharin, 41.0.

Properties

Bis(1,2-benzisothiazol-3(2H)-one 1,1-dioxidato)tetrakis(pyridine)vanadium-(II)-dipyridine forms deep red crystals (almost black). As obtained by the above procedure, the crystals are relatively large, and can be exposed to air for a few minutes without any apparent decomposition. They can be weighed in air without any problem. The compound is soluble in pyridine, ethanol, and, to a lesser extent, in warm tetrahydrofuran (THF), from which it is recrystallized as $[V(C_7H_4NO_3S)_2(C_5H_5N)_4] \cdot 2H_2O$.[2]

Single crystal X-ray analysis shows that the compound crystallizes in the orthorhombic space group *Pbna* with $a = 15.430(6)$Å, $b = 18.323(4)$Å, $c = 15.966(5)$Å, $V = 4514(2)$Å^3, and $Z = 4$. The coordination of the metal atom is octahedral, with deviations from regularity. There is a square set of four pyridine nitrogen atoms and two trans oxygen atoms from the carbonyl groups of the saccharinate ligands surrounding the vanadium atom.

The solid state KBr pellet IR spectrum shows the following main bands: 3600–3100 (s, b), 1605 (s), 1560 (s), 1480 (w), 1455 (m), 1440 (m), 1370 (m), 1290 (s), 1150 (s), 1115 (m), 1045 (m), 935 (m), and 745 (m) cm^{-1}.

C. TETRAAQUABIS(1,2-BENZISOTHIAZOL-3(2H)-ONE 1,1-DIOXIDATO)CHROMIUM(II) DIHYDRATE

$$Cr^{2+}_{aq} + 2NaC_7H_4NO_3S \longrightarrow [Cr(C_7H_4NO_3S)_2(H_2O)_4] \cdot 2H_2O + 2Na^+_{aq}$$

Procedure

To 60 mL of a stock Cr(II) solution cooled to $\sim 0\,°C$ is added a solution of sodium saccharinate at $\sim 12\,°C$. The latter is prepared by adding 36 mL of water to 24.2 g of sodium saccharinate hydrate and then adding 70 mL of absolute ethanol.

The yellowish-green solution is allowed to stand at $-20\,°C$ for 2 days. The green crystals that form are filtered and washed with two portions of 15 mL of cold water. The filtrate is almost colorless. The crystals are dried at room temperature under vacuum for 3 h. Yield: 6.45 g.

Anal. Calcd. for $Cr(C_7H_4NO_3S)_2 \cdot 6H_2O$:Cr, 10.0; saccharin, 69.5; Cl, 0.00. Found: Cr, 10.2; saccharin, 69.7; Cl, 0.00.

Properties

Tetraaquabis(1,2-benzisothiazol-3(2H)-one 1,1-dioxidato)chromium(II) di-hydrate is a quite air-stable compound. Its crystals remain unaltered for days when exposed to the air, but the powder tends to decompose a little more rapidly. Its substitution reactions are relatively fast.[3] Its solubility in water is low, but it is more soluble in pyridine. The addition of polar solvents such as ethanol, THF, or acetone transforms the compound to a metal–metal bonded dimer.[6]

Single crystal X-ray analysis shows that the compound is isomorphous with its vanadium analog, as well as the other divalent transition metal saccharinates. It crystallizes in the monoclinic space group $P2_1/c$ with $a = 8.042(4)$Å, $b = 16.032(8)$Å, $c = 7.804(4)$Å, $\beta = 100.90(4)$, $V = 9881(8)$Å^3, and $Z = 2$.

The solid state IR spectrum is almost identical to its vanadium analog.

References

1. F. A. Cotton and C. A. Murillo, *Ing, Cienc. Quim.*, **9**, 5 (1985).

2. F. A. Cotton, L. R. Falvello, R. Llusar, E. Libby, C. A. Murillo, and W. Schwotzer, *Inorg. Chem.*, **25**, 3423 (1986).

3. E. Libby, Licentiate Thesis, University of Costa Rica, 1986.

4. F. A. Cotton, G. E. Lewis, C. A. Murillo, W. Schwotzer, and G. Valle, *Inorg. Chem.*, **23**, 4038 (1984).

5. The IR spectrum is similar to that given by the other first-row transition metal M(II) saccharinates, as reported in S. Z. Haider, K. M. A. Malik, and K. J. Ahmed, *Inorg. Synth.*, **23**, 47 (1985).

6. F. A. Cotton and C. A. Murillo, unpublished results.

60. PYRIDINIUM FLUOROTRIOXOCHROMATE(VI), $(C_5H_5NH)[CrO_3F]$

$$CrO_3 + HF + C_5H_5N \longrightarrow (C_5H_5NH)[CrO_3F]$$

Submitted by MANABENDRA N. BHATTACHARJEE* and MIHIR K. CHAUDHURI*
Checked by JAY H. WORRELL† and THOMAS LI†

One of the main interests in chromium(VI) chemistry is the synthesis of new chromium(VI) compounds suitable for the effective and selective oxidation of organic substrates, especially alcohols, under mild conditions. Pyridinium

*Department of Chemistry, North-Eastern Hill University, Shillong 793 003, India. Financial support by the Department of Atomic Energy, Government of India, is gratefully acknowledged.
†Department of Chemistry, University of South Florida, Tampa, FL 33620.

fluorotrioxochromate(VI), $(C_5H_5NH)[CrO_3F]$ (PFC),[1] is one of several newly synthesized chromium(VI) compounds that has been shown to be a very effective reagent,[1,2] exhibiting some specific advantages[1,2] over other similar compounds used for the oxidation of organic substrates. The method developed for the synthesis of PFC[1] involves the reaction of chromium(VI) oxide, CrO_3, with 40% hydrofluoric acid and pyridine. The method is rapid and can be scaled up to larger quantities if desired.

■ **Caution.** *Hydrofluoric acid is toxic and causes severe skin and eye burns. Protective gloves and goggles should be worn when handling it. The synthesis should be carried out in a well-ventilated fume hood.*

Procedure

A 5.0-g (50 mmol) sample of chromium(VI) oxide, CrO_3, and 3.5 mL (70 mmol) of 40% hydrofluoric acid are added to 12 mL of water in a 50 mL polyethylene beaker. The mixture is stirred for 2–5 min, yielding a clear orange colored solution. A 4.1-mL volume (50 mmol) of distilled pyridine is added, in small portions, to this solution with stirring. The mixture is heated on a steam bath for 10–15 min to give bright orange, crystalline pyridinium fluorotrioxochromate(VI). The compound is separated by filtration (Whatman No. 41 paper) on a polyethylene funnel and is washed three times with hexane. It is dried by pressing between the folds of a filter paper and finally put under vacuum. Yield: $C_5H_5NHCrO_3F$ 8.9 g (90%).

Anal. Calcd. for $H_6FO_3C_5NCr$: C, 30.16; H, 3.04; N, 7.04; F, 9.54; Cr, 26.12. Found: C, 30.1; H, 3.07; N, 6.96; F, 9.6; Cr, 26.2.

Properties

Pyridinium fluorotrioxochromate(VI) is a bright orange, highly crystalline, diamagnetic compound. It exhibits a melting point in the 106–108 °C range.* The compound is stable at ambient temperatures and may be stored for prolonged periods in a sealed polyethylene bag.† It can be checked periodically by estimation of either chromium or C, H, and N. Pyridinium fluorotrioxochromate(VI) is soluble in water and acetone, less soluble in dichloromethane and acetonitrile, and only sparingly soluble in benzene, chloroform and hexane. The compound is an excellent oxidant[1-3] for organic substrates.

---

*The checkers observed a melting point of 102–108 °C, with an orange to orange-brown color change begining at ∼ 85 °C.
†The checkers stored the compound in plastic vials and observed no decomposition over several months.

The IR spectrum (KBr pellet) shows bands (cm^{-1}) at 952 $(v_1$, Cr—O sym), 640 $(v_2$, Cr—F$_{stretch})$, 340 $(v_3$, O—Cr—O, O—Cr—F), 980 $(v_4$, Cr—O$_{asym\,stretch})$, 375 $(v_5$, O—Cr—O), and 260 $(v_6$, O—Cr—F). Molar conductance of a $1 \times 10^{-3}\,M$ solution in water at 25 °C: $\Lambda_M = 128\,\Omega^{-1}\,cm^2\,mol^{-1}$. pH of a $1 \times 10^{-2}\,M$ solution in water is 2.45.

References

1. M. N. Bhattacharjee, M. K. Chaudhuri, H. S. Dasgupta, N. Roy, and D. T. Khathing, *Synthesis* **1982**, 588.
2. T. Nonaka, S. Kanemoto, K. Oshima, and H. Nazaki, *Bull. Chem. Soc. Jpn.*,**57**, 2019 (1984).
3. M. N. Bhattacharjee, M. K. Chaudhuri, and S. Purkayastha, *Tetrahedron*, **43**, 5389 (1987).

61. DIPOTASSIUM TRIFLUOROSULFATO-MANGANATE(III)

$$KMnO_4 + 2HCHO + K_2SO_4 + 4HF$$

$$\longrightarrow K_2[MnF_3(SO_4)] + KF + 2HCOOH + 2H_2O$$

Submitted by MANABENDRA N. BHATTACHARJEE* and MIHIR K. CHAUDHURI*
Checked by THOMAS LI† and JAY H. WORRELL†

Dipotassium trifluorosulfatomanganate(III), $K_2[MnF_3(SO_4)]$, was obtained for the first time[1] in small amounts as a by-product from a preparation of crystals of dipotassium pentafluoromanganate(III) monohydrate, $K_2[MnF_5]\cdot H_2O$, by an adaptation of Palmer's method.[2] This result represents a rare example of a manganese(III) compound having a tetragonally compressed octahedral structure.[1] No method dedicated to the preparation of $K_2[MnF_3(SO_4)]$ was available in the literature until recently.[3] It was reported in 1984 that this interesting compound could be synthesized[3] by allowing $KMnO_4$ to react with 40% HF and K_2SO_4 in the presence of formaldehyde, or from the reaction of MnO(OH) with 40% HF and K_2SO_4. Alternatively, the compound can be synthesized by the reaction of MnO(OH) with 40% HF and $K_2S_2O_8$. The procedure presented here is based on the reaction of $KMnO_4$ with aqueous hydrofluoric acid and K_2SO_4 in the presence of formaldehyde. The method is direct and can be scaled up if desired.

*Department of Chemistry, North-Eastern Hill University, Shilling 793 003, India. Financial support by the Department of Atomic Energy, Government of India, is gratefully acknowledged.
†Department of Chemistry, University of South Florida, Tampa, FL 33620.

■ **Caution.** *Hydrofluoric acid and formaldehyde are toxic and cause severe skin (HF) and eye (HF and HCHO) burns. Protective gloves and goggles should be worn, and the synthesis should be carried out in a well-ventilated fume hood.*

Procedure

A 2.0-g (12.7 mmol) sample of potassium permanganate and 2.25 g (12.93 mmol) of potassium sulfate are ground together. The mixed powder is transferred to a 200-mL polyethylene beaker and is dissolved in 40 mL of a very dilute solution of hydrofluoric acid, made up of 4 mL (80 mmol) of 40% hydrofluoric acid and 36 mL of water, by warming over a steam bath for 5–10 min. The mixture is then filtered through Whatman No. 41 filter paper on a polyethylene funnel. The filtrate is collected in a 200 mL of polyethylene beaker and again warmed over a steam bath. A 4-mL amount of 38% formaldehyde solution is added to the warm solution in small portions with constant stirring, to obtain a deep brown solution. The solution is concentrated to ~25 mL by heating* on a steam bath; pink crystalline $K_2[MnF_3(SO_4)]$ begins to appear in a few minutes. As the reaction proceeds the solution color fades. The solution is allowed to cool to room temperature for ~1 h, to give pink crystalline $K_2[MnF_3(SO_4)]$.[1,3] The compound is separated by filtration on a polyethylene funnel and is washed three times with ethanol. It is dried under vacuum. Yield: $K_2[MnF_3(SO_4)]$ 2.9 g (80.6%).

Anal. Calcd. for $K_2[MnF_3(SO_4)]$: K, 27.32; Mn, 19.2; F, 19.92; SO_4, 33.56. Found: K, 27.5; Mn, 19.5; F, 20.2; SO_4, 33.4.

Properties

The $K_2[MnF_3(SO_4)]$ is a pink crystalline compound that is unstable in water and attacks glass surfaces very slowly in the presence of moist air. The compound can be stored for prolonged periods in a sealed polyethylene envelope. Its purity can be ascertained periodically by estimation of manganese. The chemically estimated oxidation state[3] of manganese is between 2.9 and 3.1, supporting the presence of the metal in its +3 state. The μ_{eff} of the compound at 288 K is 4.0 BM, a value higher than that observed for the corresponding binary fluoromanganate(III) compound $K_2[MnF_5]\cdot H_2O$.[4] The IR spectrum (KBr pellet) shows bands (cm^{-1}) at

---

*The checkers note that if the mixture is heated rapidly on a hot plate and/or cooled rapidly, an impure product is obtained that contains a gray-pink powder mixed with the desired dark red-pink crystals.

1240(s), 1150(s) and 1020(s) (v_3), 970(s) (v_1), and 684(s), 636(s), and 605(s) (v_4) [v_1, v_3, and v_4 are all S—O modes]; 520(s) (v_{Mn-F}).

References

1. A. J. Edwards, *J. Chem. Soc. (A)*, 1971, 3074.
2. W. G. Palmer, *Experimental Inorganic Chemistry*, Cambridge University Press, Cambridge, 1950, p. 479.
3. M. N. Bhattacharjee and M. K. Chaudhuri, *Polyhedron*, 3, 599 (1984).
4. M. N. Bhattacharjee and M. K. Chaudhuri, *Inorg. Synth.*, 25, 50 (1986).

62. *cis*-DICHLOROBIS(1, 2-ETHANEDIAMINE)-PLATINUM(IV) CHLORIDE

$$2HCl + [Pt(en)_2]Cl_2 \xrightarrow[\Delta, 7h]{conc\, HCl} trans\text{-}[Pt(en \cdot HCl)_2 Cl_2]$$

$$trans\text{-}[Pt(en \cdot HCl)_2 Cl_2] + H_2O_2 \xrightarrow[\Delta, 2h]{} cis\text{-}[Pt(en)_2 Cl_2]Cl_2 + 2H_2O$$

en = 1, 2-ethanediamine

Submitted by ROBERT M. KUKSUK,* WADE A. FREEMAN,* and
PIERRE R. LEBRETON*
Checked by GEORGE B. KAUFFMAN† and MATTHEW L. ADAMS†

The synthesis of the *cis*-dichlorobis(1, 2-ethanediamine)platinum(IV) ion was first claimed by Heneghan and Bailar.[1] Later, Liu and Doyle[2] reported the optical resolution, physical (IR, NMR, and ORD) and chemical characterization of the complex ion, expressing doubt that the Heneghan and Bailar procedure worked. Liu and Ibers[3] confirmed the characterization of the resolved complex with an X-ray diffraction study, and Liu and Yoo[4] reported some of its substitution reactions. The details of the synthesis of this complex ion do not appear in refs. 2 to 4, but only in a more obscure reference.[5] Perhaps because of this, little work with this interesting complex has appeared, and some workers[6] have attempted to use the original Heneghan and Bailar synthesis, which, according to Liu,[7] yields only the *trans* isomer.

*Department of Chemistry, University of Illinois at Chicago, Chicago IL 60680.
†Department of Chemistry, California State University, Fresno, Fresno, CA 93740.

In view of the continuing interest in the substitution reactions of *cis*-dichloroplatinum(II) complexes, we now report an improved procedure for the synthesis of this related platinum(IV) complex. The synthesis uses the approach designed by Doyle,[5] but alters the details of the reaction conditions so that the procedure is more reliable.

*Procedure**

▪ **Caution.** *Concentrated HCl and 30% H_2O_2 cause burns and have irritating vapors. A well-ventilated fume hood, gloves, and safety glasses are required when these materials are used.*

A. *trans*-DICHLOROBIS(1, 2-ETHANEDIAMINE MONO-HYDROCHLORIDE)PLATINUM(II)

A 11.46-g (30.00 mmol) quantity of bis(1, 2-ethanediamine)platinum(II) chloride (Alfa) is dissolved in a minimal amount of water in a 500-mL Erlenmeyer flask. [The bis(1, 2-ethanediamine)platinum(II) chloride must first have been carefully recrystallized twice to insure its purity. Recrystallization is best accomplished by dissolving the solid in ~ 50 mL of water, gravity filtering, and allowing the filtrate to evaporate slowly. Pure [Pt(en)$_2$]Cl$_2$ gives a colorless (not brown or yellow) solution when dissolved in water. Impure [Pt(en)$_2$]Cl$_2$ drastically lowers the yield of this procedure.] A 200-mL volume of concentrated HCl is added, and the solution is heated to 89 °C for 7 h. The solution must not be boiled. The reaction mixture is allowed to cool to room temperature, then it is chilled in an ice–water bath. The solid that forms is collected by vacuum filtration, and the filtrate is set aside for platinum recovery. The solid is then washed with 100 mL of water acidified with five drops of concentrated HCl. The product dissolves in this washing, and some insoluble yellow by-product, which is [Pt(en)Cl$_2$], remains on the paper. This second filtrate is treated with 200 mL of concentrated HCl at room temperature, which precipitates the pure product. Yield: 4.65 g (30%).

Anal. Calc. for [Pt(C$_2$H$_8$N$_2$·HCl)$_2$Cl$_2$]: C, 10.46; H, 3.95; N, 12.20; Cl, 30.89. Found: C, 10.47; H, 3.92; N, 12.25; Cl, 30.89.

trans-Dichlorobis(1, 2-ethanediamine monohydrochloride)platinum(II) is a pale yellow solid. The ^{13}C NMR spectrum shows two peaks: δ 39.8 and 43.4 (ppm relative to TMS), measured against an external reference of 5% *p*-dioxane, which gives a single peak at 67.4 ppm.

Note: The synthesis may be scaled down by proportional adjustment of the amounts of reagents used.

B. *cis*-DICHLOROBIS(1, 2-ETHANEDIAMINE)PLATINUM(IV) CHLORIDE

cis-Dichlorobis(1, 2-ethanediamine)platinum(IV) chloride is prepared by the controlled oxidation of *trans*-dichlorobis(1, 2-ethanediamine monohydro-chloride)platinum(II). A 1.92-g (4.2 mmol) quantity of pure *trans*-[Pt(en·HCl)$_2$ Cl$_2$] is dissolved in a solution of three drops of concentrated HCl in 20 mL of water in a 100-mL beaker. A 1.32-mL volume of commercial 30% H$_2$O$_2$ is added. (A three- to sevenfold excess of H$_2$O$_2$ works.) The mixture is heated to 89 °C for 2 h. It must not be boiled. It is then evaporated, at room temperature, under a gentle stream of air, until some crystals appear. The addition of approximately five times the volume of the solution of ice-cold absolute ethanol precipitates the bulk of the crude product. A second crop can be obtained by repeating these steps. The crude solid is collected by vacuum filtration, and washed, first with absolute ethanol, then with diethyl ether.

The solid is dissolved in a minimal amount of a solution prepared by mixing 100 mL of water with 15 drops of concentrated HCl. The solution is cooled in an ice–water bath for ~ 15 min and vacuum filtered. This removes a red by-product that appears in some experiments. The pure product is obtained by evaporating the solution to crystallization on a steam bath, cooling it slowly to ice temperatures, and then adding five times the volume of the solution of ice-cold absolute ethanol.

Anal. Calcd. for [Pt(C$_2$H$_8$N$_2$)$_2$Cl$_2$]Cl$_2$: C, 10.51; H, 3.54; N, 12.26; Cl, 31.02. Found: C, 10.79; H, 3.54; N, 12.38; Cl, 30.67. Yield: 1.35 g (70.6%).*

Properties

cis-Dichlorobis(1, 2-ethanediamine)platinum(IV) chloride is a white (some-times off-white) solid, soluble in water. The ^{13}C NMR spectrum shows two peaks: δ 46.6 and 49.9 (ppm relative to TMS), measured against an external reference of 5% *p*-dioxane. Bands in the IR spectrum are 2988 (vs), 1640 (m), 1553 (m), 1522 (m), 1445 (m), 1304 (m), 1176 (s), 1136 (m), 1053 (s), and 988 (w) cm^{-1}. Recrystallization from water (slightly acidified with HCl) gives colorless prisms with an optic axial angle, 2V, equal to 80 °.

References

1. L. F. Heneghan and J. C. Bailar, Jr., *J. Am. Chem. Soc.*, **75**, 1840, 1953.
2. C. F. Liu and J. Doyle, *Chem. Commun.*, **219**, 412, 1967.

*The checkers report a yield of 66.2%.

3. C. F. Liu and J. Ibers, *Inorg. Chem.*, **9**, 773, 1970.
4. C. F. Liu and M. Yoo, *Inorg. Chem.*, **15**, 2415, 1976.
5. J. L. Doyle, Ph.D. Thesis, University of Michigan, 1967.
6. O. N. Adrianova, N. Sh. Gladkaya, and V. N. Vorotnikova, *Russ. J. Inorg. Chem.*, **15**, 1278 (1970).
7. C. F. Liu, personal communication.

63. BIS(1,1,1,5,5,5-HEXAFLURO-2,4-PENTANE-DIONATO)PALLADIUM AND ITS COMPLEXES WITH 2,2'-BIPYRIDINE AND BIS[2-(DIPHENYLPHOSPHINO)-ETHYL]PHENYLPHOSPHINE

Submitted by A. R. SIEDLE*
Checked by X. L. LUO and R. H. CRABTREE[†]

Bis(1,1,1,5,5,5-hexafluoro-2,4-pentanedionato)palladium has an extensive acid base chemistry. Reactions with nucleophilic ligands lead to sequential Pd—O bond cleavage and formation of complexes of the type $Pd(CF_3COCHCOCF_3)_2(\text{ligand})_n$ $(n = 1-4)$.[1] The ionic complex with 2,2'-bipyridine (bipy) may be represented as $[Pd(\text{bipy})(O,O\text{-}CF_3COCHCOCF_3)][CF_3COCHCOCF_3]$; the coordinated $[CF_3COCHCOCF_3]^-$ ligand in it is readily displaced by other ligands so that the complex is a source of $Pd(\text{bipy})^{2+}$.[2] The semichelating $CF_3COCHCOCF_3$ ligand in $[Pd[Ph_2PC_2\text{-}H_4P(Ph)C_2H_4PPh_2](CF_3COCHCOCF_3)][CF_3COCHCOCF_3]$ similarly undergoes facile displacement, and this compound provides a source of $Pd(Ph_2PC_2H_4P(Ph)C_2H_4PPh_2)^{2+}$.[3] Bis(hexafluoropentanedionato) palladium is also useful in preparing formazan palladium chelates[4] and ortho-metalated compounds such as $Pd(C_6H_4N{=}NC_6H_5)(CF_3COCHCOCF_3)$.[5,6] It reacts readily with condensed phase donors such as alumina to form a surface complex that is easily reduced to catalytically active coatings of metallic palladium.[7] Unlike its nickel(II) analog, the palladium complex has a low affinity for oxygen donors and is obtained from aqueous media in anhydrous form. The palladium starting material used here, $PdCl_2$, was selected on account of availability and relatively low cost. However, hydrated $Pd(NO_3)_2$ or the Li^+ or Na^+ salts of $PdCl_4^{2-}$ may be substituted.

*3M Corporate Research Laboratories, St. Paul, MN 55144.
†Department of Chemistry, Yale University, New Haven, CT 06520.

A.   BIS(1,,1,1,5,5,5-HEXAFLUORO-2,4-PENTANEDIONATO)-PALLADIUM

$$2Na[CF_3COCHCOCF_3] + PdCl_2 \longrightarrow Pd(CF_3COCHCOCF_3)_2 + 2NaCl$$

Procedure

▪ **Caution.** *Contact with liquid hexafluoroacetylacetone or its vapor or with PdCl$_2$ dust should be avoided. These materials are toxic and should be handled with gloves in a well-ventilated hood.*

The preparation is conveniently carried out in a 500-mL creased flask fitted with a paddle stirrer or, alternatively, in a blender. Smaller scale preparations may be conducted in an Erlenmeyer flask using a magnetic stirrer. Because PdCl$_2$ is only sparingly soluble in water, the mixture is heterogeneous and the amount of time required for completion of the reaction varies with efficiency of stirring.

A solution of Na[CF$_3$COCHCOCF$_3$] is prepared by adding 22.7 g (0.11 mol) commercial 1,1,1,5,5,5-hexafluoro-2,4-pentanedione to 110 mL of 1 *M* aqueous sodium hydroxide and stirring until the diketone dissolves. Powdered palladium(II) chloride, 9.8 g (0.055 mol) is then added in four portions with vigorous stirring. The reaction mixture thickens as the product separates, and 50 mL of water is added to reduce the viscosity and facilitate stirring. After stirring for 12–14 h, the reaction mixture is filtered through a medium porosity glass frit. The solids are sucked dry (because of the volatility of the product, the yield is reduced if this operation is prolonged) and then extracted with 50 mL of dichloromethane. The yellow dichloromethane solution is separated from a small amount of water by decantation or with a separatory funnel, filtered through paper, and evaporated to dryness under water aspirator vacuum. Sublimation at 50 °C onto a 0 °C cold finger using a mechanical pump provides 19.2 g (67%) of product as dense, yellow-orange needles. Insertion of a small plug of glass wool beneath the cold finger prevents contamination by a small amount of black, fluffy nonvolatile residue. Prolonged sublimation under dynamic vacuum results in some loss of the product into the cold trap. (The checkers ran the reaction on one fifth this scale and, using a water aspirator vacuum, obtained a yield of 72%.)

Anal. Calcd. for C$_{10}$H$_2$F$_{12}$O$_4$Pd: C, 23.1; H, 0.4; F, 43.9; Pd, 20.4. Found: C, 23.0; H, 0.5; F, 43.8; Pd, 20.7.

Properties

The compound is readily soluble in hydrocarbons such as hexane and toluene, in chlorinated hydrocarbons such as dichloromethane and chloroform, and in tetrahydrofuran, acetonitrile, and methanol. Solubility in water is very low. Solutions in aliphatic alcohols slowly decompose with deposition of metallic palladium.[7] Bis(1,1,1,5,5,5-hexafluoro-2,4-pentanedionato)-palladium is quite volatile and is best purified by sublimation. The melting point, determined by differential scanning calorimetry, is 99.6 °C, but, because of the volatility of this material, melting points measured in capillaries are not reliable criteria of purity and this is better determined by NMR and elemental analyses. The ^{1}H and ^{19}F NMR spectra [in CDCl$_3$ relative to internal (CH$_3$)$_4$Si and CFCl$_3$] comprise singlets at 6.42 and -73.6 ppm, respectively. The IR spectrum (Nujol mull) contains strong bands at 1600, 1450, 1265, 1230, 1210, 1175, 1155, 1105, and 810 cm^{-1}. The crystal structure has been reported.[1]

B. (2,2′-BIPYRIDINE)(1,1,1,5,5,5-HEXAFLUORO-2,4-PENTANEDIONATO)PALLADIUM(II) 1,1,1,5,5,5-HEXA-FLUORO-2,4-DIOXO-3-PENTANIDE

$$(CF_3COCHCOCF_3)_2Pd + C_{10}H_8N_2$$
$$\longrightarrow [Pd(C_{10}H_8N_2)(CF_3COCHCOCF_3][CF_3COCHCOCF_3]$$

Procedure

A solution of 0.26 g (0.5 mmol) bis(hexafluoropentanedionato)palladium in 8 mL of toluene is added to 0.08 g (0.5 mmol) 2,2′-bipyridine in 3 mL of the same solvent. The product separates as yellow crystals that are collected on a filter, washed with toluene then pentane, and vacuum dried using a mechanical pump. Yield: 0.32 g (95%), mp 193 °C (dec).

Anal. Calcd. for C$_{20}$H$_{10}$F$_{12}$N$_2$O$_4$Pd: C, 35.5; H, 1.5; N, 4.1. Found: C, 35.2; H, 1.4; N, 4.0.

Properties

The compound is soluble in dichloromethane, methanol, and acetonitrile, slightly soluble in chloroform and insoluble in water and aromatic hydrocarbons. At 24 °C, the 94.2 MHz ^{19}F NMR spectrum in CD$_3$CN comprises a singlet at -74.7 ppm (w/2 = 240 Hz). The 100 MHz NMR spectrum in acetone-d_6 contains peaks at 8.96 (dd, 1.4, 8.3), 8.56 (dt, 1.5, 7.9), 8.56 (dd, 1.5, 5.8), and 8.03 (ddd, 1.3, 5.7, 8.0 Hz) due to the bipy ligand and a broad singlet at 6.1 ppm due to the methine protons in the exchanging CF$_3$COCHCOCF$_3$ groups. The IR spectrum (Nujol mull) contains strong bands at 1670, 1630, 1610, 1455, 1275, 1215, 1155, 1130, and 780 cm^{-1}. The crystal structure has been reported.[2]

C. [BIS[2-(DIPHENYLPHOSPHINO)ETHYL]PHENYL-PHOSPHINE](1,1,1,5,5,5-HEXAFLUORO-2,4-PENTANEDIONATO)PALLADIUM(II) 1,1,1,5,5,5-HEXAFLUORO-2,4-DIOXO-3-PENTANIDE

$$(CF_3COCHCOCF_3)_2Pd + (Ph_2PC_2H_4)_2PPh$$

$$\longrightarrow [Pd[(Ph_2PC_2H_4)_2PPh](CF_3COCHCOCF_3)][CF_3COCHCOCF_3]$$

Procedure

A solution of 0.52 g (1 mmol) Pd(CF$_3$COCHCOCF$_3$)$_2$ in 3 mL of toluene is added with stirring to 0.53 g (1 mmol) (Ph$_2$PC$_2$H$_4$)$_2$PPh (triphos) in 10 mL of warm toluene. Methylcyclohexane, 15 mL, is added to the resulting orange solution. After standing for 1 h at room temperature, the reaction mixture is cooled in a wet ice bath, then filtered to provide 0.68 g of product as fibrous, cream colored needles. These are sufficiently pure for synthetic purposes. Analytically pure material, mp 192 °C, results from recrystallization by slow rotary evaporation of a dichloromethane–hexane solution of the compound. The yield of the recrystallized complex is 0.60 g (57%). (The checkers ran this reaction under argon on one half the above scale and obtained a 71% yield after recrystallization.)

Anal. Calcd. for C$_{44}$H$_{35}$F$_{12}$O$_4$P$_3$Pd: C, 50.1; H, 3.3; P, 8.8; Pd, 10.0. Found: C, 50.3; H, 3.2; P, 8.5; Pd, 10.3.

Properties

The complex is soluble in dichloromethane, acetonitrile, acetone and methanol. The coordinated semichelating CF$_3$COCHCOCF$_3$ group undergoes an intramolecular rearrangement and also exchanges with the noncoordinated [CF$_3$COCHCOCF$_3$]$^-$. Therefore, the ^{1}H and ^{19}F NMR spectra are temperature and field dependent. At 27 °C, the 400 MHz ^{1}H spectrum in CDCl$_3$ contains two broad singlets at 5.62 and 5.23 ppm of unit area due to the methine protons in the nonequivalent CF$_3$COCHCOCF$_3$ groups (the checkers observed 5.62 and 5.19 ppm at 250 MHz). The ^{31}P NMR spectrum shows peaks at 108.2 (d) and 48.6 (t, $J_{PP} = 9.5$ Hz) (external 85% H$_3$PO$_4$ reference). The IR spectrum (Nujol mull) contains strong bands at 1670, 1550, 1535, 1255, 1210, 1185, 1135, and 480 cm^{-1}.

References

1. A. R. Siedle, R. A. Newmark, and L. H. Pignolet, *Inorg. Chem.*, **22**, 2281 (1983).
2. A. R. Siedle, R. A. Newmark, A. A. Kruger, and L. H. Pignolet, *Inorg. Chem.*, **20**, 3399 (1981).
3. A. R. Siedle, R. A. Newmark, and L. H. Pignolet, *J. Am. Chem. Soc.*, **103**, 4947 (1981).
4. A. R. Siedle and L. H. Pignolet, *Inorg. Chem.*, **19**, 2052 (1980).
5. A. R. Siedle, *J. Organometal. Chem.*, **208**, 115 (1981).
6. M. C. Etter and A. R. Siedle, *J. Am. Chem. Soc.*, **105**, 641 (1983).
7. A. R. Siedle, P. M. Sperl, and T. W. Rusch, *Appl. Surf. Sci.*, **6**, 149 (1980).

Chapter Nine

LIGANDS AND OTHER MAIN GROUP COMPOUNDS

64. HYBRID TERTIARY PHOSPHINE AMINE AND AMIDE CHELATE LIGANDS: *N*-[2-(DIPHENYLPHOSPHINO)-PHENYL]BENZAMIDE AND 2-(DIPHENYLPHOSPHINO)-*N*-PHENYLBENZAMIDE

Submitted by DAVID HEDDEN* and D. MAX ROUNDHILL*
Checked by BRUCE N. STORHOFF[†] and M. BRIAN ARNOLD[†]

Functionalized tertiary aryl phosphines play an important role in transition metal coordination chemistry. These compounds have been used as ligands in synthesis, catalysis, mechanistic studies, and in the study of coordination compounds as structural models.[1] In this contribution the syntheses of two new types of these ligands, tertiary aryl phosphines functionalized by an amide group, are detailed. The published coordination chemistry of these compounds includes the study of intramolecular N—H oxidative addition, the synthesis of chelates stabilized amido complexes, and the preparation of complexes with both five- and six-membered chelate rings.[2]

The starting materials for these ligand syntheses are known phosphines that can be modified by standard synthetic methodology. Furthermore, these synthetic routes are general methods since the reagent that modifies the nitrogen group of the starting phosphine (synthesis in Section A), or which introduces the nitrogen functionality into the ligand (synthesis in section B)

*Department of Chemistry, Tulane University, New Orleans, LA 70118.
†Department of Chemistry, Ball State University, Muncie, IN 47306.

can be replaced by a homolog reagent. In this manner, the steric and/or electronic properties of the nitrogen group can be controlled.

A.   *N*-[2-(DIPHENYLPHOSPHINO)PHENYL]BENZAMIDE

$$\text{(ring)}\underset{P(C_6H_5)_2}{\overset{NH_2}{<}} + C_6H_5COCl + C_5H_5N$$

$$\xrightarrow{\text{THF}} \text{(ring)}\underset{P(C_6H_5)_2}{\overset{NHC(O)C_6H_5}{<}} + C_5H_5NHCl$$

tetrahydrofuran = THF (solvent)

Procedure

▪ **Caution.** *Tetrahydrofuran is extremely flammable and forms explosive peroxides; only fresh peroxide-free material should be used. Pyridine, benzoyl chloride, and dichloromethane are harmful if inhaled. Benzoyl chloride causes severe skin and eye burns. All manipulations should be carried out in a well-ventilated fume hood, protective gloves and goggles should be worn.*

Tetrahydrofuran is dried over sodium benzophenone and a fresh sample is distilled under nitrogen into the reaction vessel immediately before use. Pyridine is sequentially dried over powdered KOH and CaH_2 for a total of 12 h. The CaH_2-dried sample is distilled under nitrogen into a flask containing activated 4 Å molecular sieves (Aldrich). The flask is stoppered under a nitrogen blanket. Benzoyl chloride is distilled under nitrogen and is used immediately.

2-(Diphenylphosphino)benzenamine[3] (2.46 g, 8.9 mmol) and pyridine (2.12 g, 26.7 mmol) are dissolved in dry THF (10 mL) contained in a 25-mL two-necked round-bottomed flask fitted with a Suba-Seal rubber septum (Strem), an oil bubbler connected to a nitrogen source, and a magnetic stirring bar. Freshly distilled benzoyl chloride (1.25 g, 8.9 mmol) is rapidly added to the stirred solution via a syringe through the rubber septum. A white precipitate of pyridine hydrochloride is formed immediately. The suspension is stirred under nitrogen for 30 min. The white precipitate is removed by filtration in air and then washed with THF (4 × 20 mL). Removal of the solvent from the combined washings and filtrate using a rotary evaporator gives a viscous white oil. Unreacted pyridine is removed by washing the oil with water (6 × 25 mL) using a separatory funnel.* The remaining oil is dissolved in

---

*The checkers dissolved the oil in 50 mL of dichloromethane and then washed the solution with 2 × 25 mL of 5% (aq) HCl. This procedure reduced the pyridine odor.

dichloromethane (15 mL) and the solution is dried over anhydrous $MgSO_4$ for 12 h. After filtering off the solids and washing them with dichloromethane (20 mL), the volume of the solution is reduced to ~ 3 mL on a rotary evaporator. Hexane (100 mL) is added, and the cloudy solution is stored overnight at $-10\,°C$. The resulting white needles are collected by filtration, lightly washed with hexane (10 mL), and dried *in vacuo* (1 torr, 24 h, 25 °C). Yield: 2.81 g (83%); mp 107 to 108 °C.

Anal. Calcd. for $C_{25}H_{22}NOP$: C, 78.7; H, 5.29; N, 3.67; P, 8.12. Found: C, 78.8; H, 5.47; N, 3.67; P, 8.22.

The previous procedure has been scaled up by a factor of 10 with no decrease in yield or product purity.

Properties

The crystalline solid has been stored in air for several months without any detectable decomposition. It displays the solubility properties expected for a tertiary aryl phosphine: It is soluble in ethers, chloroform, toluene, and acetone, but insoluble in hexane and ethanol. IR(Nujol mull): $v_{(NH)} = 3350\,cm^{-1}$, $v_{(CO)} = 1680\,cm^{-1}$. ^1H NMR (CDCl$_3$): $\delta 8.80$ (NH, 1H, broad): 8.53–8.45, 7.7–6.9 (ArH, 19H, multiplet). ^{31}P NMR (CDCl$_3$): $\delta - 20.3$.

This synthesis yields exclusively the **Z** rotational isomer of the amide as determined by IR spectroscopy.[4] Conversion to the **E** isomer cannot be induced by either deprotonation–protonation of the NH group, or by heating *in vacuo* at 220 °C.

N-[2-(Diphenylphosphino)phenyl]acetamide has been synthesized by substituting benzoyl chloride by acetyl chloride. It should also be possible to use other types of *N*-acylating reagents such as acid anhydrides.[5]

B. 2-(DIPHENYLPHOSPHINO)-*N*-PHENYLBENZAMIDE

Procedure

■ **Caution.** *Chloroform is a suspected carcinogen and may cause adverse reproductive effects. Aniline may cause cyanosis and may be fatal if absorbed* *through the skin. Dicyclohexylcarbodiimide may cause allergic skin reactions. Protective gloves and goggles should be worn, and all manipulations should be carried out in well-ventilated fume hood.*

A 250-mL two-necked round-bottomed flask is equipped with a Teflon coated magnetic stirring bar and a 125-mL pressure equalizing dropping funnel that is capped with an oil bubbler. The oil bubbler is connected to a nitrogen source. Chloroform (50 mL of reagent grade) is then placed in the flask and is deoxygenated with nitrogen. Under a nitrogen flush, 2-(diphenylphosphino)benzoic acid (6.12 g, 20 mmol)[6] and freshly distilled aniline (0.93 g, 20 mmol) are added to the flask through the open neck. The flask is immersed in an ice bath, and the solids are dissolved with stirring to give a yellow solution. While the mixture in the flask is cooling, the dropping funnel is charged with a nitrogen saturated solution of N, N'-dicyclohexylcarbodiimide (Aldrich) (4.12 g, 20 mmol) in $CHCl_3$ (50 mL). The solution is added dropwise over a 30-min period to the stirred, cooled solution in the flask. A white precipitate, N, N'-dicyclohexylurea, forms. The reaction mixture is allowed to warm to room temperature, and it is then stirred under nitrogen for 2 h. The suspension is vacuum filtered through a medium porosity glass frit, and the white solids left on the frit are washed with $CHCl_3$ (2 × 20 mL). Rotary evaporation of the combined filtrate and washings gives a yellow oil. Product isolation is achieved by column chromatography on silica gel (230–400 mesh silica gel, 6 × 40-cm column).* Elution with CH_2Cl_2 gives two colorless, mobile bands. The eluant is collected in 10-mL fractions. The progress of the chromatogram is monitored by thin layer chromatography (TLC) (CH_2Cl_2 eluant; silica gel $60F_{254}$ TLC plates (EM Reagents); Band 1: $R_f = 0.6$, Band 2: $R_f = 0.35$). The leading band contains the product. The fractions that test positive for the product are combined, and solvent is removed by rotary evaporation to give the product as a white powder. Yield: 3.41 g (56% based on reacted 2-(diphenylphosphino)-benzoic acid, see below),† mp 179–180 °C. An analytical sample is prepared by the addition of hexane to a saturated CH_2Cl_2 solution of the compound. The fluffy, white needles are isolated by filtration and dried *in vacuo* (1 torr, 25 °C, 24 h).

---

*The checkers isolated the product by standard flash separation chromatography on Bakerflex flash-grade silica. Unreacted 2-(diphenylphosphino)benzoic acid was obtained as the first band, and the product was obtained cleanly as the second band. An unidentified phosphorus compound was detected in the column wash by ^{31}P NMR.

†The checkers obtained a yield of 1.44 g (19%), and observed only a trace of N, N'-dicyclohexylurea precipitate during addition of the carbodiimide. They attribute the low yield to contamination of their nitrogen gas with water, a condition discovered after the experiments were completed. The submitters note that, over the 30-min addition period, water in the nitrogen gas would hydrolyze the moisture sensitive N, N'-dicyclohexylcarbodiimide and would thus reduce the yield.

Anal. Calcd. for $C_{25}H_{20}NOP$: C, 78.7; H, 5.29; N, 3.67; P, 8.12. Found: C, 78.8; H, 5.34; N, 3.66; P, 8.06.

The (amide I) second chromatographic band is discarded. The yellow band that remains at the top of the column is removed by elution with CH_3CN. Solvent removal gives 1.2 g unreacted 2-(diphenylphosphino)benzoic acid.

The above procedure has been scaled up by a factor of 10 without any decrease in yield or product purity when the product mixture is chromatographed in portions.

Properties

The **E** rotational isomer of the amide is the sole product of this reaction, as determined by IR spectroscopy.[4] IR (Nujol mull): $v_{(NH)} = 3240 \, cm^{-1}$; $v_{(CO)} = 1650 \, cm^{-1}$. ^1H NMR (CDCl$_3$): $\delta 7.8$–6.8 (ArH and NH, 20H, multiplet). ^{31}P NMR (CDCl$_3$): $\delta - 9.8$.

Sublimation (1 torr, 250 °C) of the **E** isomer results in a thermally induced **E → Z** isomerization. For the **Z** isomer: mp 145–146 °C.

Anal. Calcd. for $C_{25}H_{20}NOP$: C, 78.7; H, 5.29; N, 3.67; P, 8.12. Found: C, 78.6; H, 5.35; N, 3.64; P, 7.99.

IR (Nujolmull): $v_{(NH)} = 3360 \, cm^{-1}$; $v_{(CO)} = 1650 \, cm^{-1}$.

Both isomers are air-stable crystalline solids that are soluble in ethers, chloroform, benzene, and acetone.

References

1. T. B. Rauchfuss in *Homogenous Catalysis with Metal Phosphine Complexes*, L. H. Pignolet (ed.) Plenum, New York, 1983, pp. 239–256.
2. D. Hedden, D. M. Roundhill, W. C. Fultz, and A. L. Rheingold, *J. Am. Chem. Soc.*, **106**, 5014 (1984). D. Hedden and D. M. Roundhill, *Inorg. Chem.*, **24**, 4152 (1985); **25**, 9 (1986). D. Hedden, D. M. Roundhill, W. C. Fultz, and A. L. Rheingold, *Organometallics*, **5**, 336 (1986). S. Park, D. Hedden, A. L. Rheingold, and D. M. Roundhill, *Organometallics*, **5**, 1305 (1986).
3. M. K. Cooper, and J. M. Downes, *Inorg. Chem.*, **17**, 880 (1978); M. K. Cooper, J. M. Downes, and P. A. Duckworth, *Inorg. Synth.*, 25, XXX (1988).
4. B. C. Challis and J. A. Challis, in *Comprehensive Organic Chemistry*, Vol. 2, D. H. R. Barton and W. D. Ollis (eds.), Pergamon, Oxford, 1979, pp. 990–991.
5. M. E. Wilson, R. G. Nuzzo, and G. M. Whitesides, *J. Am. Chem. Soc.*, **100**, 2269 (1978).
6. J. E. Hoots, T. B. Rauchfuss, and D. A. Wrobleski, *Inorg. Synth.*, **21**, 175 (1982).

65.   1,1-DIMETHYL-*N*-(TRIMETHYLSILYL)ETHANAMINE

Submitted by A. J. NIELSON*
Checked by B. L. HAYMORE[†]

$$Me_3SiCl + 2Me_3CNH_2 \xrightarrow[0°C]{petroleum\ ether} Me_3SiNHCMe_3 + Me_3CNH_3Cl$$

N-(Trimethylsilyl)alkanamines (Me_3SiNHR) are important reagents for the synthesis of alkylimido complexes ($M≡NR$) of the earlier transition metals.[1] In particular, they are useful in the preparation of bis(organoimido) complexes of tungsten(VI).[2] The silylamines have been prepared by alcohol elimination[3] and transamination[4] reactions, but the most convenient method is the reaction between chlorotrimethylsilane and a primary alkylamine,[5] although the yields have generally been low. The following procedure for the synthesis of 1,1-dimethyl-*N*-(trimethylsilyl)ethanamine, employing strictly anhydrous conditions, gives a yield in excess of 80%. The method is applicable to a variety of *N*-(trimethylsilyl)alkanamines.

Starting Materials and General Procedure

Petroleum ether (bp fraction 40–60 °C) is dried over sodium wire, and chlorotrimethylsilane (Aldrich) is used without further purification. 1,1-Dimethylethanamine is dried over and distilled from calcium hydride and stored and handled under nitrogen under strictly anhydrous conditions. Commercial oxygen-free dry nitrogen is used. When solutions are transferred between flasks, or between flask and dropping funnel, a stainless steel transfer tube is used. Each flask is fitted with a gas inlet tap and a serum cap through which the transfer tube passes. The nitrogen supply to the receiving vessel is turned off, and a vent needle is passed through the septum. With the transfer tube placed below the level of liquid, a positive nitrogen pressure is used to force the solution into the receiving flask. All glassware is oven dried and completely flushed with nitrogen before use.

Procedure

■   **Caution.**   *Chlorotrimethylsilane   and   1,1-dimethylethanamine   are extremely flammable, high vapor pressure, toxic irritants. They can cause severe*

*Department of Chemistry, University of Auckland, Private Bag, Auckland, New Zealand.
†Monsanto Co., 800 N. Lindbergh Boulevard, St. Louis, MO 63167.

skin and eye burns. Chlorotrimethylsilane reacts violently with water. The synthesis should be carried out in well-ventilated fume hood, and protective gloves and goggles should be worn.

Petroleum ether (400 mL) is placed in a 1-L three-necked round-bottomed flask, fitted with a gas inlet tap, a 250-mL dropping funnel, and a mechanical stirrer. 1, 1-Dimethylethanamine [135 mL (94 g), 1.28 mmol] is added by syringe to the petroleum ether; the flask is then flushed with N_2 and cooled with an ice bath. Chlorotrimethylsilane [75 mL (64.2 g), 0.59 mmol] in petroleum ether (100 mL) is transferred to the dropping funnel, and the pressure is equalized to the flask. The silylchloride is added dropwise to the 1, 1-dimethylethanamine–petroleum ether solution over a period of 2 h, while the mixture is rapidly stirred and the ice bath maintained. A dense precipitate of 1, 1-dimethylethanamine hydrochloride forms during the reaction. After the addition is complete, stirring is continued for a further 4 h with the ice bath removed, and the reaction mixture is then allowed to stand overnight under N_2. The solution is quickly filtered in the air through a sintered filter funnel into another 1-L three-necked round-bottomed flask equipped with a gas inlet tap and stopper. The amine hydrochloride is washed twice with petroleum ether (100 mL), and the flask is then stoppered and flushed with nitrogen. (Approximately 60 g of the amine hydrochloride is obtained after drying in an oven at 100 °C, indicating that the reaction has gone to > 90% completion.) The flask is set up for distillation through an 18-in. fractionating column, keeping all the apparatus flushed with nitrogen, and the petroleum ether is distilled off. (*Note:* the distillation should not be discontinued until the temperature of the solution reaches 90 °C because some of the petroleum ether fraction is retained at this temperature.) The contents of the flask are transferred under N_2 to a 250-mL two-necked round-bottomed flask fitted with a gas inlet tap and rubber serum cap. The flask is set up for distillation through an 18-in. fractionating column, which is well lagged or heated, and the fraction is collected when the thermometer reads 110 to 120 °C but no higher than 125 °C. The yield of 1, 1-dimethyl-N-(trimethylsilyl)ethanamine is 93 mL (82%).* (*Note:* Depending on the efficiency of the column some product will collect with the fraction obtained up to 110 °C. This may be recovered by refractionation, bringing the yield to ~ 90%.)

Properties

The product obtained from the above procedure is pure enough for most applications, but it should be redistilled through a fractionation column, and

---

* The checker obtained a yield of 72% using glassware that had not been oven dried. Dried glassware improved the yield to 76%. Dried glassware and freshly distilled Me_3SiCl gave 89% yield.

the distillate collected at 120–122 °C, if extremely pure amine is required. 1,1-Dimethyl-*N*-(trimethylsilyl)ethanamine is hydrolytically sensitive and should be stored under N_2 in a Schlenk tube fitted with a rotoflow cap and transferred with a syringe. The ^{1}H NMR spectrum shows equal intensity methyl group singlets at 0.0 and 1.1 ppm from TMS. Physical data for a series of *N*-(trimethylsilyl)alkanamines are shown in Table I.

TABLE I   Physical Data for *N*-(Trimethylsilyl)alkanamines[5,6]

	bp(°C)	N_D^{20}	d^{20}
$Me_3SiNHCMe_3$	121–122	1.4076	0.754
$Me_3SiNHCHMe_2$	102–103	1.3950	0.733
$Me_3SiNHEt$	90–91	1.3929	0.730
$Me_3SiNHMe$	71	1.3901	0.739

References

1. W. A. Nugent and B. L. Haymore, *Coord. Chem. Rev.*, **31**, 123 (1980); T. C. Jones, A. J. Nielson, and C. E. F. Rickard, *J. Chem. Soc. Chem. Commun.*, **1984**, 205.

2. A. J. Nielson and B. L. Haymore, *Inorg. Synth.*, **27**, 300 (1990)

3. K. Ruhlmann, *J. Prakt. Chem.*, **16**, 172 (1962); E. W. Abel, *J. Chem. Soc.*, **1961**, 4933.

4. S. H. Langer, S. Connell, and I. Wender, *J. Org. Chem.*, **23**, 50 (1958).

5. E. W. Abel and G. R. Willey, *J. Chem. Soc.*, **1964**, 1528; R. O. Sauer and R. H. Hasek, *J. Am. Chem. Soc.*, **68**, 241 (1946); E. W. Abel, D. A. Armitage, and G. R. Willey, *Trans. Faraday Soc.*, **60**, 1527 (1964); A. W. Jarvie and D. J. Lewis, *J. Chem. Soc.*, **1963**, 1073; R. Fessenden and J. S. Fessenden, *Chem. Rev.*, **61**, 361 (1961).

6. G. R. Willey, *J. Phys. Chem.*, **71**, 4294 (1967).

66.   ETHYNYLPENTAFLUORO-λ^6-SULFANE ($SF_5C \equiv CH$)

$$SF_5Br + HC \equiv CH \xrightarrow{57\,°C} SF_5CH = CHBr$$

$$SF_5CH = CHBr + KOH \xrightarrow{\text{petrolium ether}} SF_5C \equiv CH + KBr + H_2O$$

Submitted by ROBIN J. TERJESON,* JO ANN CANICH,* and GARY L. GARD*
Checked by MARK R. COLSMAN,† MARK A. UREMOVICH,† and STEVEN H. STRAUSS†

It is known that the introduction of SF_5 groups into molecular systems can bring about significant changes in their physical, chemical, and biological

*Department of Chemistry, Portland State University, Portland, OR 97207.
†Department of Chemistry, Colorado State University, Fort Collins, CO 80523.

properties. These properties are manifested by various applications, such as solvents for polymers, perfluorinated blood substitutes, surface-active agents, fumigants, and as thermally and chemically stable systems.[1] The syntheses of these compounds are the subjects of ongoing studies. One compound of particular interest is ethynylpentafluoro-λ^6- sulfane (SF$_5$C≡CH), which has been used as the starting reagent for the syntheses of a number of novel and interesting SF$_5$ derivatives.[2,3]

The two-step procedure described here represents a convenient, facile synthesis that produces SF$_5$C≡CH in high yields.[4]

Procedure

■ **Caution.** *Sulfur bromide pentafluoride is an extremely toxic and moisture sensitive compound. This material should be handled in a well-ventilated fume hoods or on vacuum lines. Acetylene is a reactive gas, and care must be taken to avoid the presence of oxygen and pressure build-up; in closed reaction vessels, the total pressure must be under 30 atm.*

A.   SF$_5$CH=CHBr

Into an evacuated 150-mL Hoke stainless steel vessel equipped with a Whitey stainless steel value, SF$_5$Br* (5.69 g, 27.5 mmol) and CH≡CH (1.00 g, 38.5 mmol; Airco, technical grade, used as received) are condensed at − 196 °C. The mixture is heated at 55 ± 2 °C for 3.8 days. The pure, hydrolytically and thermally stable product, SF$_5$CH=CHBr (5.13 g, 22.0 mmol) is obtained by distillation in a Kontes (14/20), all glass apparatus, at atmospheric pressure (bp 86 ± 2 °C). Yield: 80%.[4,†]

B.   SF$_5$C≡CH

The apparatus that is used for the dehydrobromination of SF$_5$CH=CHBr consists of a 250-mL three-necked round-bottomed Pyrex glass vessel, equipped with a Teflon stirring bar, an addition funnel (125 mL), a nitrogen

*SF$_5$Br can be prepared by the interaction of BrF (45 mmol) with SF$_4$ (45 mmol) in the presence of CsF (13 mmol) in a 30-mL Hoke stainless steel vessel at 80–90 °C for 24 h. The reaction products are separated under dynamic vacuum by a trap-to-trap system with traps cooled to −95 and −112 °C. The SF$_5$Br is collected in the −112 °C trap (Yield: ∼75%). The compound BrF is prepared by equilibration of BrF$_3$ (16 mmol) and Br$_2$ (16 mmol) in a 150-mL stainless steel vessel at room temperature.[5] The chemicals SF$_4$ and CsF may be purchased from Peninsular ChemResearch (PCR), BrF$_3$ from Ozark-Mahoning and Br$_2$ from Mallinckrodt.

†The checkers vacuum distilled the product and collected it in a −35 °C trap. Yield: 65%. They noted the presence of a small amount of nonvolatile solid when the product is frozen at −196 °C and then warmed to room temperature.

inlet tube, and a West (Kontes) reflux condenser. A 100-mL Pyrex glass vacuum trap (cooled to $-196\,°C$) is connected to the condenser and protected from the atmosphere by a mercury bubbler. In the reaction vessel, 80 mL of petroleum ether (90–120 °C fraction; Baker)* is heated to reflux, and KOH (11.5 g, 205 mmol; Baker) is added. The compound, $SF_5CH{=}CHBr$ (20.0 g, 85.8 mmol) is added slowly over 0.8 h, and additional KOH (17.1 g, 305 mmol) is added during this period (the KOH turns brown and sludgelike). The mixture is allowed to heat at reflux under slow nitrogen flow for an additional 2 h. The product is transferred under vacuum (< 1 torr) at $-78\,°C$ to another 100-mL vacuum trap cooled to $-196\,°C$. The product, $SF_5C{\equiv}CH$, is collected in a 49% yield (6.38 g, 42.0 mmol).[4,†] The IR spectrum agrees with that previously reported.[2]

Properties

Ethynylpentafluoro-λ^6-sulfane is a hydrolytically and air stable colorless gas with a boiling point of 6 °C.[2]

The IR spectrum[4] contains the following bands (cm^{-1}): 3338 (ms), 2118 (m), 1613 (vw), 1506 (vw), 1344 (w with sh at 1338), 893 (vs, b), 730 (w), 720 (w), 674 (ms), 628 (m), 621 (m), and 590 (ms). The very strong band centered at 893 cm^{-1} is characteristic of the S—F stretching vibration; the absorption band at 620 cm^{-1} is attributed to the S—F deformation mode.[6] The acetylenic (—C≡C—) and the C—H absorptions occur at 2118 and 3338 cm^{-1}, respectively. The ^{19}F NMR spectrum: $\phi 71.8$ (SF), and $\phi 80.3$ (SF$_4$) are multiplets with $J_{SF_4-F} = 151.8$ Hz. The ^1H NMR spectrum: $\delta 2.67$ (pentet; $J_{F-H} = 3.15$ Hz).[4] The positive ion mass spectrum for $SF_5C{\equiv}CH$ has been determined: $(SF_5C_2H)^+$, $(SF_4C_2H)^+$, $(SF_5)^+$, $(SF_3)^+$, and $(C_2HF)^+$.[2]

References

1. See, for example, G. L. Gard and C. W. Woolf, U.S. Patent 3,448,121, (1969). G. L. Gard, J. Bach, and C. W. Woolf, British Patent 1,167,112, (1969). E. E. Gilbert and G. L. Gard, U.S. Patent 3,475,453, (1969). R. E. Banks and R. N. Haszeldine, British Patent 1,145,263. Y. Michimasa, *Chem. Abstr.*, **82**, 175255g (1975). W. A. Sheppard, U.S. Patent 3,219,690, (1965).

2. F. W. Hoover and D. D. Coffman, *J. Org. Chem.*, **29**, 3567 (1964).

3. A. D. Berry, R. A. DeMarco, and W. B. Fox, *J. Am. Chem. Soc.*, **101**, 737 (1979).

*The checkers used octane.

†The checkers used 3 g of $SF_5CH{=}CHBr$ in the dehydrobromination step; their yield was 25%. They note that the use of a $-196\,°C$ trap in the gaseous nitrogen stream causes condensation of liquid nitrogen as well as the product. They attribute their low yield to loss of products as the liquid nitrogen in the trap boils away during the vacuum transfer, and they suggest using a $-131\,°C$ (pentane slush) trap to avoid this problem.

4. J. M. Canich, M. M. Ludvig, W. W. Paudler, G. L. Gard, and J. M. Shreeve, *Inorg. Chem.*, **24**, 3668 (1985).
5. K. O. Christe, E. C. Curtis, and C. J. Schack, *Spectrochim. Acta*, **33A**, 69 (1977).
6. H. L. Cross, G. Cushing, and H. L. Roberts, *Spectrochim. Acta*, **17**, 344 (1961).

67. IODINE AND BROMINE POLYSULFUR HEXAFLUOROARSENATE(V) AND HEXAFLUOROANTIMONATE(V)

Submitted by M. P. MURCHIE,* J. PASSMORE,* and C.-M. WONG*
Checked by T. GRELBIG[†] and K. SEPPELT[†] (Sections A and B)
M. WITT[‡] and H. W. ROESKY[‡] (Sections C and D)

Neutral binary sulfur iodides have low stability and are in fact only stable at low temperatures, for example, S_2I_2 (refs. 1, 2) at 183 K and SI_2 at 9 K.[3] Bromides are more stable, and S_xBr_2 compounds ($x = 2$ to ~ 10) are known, of which only S_2Br_2 has been isolated as a pure compound and is well characterized.[4] Binary sulfur–bromine salts include $[SBr_3][MF_6]$ (M = As and Sb)[5] and the unstable NH_4SBr.[4] A number of iodine polysulfur and bromine polysulfur cations have been prepared quantitatively in recent years, as salts of very weakly basic $[AsF_6]^-$ and $[SbF_6]^-$ anions. The iodine containing salts are more stable than their bromine counterparts. In addition, the $[SbF_6]^-$ salts are more stable than those of $[AsF_6]^-$, although they are more difficult to prepare because of the difficulties of handling SbF_5. Below, the preparations of $[S_7I][MF_6]$ (M = As and Sb),[6] $[(S_7I)_2I][SbF_6]_3 \cdot 2AsF_3$,[7] $[S_7Br][MF_6]$,[8] $[(S_7I)_4S_4][AsF_6]_6$,[7] and $[(S_7Br)_4S_4][AsF_6]_6$ (ref. 8) are described. These syntheses involve the oxidation of mixtures of sulfur and halogen (iodine or bromine) with either arsenic pentafluoride or antimony pentafluoride in an appropriate solvent (arsenic trifluoride or sulfur dioxide).

General Procedures

The apparatus used in the preparations consists of a two-bulbed reaction vessel equipped with a Teflon-stemmed Pyrex valve (J. Young) and a coarse sintered glass frit between the bulbs.[9] In the case of compounds containing

*Department of Chemistry, University of New Brunswick, Fredericton, New Brunswick, Canada E3B 6E2.
[†]Freie Universität Berlin, Institut für Anorganische und Analytische Chemie, Fabeckstrasse 34-36, 1000, Berlin 33, Federal Republic of Germany.
[‡]Institut für Anorganische Chemie der Universität Göttingen, Tammannstrasse 4, D-3400 Göttingen, Federal Republic of Germany.

antimony, a medium glass frit can be used if high purity of products is needed. The transfers become more difficult with the medium frit, and total yields will be less but the purity will likely be higher. Volatile reagents are manipulated in a preconditioned metal vacuum line.[9] Air sensitive, nonvolatile materials are manipulated in a good dry box [e.g., Vacuum Atmospheres Dri-Lab with Dri-Train (HE-493)].

Sulfur dioxide (Matheson) is dried and stored over CaH_2. Sulfuryl chloride fluoride (Aldrich) and arsenic trifluoride (Ozark-Mahoning) are stored over 4 A molecular sieves and NaF, respectively.

In order to avoid by-products that cannot be separated from the salts, it is absolutely necessary to maintain the correct stoichiometry, unless a slight excess of one of the reactants is specified in the procedures. (The checkers recommend monitoring the drop in vapor pressure on pumping, for example, by a Pirani vacuum guage, as small amounts of SO_2 are retained by the crystalline solids. However, very extended pumping on some salts can lead to some decomposition.)

- **Caution.** *All reactions should be conducted in a well-ventilated fume hood with appropriate precautions. Arsenic and antimony pentafluoride and arsenic trifluoride are very poisonous and hydrolyze readily to form HF.[10,11] Bromine is corrosive and is harmful to the skin and mucous membranes. Sulfur dioxide is poisonous and can generate 3 to 4 atmospheres pressure at room temperature. Well-constructed glass vessels or metal systems must be employed to prevent pressure bursts. The use of rubber gloves, safety glasses, and face shields is recommended.*

Unfortunately, there is no very easy way to identify these salts. An IR spectrum establishes the presence of $[AsF_6]^-$ (699 cm^{-1}) and $[SbF_6]^-$ (669 cm^{-1}).[12] Ideally, Raman spectroscopy would be the identification tool of choice, but these compounds decompose in the laser beam even at liquid nitrogen temperature. The product weights themselves are a good identification of the product formed, as are elemental analyses. However, the latter can be unreliable, particularly for the antimony containing salts. Cell dimensions obtained from single crystals provide an unambiguous method of identification, although it is rather inconvenient. The reader is referred to the original publications for these data.

A. IODO-*CYCLO*-HEPTASULFUR HEXAFLUOROARSENATE(V), $[S_7I][AsF_6]$ AND IODO-*CYCLO*-HEPTASULFUR HEXAFLUOROANTIMONATE(V), $[S_7I][SbF_6]$ (REF. 6)

$$\tfrac{14}{8}S_8 + I_2 + 3AsF_5 \longrightarrow 2[S_7I][AsF_6] + AsF_3$$

$$\tfrac{42}{8}S_8 + 3I_2 + 10SbF_5 \longrightarrow 6[S_7I][SbF_6] + (SbF_3)_3 \cdot SbF_5$$

Procedure [S₇I][AsF₆]

Procedure $[S_7I][AsF_6]$

Elemental sulfur (Fisher) (2.16 g, 8.45 mmol) is introduced into the two-bulbed vessel by removing the Teflon value stem. The assembled vessel is connected to the vacuum line and evacuated ($\sim$ 1 h). Elemental iodine (BDH Chemicals) (2.23 g, 8.77 mmol) is introduced into the vessel in a similar manner to that described for sulfur, and the vessel is dried by gentle heating. Arsenic trifluoride (6.5 mL) is condensed into the bulb containing the sulfur–iodine mixture by cooling the bulb in a liquid nitrogen bath ($-196\,°C$). Arsenic penta-fluoride (Ozark-Mahoning) (1.89 g, 11.1 mmol), the limiting reagent, is condensed onto the mixture in aliquots from a premeasured volume at a known pressure.[13,14] On warming to room temperature, an intense orange-brown solution over a dark solid is noted. After 72 h, the solution is transferred through the sintered glass frit into the second bulb. Approximately 50% of the AsF_3 is condensed back onto the solid remaining in the first bulb by cooling the first bulb at $-196\,°C$. On warming to room temperature, some of the solid dissolves in the AsF_3, and the solution is again poured through the frit into the second bulb. This process is repeated ($\sim$ 5 times) until all soluble material has been transferred to the second bulb. The volatile materials are then slowly removed under dynamic vacuum, with the solution being held at $0\,°C$, producing a dark brown microcrystalline solid.

Anal. Calcd. weight of $[S_7I][AsF_6]$, based on AsF_5: 4.00 g. Found: 3.89 g. Calcd. weight of unreacted sulfur and iodine: 1.79 g. Found: 1.78 g.

Procedure $[S_7I][SbF_6]$

Sulfur (2.98 g, 11.6 mmol) and iodine (1.89 g, 7.46 mmol) are added to the glass vessel in a manner similar to that described for $[S_7I][AsF_6]$. Sulfur dioxide (5.5 mL) is condensed onto the sulfur–iodine mixture at $-196\,°C$. This results in the formation of a transparent purple solution over unreacted sulfur and iodine at room temperature. The solubility of iodine in SO_2 at room temperature is $\sim 0.005\,g\,g^{-1}$, and sulfur is essentially insoluble. Antimony pentafluoride (4.67 g, 21.5 mmol),* the limiting reagent, is transferred to a separate vessel, and then quantitatively transferred into the reaction vessel containing the mixture, which is held at $-196\,°C$. On warming to room temperature, an intense orange-brown solution over a dark solid is obtained. After 48 h with stirring, the soluble and insoluble materials are separated by filtration and extraction with SO_2 as described for $[S_7I][AsF_6]$.

---

*Antimony pentafluoride (Ozark-Mahoning) is triply distilled *in vacuo* in a rigorously dried glass apparatus and stored in a Pyrex round-bottomed flask fitted with a stainless steel valve [Whitey (1KS4)] and Teflon compression fittings (Swagelok).[15]

In order to facilitate the formation of a crystalline product, SO_2ClF (1.45 g) is condensed onto the solution in the second bulb at $-196\,°C$. After warming to room temperature, the solvent (SO_2/SO_2ClF) is slowly condensed back onto the insoluble materials producing a dark brown crystalline solid. This is achieved by cooling the bulb containing the insoluble material with running tap water ($\sim 11\,°C$) while the bulb containing the solvent–solute mixture is held at room temperature ($\sim 20\,°C$) overnight. The volatile materials are then removed under dynamic vacuum.

Anal. Calcd. weight of $[S_7I][SbF_6]$, based on SbF_5: 7.58 g. Found: 7.58 g. Calcd. weight of insoluble materials [i.e., $(SbF_3)_3\cdot SbF_5$ and unreacted sulfur and iodine]: 1.96 g. Found: 2.01 g.

Properties

Solutions of $[S_7I][MF_6]$ (M = As and Sb) in SO_2 or AsF_3 are stable for several weeks at room temperature. The solids, in sealed glass tubes under an atmosphere of dry nitrogen, are indefinitely stable (> 1 year) at $-20\,°C$. Solid $[S_7I][AsF_6]$ decomposes after several months at $5\,°C$, and more quickly (several weeks) at room temperature, as indicated by loss of crystallinity and evolution of iodine. Solid $[S_7I][SbF_6]$ appears to be stable under the same conditions.

B. μ-IODO-BIS(4-IODO-*CYCLO*-HEPTASULFUR) TRIS(HEXAFLUOROANTIMONATE) BIS(ARSENIC TRIFLUORIDE), $[(S_7I)_2I][SbF_6]_3\cdot 2AsF_3$ (REF. 7)

$$\tfrac{28}{8}S_8 + 3I_2 + 10SbF_5 \xrightarrow{AsF_3} 2[(S_7I)_2I][SbF_6]_3\cdot xAsF_3 + (SbF_3)_3\cdot SbF_5$$

Sulfur (1.01 g, 3.96 mmol) and a slight excess of iodine are added to the glass reaction vessel. The excess iodine is slowly removed on evacuation leaving the required amount of I_2 (0.88 g, 3.47 mmol). Arsenic trifluoride (5 mL) and antimony pentafluoride (2.49 g, 11.25 mmol) are condensed onto the sulfur–iodine mixture in a similar manner to that described for $[S_7I][AsF_6]$ and $[S_7I][SbF_6]$. After stirring for 72 h, a brown solution containing solids is obtained. The products are separated by filtration and extraction as previously described, and the volatile materials are removed. This produces a white insoluble solid, identified as $(SbF_3)_3\cdot SbF_5$ by Raman spectroscopy,[16] and a soluble microcrystalline red-orange solid identified as $[(S_7I)_2I][SbF_6]_3\cdot 2AsF_3$.[7]

Anal. Calcd. weight of $[(S_7I)_2I][SbF_6]_3 \cdot 2AsF_3$, based on sulfur: 3.48 g. Found: 3.67 g. Calcd. weight of $(SbF_3)_3 \cdot SbF_5$: 0.86 g. Found: 0.66 g.

The discrepancy in weights may be due to some transfer of $(SbF_3)_3 \cdot SbF_5$ through the coarse frit. If more nearly pure material is required a medium sintered frit is recommended. The $[(S_7I)_2I][SbF_6]_3 \cdot 2AsF_3$ is stored under an atmosphere of dry nitrogen in a glass sample tube at $-20\,°C$.

Properties

Solid $[(S_7I)_2I][SbF_6]_3 \cdot 2AsF_3$ slowly (~ 0.1 g/day) loses AsF_3 and iodine on prolonged pumping at room temperature, with loss of crystallinity. Solutions of $[(S_7I)_2I][SbF_6]_3$ in AsF_3 are stable for several weeks at room temperature.

C. BROMO-*CYCLO*-HEPTASULFUR HEXAFLUORARSENATE(V), $[S_7Br][AsF_6]$ AND BROMO-*CYCLO*-HEPTASULFUR HEXAFLUOROANTIMONATE(V), $[S_7Br][SbF_6]$ (REF. 8)

$$\tfrac{14}{8}S_8 + Br_2 + 3AsF_5 \longrightarrow 2[S_7Br][AsF_6] + AsF_3$$

$$\tfrac{42}{8}S_8 + 3Br_2 + 10SbF_5 \longrightarrow 6[S_7Br][SbF_6] + (SbF_3)_3 \cdot SbF_5$$

Procedure $[S_7Br][AsF_6]$

Arsenic pentafluoride (2.40 g, 14.1 mmol) is condensed onto a mixture of sulfur (2.79 g, 10.9 mmol) and AsF_3 (5 mL) in a glass reaction vessel as described for $[S_7I][AsF_6]$. After 12 h, bromine (0.77 g, 4.81 mmol), which has been stored over P_2O_5, is quantitatively condensed onto the solution at $-196\,°C$ from a vessel containing the preweighted quantity of bromine. [*Note*: Reactions carried out by adding bromine before the AsF_5 give material of very poor quality. In addition, an exact stoichiometry for AsF_5 and Br_2 is required, to avoid contamination with S_2Br_2 and $[(S_7Br)_4S_4][(AsF_6)_6]$. The solution is warmed to room temperature and stirred for 2 h, yielding a red-brown solution containing solids. The products are separated by filtration, and the volatile materials are very slowly (1 h) removed, producing a noncrystalline soluble red-brown solid and unreacted sulfur.

Anal. Calcd. weight of $[S_7Br][AsF_6]$, based on AsF_5: 4.64 g. Found: 4.64 g.

Procedure $[S_7Br][SbF_6]$

Typically, SbF_5 (3.04 g, 14.0 mmol) is condensed onto sulfur (3.03 g, 11.8 mmol) in SO_2 (6 mL), and the mixture is stirred for 1 h. A preweighted quantity of bromine (0.73 g, 4.56 mmol) is then transferred quantitatively to the mixture, and the resulting red-brown solution is stirred for a further 2 h (an exact stoichiometry for SbF_5 and Br_2 is required). The soluble and insoluble products are separated, and SO_2ClF (3.01 g) is added to the solution to facilitate crystal formation as described for $[S_7Br][SbF_6]$. The volatile materials are slowly ($\sim$ 12 h) removed under dynamic vacuum at 0 °C, yielding a crystalline red-brown soluble solid and a mixture of sulfur and $(SbF_3)_3 \cdot SbF_5$.

Anal. Calcd. weight of $[S_7Br][SbF_6]$, based on SbF_5: 4.54 g. Found: 4.51 g. Calcd. weight of insoluble solid [i.e., unreacted sulfur and $(SbF_3)_3 \cdot SbF_5$]: 2.20 g. Found: 2.28 g.

Properties

Solid $[S_7Br][AsF_6]$, under an atmosphere of dry nitrogen in a sealed glass tube, decomposes at room temperature after several days, while $[S_7Br][SbF_6]$ is stable under the same conditions for several weeks. *In situ* samples of both $[S_7Br][AsF_6]$ and $[S_7Br][SbF_6]$ in concentrated (SO_2) solution deposit a yellow solid (presumably S_8) with the evolution of a dark red liquid (presumably S_xBr_2) on standing at room temperature for several weeks. Both pure $[S_7Br][AsF_6]$ and $[S_7Br][SbF_6]$ in the solid phase can be isolated only from fresh solutions. As solids, they are stable for at least several months at -20 °C.

D. $[(S_7I)_4S_4][AsF_6]_6$ AND $[(S_7Br)_4S_4][AsF_6]_6$ CONTAINING THE IODO-*CYCLO*-HEPTASULFUR(1 +), BROMO-*CYCLO*-HEPTASULFUR(1 +), AND TETRASULFUR(2 +) CATIONS

$$4S_8 + 2X_2 + 9AsF_5 \xrightarrow{SO_2} [(S_7X)_4S_4][AsF_6]_6 + 3AsF_3 \qquad (X = I \text{ or } Br)$$

Procedure $[(S_7I)_4S_4][AsF_6]_6$ *(ref. 7)*

Arsenic pentafluoride (3.12 g, 18.4 mmol) is condensed onto a mixture of sulfur (2.04 g, 7.95 mmol) and iodine (1.01 g, 3.98 mmol) in SO_2 (3 mL) in a two-bulbed glass reaction vessel. (The $I_2:S_8$ ratio of 1:2 is necessary together with a slight excess of AsF_5.) After 24 h, a brown solution over brown needle-like crystals is obtained. The volatile material is removed by pumping slowly under dynamic vacuum (1 h), producing 5.25 g of the needlelike crystals. Calcd. weight for $[(S_7I)_4S_4][AsF_6]_4$: 5.30 g.

Procedure $[(S_7Br)_4S_4][AsF_6]_6$ *(ref. 8)*

Arsenic pentafluoride (2.18 g, 12.8 mmol) is condensed onto a mixture of sulfur (1.34 g, 5.22 mmol) and SO_2 (6 mL) in a two-bulbed glass vessel. After 1 h, a preweighted quantity of bromine (0.48 g, 2.92 mmol) and SO_2ClF (2.95 g) are added to the solution. (A slight excess of AsF_5 and Br_2 over S_8 seems to lead to a purer product.) On warming to room temperature, a dark green-brown solution over a dark precipitate is obtained. After 12 h, the solution becomes dark red with a trace of green. The solvent is slowly (4 days) condensed into the other side of the reaction vessel with a temperature gradient of 10 °C between the two bulbs. This results in the formation of red needlelike crystals. The volatile material is then removed under dynamic vacuum.

Anal. Calcd. weight of $[(S_7Br)_4S_4][AsF_6]_6$, based on sulfur: 3.25 g. Found: 3.44 g.

Properties

Crystalline $(S_7I)_4S_4(AsF_6)_6$ and $(S_7Br)_4S_4(AsF_6)_6$, stored in glass sample tubes under an atmosphere of dry nitrogen, show no signs of decomposition after several weeks at room temperature.

References

1. G. Vahl and R. Minkwitz, *Z. Anorg. Allg. Chem.*, **443**, 217 (1978).
2. K. Manzel and R. Minkwitz, *Z. Anorg. Allg. Chem.*, **441**, 165 (1978).
3. M. Feuerhahn and G. Vahl, *Inorg. Nucl. Chem. Lett.*, **16**, 5 (1980).
4. P. S. Magee, in *Sulfur in Organic and Inorganic Chemistry*, Vol. 1, A. Senning (ed.), Dekker, New York, 1971, p. 261.
5. J. Passmore, E. K. Richardson, and P. Taylor, *Inorg. Chem.*, **17**, 1681 (1978).
6. J. Passmore, G. Sutherland, P. Taylor, T. K. Whidden, and P. S. White, *Inorg. Chem.*, **20**, 3839 (1981).
7. J. Passmore, G. Sutherland, and P. S. White, *Inorg. Chem.*, **21**, 2717 (1982).
8. J. Passmore, G. Sutherland, T. K. Whidden, P. S. White, and C.-M. Wong, *Can. J. Chem.*, **63**, 1209 (1985).
9. M. P. Murchie and J. Passmore, *Inorg. Synth.*, **24**, 76 (1986).
10. F. A. Hohorst and J. M. Shreeve, *Inorg. Synth.*, **11**, 143 (1968).
11. R. Y. Eagers, *Toxic Properties of Inorganic Fluorine Compounds*, Elsevier, Amsterdam, 1969.
12. G. M. Begun and A. C. Rutenberg, *Inorg. Chem.*, **6**, 2212 (1967).
13. P. A. W. Dean, R. J. Gillespie, and P. K. Ummat, *Inorg. Synth.*, **15**, 213 (1974).
14. C. L. Chernick, in *Noble Gas Compounds*, H. H. Hyman (ed.), The University of Chicago Press, Chicago, 1963, p. 35.

15. J. Passmore and P. Taylor, *J. Chem. Soc. Dalton Trans.*, **1976**, 804.
16. W. A. S. Nandana, J. Passmore, D. C. N. Swindells, P. Taylor, P. S. White, and J. E. Vekris, *J. Chem. Soc. Dalton Trans.*, **1983**, 619.

68. TRIMETHYLBORANE

$$\text{AlMe}_3(\overbrace{\text{hexane}}) + \text{O}(n\text{-Bu})_2 \longrightarrow \text{AlMe}_3:\text{O}(n\text{-Bu})_2$$

$$\text{AlMe}_3:\text{O}(n\text{-Bu})_2 + \text{BBr}_3 \xrightarrow[\text{O}(n\text{-Bu})_2]{\text{hexane}} \text{BMe}_3 + \text{AlBr}_3:\text{O}(n\text{-Bu})_2$$

Submitted by WILLIAM S. REES, JR.,* MICHAEL D. HAMPTON,[†]
STEPHEN W. HALL,* and JERRY L. MILLS*
Checked by PHILLIP NIEDENZU[‡] and SHELDON G. SHORE[‡]

Trimethylborane is a volatile pyrophoric compound that is frequently employed as a Lewis acid. This compound was first prepared by Stock and Zeidler by direct gas phase combination of BCl_3 with $ZnMe_2$ (ref. 1) at room temperature. The reaction resulted in high yields of BMe_3 but had to be performed on a very small scale to avoid explosions. Other attempts to produce BMe_3 on a laboratory scale have involved extreme conditions and resulted in low yields.[2-4] Preparations involving temperatures up to 800 °C have been reported,[5,6] as have methods requiring a catalytic bed of heated (325–350 °C) Zn or Al.[7,8] The Grignard method, which produces moderate yields, is not particularly compatible with vacuum line techniques, where considerable difficulty is had in separating the BMe_3 from the conventional Grignard solvents, diethyl ether or tetrahydrofuran (THF).[9]

The preparation reported here produces high yields (98–99% based on BBr_3) of pure BMe_3 at room temperature in a short time. The need for refluxing, heated catalysts, or extreme conditions is eliminated.

■ **Caution.** *The reactants and products of this synthesis are spontaneously flammable in the presence of oxygen and/or moisture. Trimethylborane is highly toxic. Thus, the reactants BBr_3 and $AlMe_3$, as well as the product BMe_3, must be handled with great care. These compounds must be manipulated only in the dry, inert atmosphere of a glove bag, dry box, or vacuum line. Further, combination of the reactants $AlMe_3$ and BBr_3 neat or in a less strongly complexing solvent than $O(n\text{-Bu})_2$ may result in explosion.*

*Department of Chemistry and Biochemistry, Texas Tech University, Lubbock, TX 79409.
[†]Department of Chemistry, University of Central Florida, Orlando, FL 32816.
[‡]Department of Chemistry, The Ohio State University, Columbus, OH 43210.

Procedure

The vacuum line used in this synthesis is a standard, calibrated, high vacuum system equipped with a two stage mechanical forepump and a dual stage mercury diffusion pump. The operation pressure in this system is 10^{-5} torr or lower.[10]

Boron tribromide reacts with most brands of stopcock grease, necessitating the use of reaction vessels which have Teflon stopcocks.

In the dry nitrogen (extra dry grade, 99.5% purity) or argon (high purity grade, 99.995% purity) atmosphere of a glove bag, 10.0 mL of dibutyl ether,* previously dried over either sodium napthalene or sodium benzophenone[10] is introduced by syringe into a 1.0-L reaction bulb containing a Teflon-coated magnetic stirring bar (Fig. 1).

A 7.8-mL volume of 2.0 M trimethylaluminum (15.6 mmol) in hexane† (Aldrich or Texas Alkyls) is carefully syringed into the reaction bulb in the glove bag. (Vapor transfer of $AlMe_3$ on a vacuum line is unsatisfactory because of the low volatility and long monomer–dimer, liquid–vapor equilibration time of this compound.) The stopcock of the reaction bulb is then closed, and the bulb is removed from the glove bag and is placed on the vacuum line. ■ **Caution.** *Some trimethylaluminum may remain in the syringe, so it must be handled with extreme care. The syringe should be rinsed with butanol before further cleaning.* The solution is then stirred for 5 min to effect dissolution of the trimethylaluminum. ■ **Caution.** *Trimethylaluminum must be dissolved in the dibutyl ether before the BBr_3 is added. If $AlMe_3$ is added instead to a complex of BBr_3 with the solvent dibutyl ether, the reaction is so exothermic that an explosion may result.* After the trimethylaluminum is dissolved, the solution is freeze–thaw degassed three times by standard procedures.[10]

Following the third degassing step, the reaction vessel cold finger is cooled to $-196\,°C$. Then, exactly 10 mmol of BBr_3‡ (Aldrich) is vapor transferred

---

*Dibutyl ether is the only solvent suitable for this reaction. Dibutyl ether has relatively low volatility and complexes with trimethylaluminum without solvent decomposition. Dibutyl ether complexes with BBr_3 sufficiently strongly to slow the rate of reaction of BBr_3 with $AlMe_3$ to a safe rate. Boron tribromide reacts explosively with trimethylaluminum in diethyl ether or anisole solutions because of the weak complexation between these solvents and BBr_3. Tetrahydrofuran and p-dioxane undergo decomposition reactions with trimethylaluminum, and thus are unsuitable as solvents in this synthesis.

†When dissolved in hexane or other hydrocarbon solvents, trimethylaluminum is much less pyrophoric. Complete safety and handling information is available from Texas Alkyls.

‡Born tribromide is used as received. It is photosensitive, but can be purified by storage over elemental mercury, followed by storage in the dark. The use of the boron halides BF_3 and BCl_3 results in much lower yields and difficulties with purification, and they are therefore considered to be unsatisfactory for this reaction.

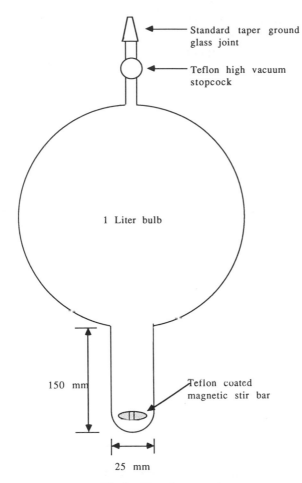

Fig. 1   Reaction vessel.

into the reaction vessel using standard high-vacuum line procedures.[10] The liquid nitrogen is then removed from around the cold finger, which is allowed to warm slowly to room temperature ($\sim$ 20 min). The solution is stirred at ambient temperature for 30 min.

With the reaction bulb finger cooled to 0 °C, the trimethylborane product, together with hexane, is transferred under static vacuum into the vacuum line. All other species should remain in the reaction bulb because of low vapor pressures.

The BMe$_3$ is purified by passage through a $-$ 112 °C trap (CS$_2$ slush) into a $-$ 196 °C trap (liquid nitrogen) under a dynamic vacuum. Hexane is retained in the $-112\,°C$ trap, and the BMe_3 is retained in the liquid nitrogen-cooled trap. Typical yields are 98–99%, based on the quantity of BBr_3 used. If an excess of BBr_3 is used, trace quantities of BBr_2Me and $BBrMe_2$ may result. The use of a stoichiometric quantity of BBr_3 in the reaction results in pure BMe_3, as indicated by IR spectroscopy, ^{11}B and ^1H NMR spectroscopy, and vapor pressure data.

Properties

Trimethylborane is a colorless liquid that is air and moisture sensitive. It is stable for long periods of time in vacuum. Its vapor pressure curve is given by $\log P = -1393.3/T + 1.75 \log T - 0.007735T + 6.1385$, where T is in Kelvins.[1] The IR spectrum of BMe_3 exhibits a very strong absorbance at 1330, strong absorbances at 3010 and 1185 cm^{-1}, a peak of medium intensity at 1210 cm^{-1}, and weak absorbances at 1010 and 990 cm^{-1}.

The ^{11}B NMR spectrum of BMe_3 exhibits a singlet at 85.98 ppm downfield from BF_3 etherate,[11] The ^1H NMR spectrum exhibits a singlet at 0.53 ppm relative to TMS.[12]

References

1. A. Stock and R. Zeidler, *Berichte*, **1921**, 531.
2. J. Iyoda and I. Shilhara, *Bull. Chem. Soc. Jpn.*, **32**, 304 (1959).
3. M. W. Hunt (Continental Oil Co.), U.S. Patent 3,000, 962 (1961).
4. J. Casanova, H. R. Kiefer, and R. E. Williams, *Org. Prep. Proced.*, **1**, 57 (1969).
5. S. Witz, J. L. Sheperd, and E. T. Hormets (Aerojet General Corp.), U.S. Patent 3,329,485 (1967).
6. W. Sundermeyer and W. Verbeck (Th. Goldschmidt, A.G.), W. German Patent 1,239,687 (1967).
7. D. T. Hurd (G.E. Company), U.S. Patent 2,446,008 (1948).
8. British Thompson-Houston Co., Ltd., Brit. Patent 618,358 (1949).
9. D. C. Mente and J. L. Mills, *Inorg. Chem.*, **14**, 1862 (1975).
10. D. F. Shriver and M. A. Drezdzon, *The Manipulation of Air-Sensitive Compounds*, 2nd ed., Wiley, New York, 1986.
11. G. R. Eaton and W. N. Lipscomb, *NMR Studies of Boron Hydrides and Related Compounds*, Benjamin, New York, 1969.
12. B. A. Amero and E. P. Schram, *Inorg. Chem.*, **15**, 2842 (1976).

SOURCES OF CHEMICALS AND EQUIPMENT

List of addresses for the suppliers mentioned in the Syntheses.

Ace Scientific Supply Co., 40-A Cotters Lane, East Brunswick, NJ 08816

Airco Special Gases, 575 Mountain Ave, Murray Hill, NJ 07974

Air Products and Chemicals, Inc., Speciality Gas Department, Hometown Facility, P.O. Box 351, Tamaqua, PA 18252

Aldrich Chemical Co., 940 West St., Paul Avenue, Milwaukee, WI 53233

Alfa Products, 152 Andover St., Danvers, MA 01923

Analtech Inc., P.O. Box 7558, Newark, DE 19714

Apiezon Products—see Biddle Instruments

J. T. Baker Chemical Co., 222 Red School Lane, Phillipsburg, NJ 08865

BASF Corp., Chemicals Div., 100 Cherry Hill Rd., Parsippany, NJ 07054

BDH Chemicals Ltd.—see Gallard-Schlesinger

Biddle Instruments, 510 Township Line Rd., Blue Bell, PA 19422

Cerac Inc., 407 N. 13th St., P.O. Box 1178, Milwaukee, WI 53201

Eastman Kodak Co., Laboratory and Research Products Division, Rochester, NY 14650

EM Science, 111 Woodcrest Rd., Cherry Hill, NJ 08034

Fisher Scientific Co., 711 Forbes Ave., Pittsburgh, PA 15219

Gallard–Schlesinger Chemical Manufacturing Co., 584 Mineola Ave., Carle Place, NY 11514

Hoke Inc., 1 Tenakill Park, Cresskill, NJ 07626

Johnson Matthey Aesar Group, Eagles Landing, P.O. Box 1087, Seabrook, NH 03874

Kontes, Spruce St., P.O. Box 729, Vineland, NJ 08360

Lithium Corp. of America, 449 N. Cox Rd., P.O. Box 3925, Gastonia, NC 28053

Mallinckrodt, Inc., Science Products Div., P.O. Box 5840, St. Louis, MO 63134

Merck & Co., Inc., P.O. Box 2000, Rahway, NJ 07065

E. Merck, Darmstadt—see EM Science

Matheson Gas Products, 30 Seaview Dr., Secaucus, NJ 07094

Ozark-Mahoning Co., 1870 South Boulder, Tulsa, OK 74119

PCR, Inc., P.O. Box 1466, Gainesville, FL 32602

Rainin Instrument Co., Mack Rd., Woburn, MA 01801-4628

Rare Earth Products, Widnes, Cheshire, England

Research Chemicals, Phoenix, AZ

Sigma Chemical Co., P.O. Box 14598, St. Louis, MO 63178

Strem Chemicals, Inc., 7 Mulliken Way, P.O. Box 108, Newburyport, MA 01950

Swagelok Co., 31400 Aurora Rd., Solon OH 44139

Texas Alkyls, P.O. Box 600, Deer Park, TX 77536

Vacuum/Atmospheres Co., 4652 W. Rosecrans Ave., Hawthorne, CA 90250

Ventron—see Alfa Products

Wheaton Scientific, 1301 N. Tenth St., Millville, NJ 08332

Whitey Co., 318 Bishop Rd., Highland heights, OH 44143

J. Young (Scientific Glassware) Ltd., 11 Colville Rd., London W3 8BS, England

INDEX OF CONTRIBUTORS

Prepared by THOMAS E. SLOAN*

*Chemical Abstracts Service, Columbus, OH.

SUBJECT INDEX

Prepared by THOMAS E. SLOAN*

Names used in this Subject Index for Volumes 26–30 are based upon IUPAC *Nomenclature of Inorganic Chemistry*, Second Edition (1970), Butterworths, London; IUPAC *Nomenclature of Organic Chemistry*, Sections A, B, C, D, E, F, and H (1979), Pergamon Press, Oxford, U.K.; and the Chemical Abstracts Service *Chemical Substance Name Selection Manual* (1978), Columbus, Ohio. For compounds whose nomenclature is not adequately treated in the above references, American Chemical Society journal editorial practices are followed as applicable.

Inverted forms of the chemical names (parent index headings) are used for most entries in the alphabetically ordered index. Organic names are listed at the "parent" based on Rule C-10, *Nomenclature of Organic Chemistry,* 1979 Edition. Coordination compounds, salts and ions are listed once at each metal or central atom "parent" index heading. Simple salts and binary compounds are entered in the usual uninverted way, e.g., *Sulfur oxide* (S_8O), *Uranium(IV) chloride* (UCl_4).

All ligands receive a separate subject entry, e.g., *2,4-Pentanedione,* iron complex. The headings *Ammines, Carbonyl complexes, Hydride complexes,* and *Nitrosyl complexes* are used for the NH_3, CO, H, and NO ligands.

Acetaldehyde, iron complex, 26:235
Acetic acid, chromium, molybdenum, and
    tungsten complexes, 27:297
  palladium complex, 26:208
  rhodium complex, 27:292
  tungsten complex, 26:224
———, chloro-, ruthenium complex, 26:256
———, trichloro-, ruthenium complex,
    26:256
———, trifluoro-, ruthenium complex,
    26:254
  tungsten complex, 26:222
Acetone, iridium complex, 26:123
———, molybdenum and tungsten com-
    plex, 26:105
Acetonitrile, molybdenum and tungsten
    complexes, 26:122, 133
  osmium complex, 26:290
  palladium complex, 26:128
  ruthenium complex, 26:356
  ruthenium(II) complex, 26:69

Acetylene, diphenyl-, molybdenum com-
    plex, 26:102
Acyl isocyanide, chromium complexes,
    26:31
Ammines, ruthenium, 26:66
Ammoniodicobaltotetracontatungstotetra-
    arsenate (23−), [(NH₄)As₄W₄₀O₁₄₀-
    [Co(H₂O)]₂]²³⁻, tricosaammonium, non-
    adecahydrate, 27:119
Ammonium ammoniodicobaltotetraconta-
    tungstotetraarsenate(23−), ((NH₄)₂₃-
    [(NH₄)As₄W₄₀O₁₄₀[Co(H₂O)]₂),
    nonadecahydrate, 27:119
Ammonium [(1*R*)(*endo, anti*)-3-bromo-
    1,7-dimethyl-2-oxobicyclo[2.2.1]hep-
    tane-7-methanesulfonate, 26:24
Ammonium dihydrogen pentamolyb-
    dobis[(4-aminobenzyl)phosphonate]
    (4−), [(NH₄)₂ H₂[Mo₅O₁₅-
    (NH₂C₆H₄CH₂PO₃)₂]], pentahydrate,
    27:126

*Chemical Abstracts Service, Columbus, OH.

FORMULA INDEX

Prepared by THOMAS E. SLOAN*

The Formula Index, as well as the Subject Index, is a Cumulative Index for Volumes 26–30. The Index is organized to allow the most efficient location of specific compounds and groups of compounds related by central metal ion or ligand grouping.

The formulas entered in the Formula Index are for the total composition of the entered compound, e.g., F_6NaU for sodium hexafluorouranate(V). The formulas consist solely of atomic symbols (abbreviations for atomic groupings are not used) and arranged in alphabetical order with carbon and hydrogen always given last, e.g., $Br_3CoN_4C_4H_{16}$. To enhance the utility of the Formula Index, all formulas are permuted on the symbols for all metal atoms, e.g., $FeO_{13}Ru_3C_{13}H_{13}$ is also listed at $Ru_3FeO_{13}C_{13}H_{13}$. Ligand groupings are also listed separately in the same order, e.g., $N_2C_2H_8$, 1,2-Ethanediamine, cobalt complexes. Thus individual compounds are found at their total formula in the alphabetical listing; compounds of any metal may be scanned at the alphabetical position of the metal symbol; and compounds of a specific ligand are listed at the formula of the ligand, e.g., NC for Cyano complexes.

Water of hydration, when so identified, is not added into the formulas of the reported compounds, e.g., $Cl_{0.30}N_4PtRb_2C_4 \cdot 3H_2O$.

*Chemical Abstracts Service, Columbus, OH.